physics
019
042

NEURAL
NETWORKS FOR
COMPUTING

AIP CONFERENCE PROCEEDINGS 151

RITA G. LERNER
SERIES EDITOR

heading

"NEURAL NETWORKS FOR COMPUTING"

SNOWBIRD, UT 1986

EDITOR:

JOHN S. DENKER
AT&T BELL LABORATORIES

AMERICAN INSTITUTE OF PHYSICS   NEW YORK 1986

L.C. Catalog Card No. 86-072481
ISBN 0-88318-351-X
DOE CONF-8604173

Printed in the United States of America

Second Printing, 1987

CONTENTS

Neural Networks for Computing
(Snowbird 1986)

The conference on "Neural Networks for Computing" was held at Snowbird, Utah on April 13th through 16th, 1986. This was a successor to the conference of the same name held at Santa Barbara a year earlier. The Santa Barbara meeting had about 60 participants, which was about right. We figured that if we made preparations to have two and a half times that many this year, we would be safe. Nevertheless, we were unable, alas, to admit all the worthy applicants, which is an indication of the growth of the field. Taking into account party-crashers and spooks, we wound up with about 160 people in the meeting room.

A more important measure of growth in the field is the advance in the level of the work. Ideas that were (quite rightly) considered the pinnacle of cleverness last year were taken for granted this year. At the previous meeting there was a feeling that collective-computation ideas had a lot of potential, even if we didn't know exactly where it all would lead, while at this meeting there was excitement of a different kind, because you could see that things were already moving from the potential to the actual.

I would like to thank the other members of the organizing committee: Gérard Dreyfus, John Hopfield, Rich Howard, Christof Koch, Alan Lapedes, Demetri Psaltis, Terry Sejnowski, especially the chief organizer, Larry Jackel, and our indefatigable conference secretary, Carol O'Rourke. I would also like to extend a tribute to the staff at Snowbird for making our days there so pleasant.

A note from the editor to the contributors: to those of you who submitted papers on time, complete, in the correct format, and with copyright releases—my sincerest thanks! I wish there were more of you—you were in the minority. And to those of you who struggled and stayed within the length limits, I convey the apologies of those who had to run over. There were several cases where it seemed more sensible for a major group to submit one long paper rather than a stack of smaller ones.

The program of the meeting listed about 40 talks and about 50 posters, to which was added quite a lot of audience participation and a number of unscheduled presentations. As expected, not all of these participants chose to contribute to the proceedings, so reading this book will be indeed a poor substitute for actually attending the meeting.

Best wishes to you all!

John S. Denker
AT&T Bell
Laboratories

NEURAL NETWORKS FOR COMPUTING?

Yaser S. Abu-Mostafa
California Institute of Technology
Pasadena, CA 91125

ABSTRACT

In this paper, we address the capabilities and cost-effectiveness of the current models of neural networks for carrying out general computation. We show that neural networks are formally capable of performing any conventional computation. However, we show that for hard 'algorithmic' problems, the speed with which the solution is achieved comes at the cost of an excessive size for the network.

1. INTRODUCTION

The basic operation of most models of neural networks is that of nearest-neigbour search or associative memory [1]. The efficiency or cost-effectiveness of this operation is measured by information capacity and radius of attraction that can be provided by a network of N neurons [2,3]. In addition to this basic operation, there are several examples of the use of neural networks in other computational tasks, e.g., [4]. Two questions are addressed here:
1. Which problems can be solved by neural networks?
2. Which problems can be *efficiently* solved by neural networks?
The answers to these questions determine the capabilities and limitations of neural networks as 'general-purpose' computers.

Whereas the information properties are determined by the statics of a system, the computation properties are determined by the dynamics, since the motion from the input state to the output state is what defines a computation. The ability to move from an arbitrary point to another relates to the first question and is discussed in section 2, while the *cost* of such process relates to the second question and is discussed in section 4. Section 3 addresses two common problems in isolating the effect of actual computation from the necessary pre- and post-processing.

We shall use the binary convention -1 and $+1$ instead of 0 and 1. The neural network model is described in the appendix. The network consists of N neurons, and the weight (transition) matrix W is a symmetric, zero-diagonal, real $N \times N$ matrix whose entries are denoted by w_{ij}.

2. COMPUTING CAPABILITIES

We show that neural networks are formally capable of solving any conventional computational problem. For simplicity, we represent the input and output of the problem in question by binary strings $\mathbf{x} = x_1 \cdots x_M$ and $\mathbf{y} = y_1 \cdots y_K$, which can be done for any finite problem. The problem can then be characterized as K Boolean functions $y_1 = f_1(\mathbf{x}), \cdots, y_K = f_K(\mathbf{x})$. Therefore, it suffices to show that any Boolean function can be simulated by the appropriate neural network.

This work was supported in part by Caltech's Program in Advanced Technologies, sponsored by Aerojet General, General Motors, GTE, and TRW.

It is well known in switching theory [5] that two-input NAND gates form a complete basis for Boolean functions. A NAND gate is one whose output is -1 iff all of its inputs are $+1$'s, and being a complete basis means that any Boolean function can be simulated by a combinational circuit all of whose gates are exclusively two-input NAND gates. A NAND gate, in turn, can be simulated by a two-input threshold function with weights $w_1 = w_2 = -2$ and threshold $t = -3$ (figure 1). Since the operation of the neurons in a neural network is that of an arbitrary threshold function, we have a crude but guranteed way of simulating any Boolean function with a neural network.

FIGURE
1

$$w_1 = -2$$

$$\equiv$$

NAND gate

$$t = -3$$

$$w_2 = -2$$

To do this, we take the combinational circuit with NAND gates and replace each gate with a neuron whose threshold function is that of figure 1. To supply the inputs, we provide a neuron with no input synapses and zero internal threshold for each x_m. The initial state of each input neuron is set according to the value of x_m. The output is collected from the final neuron after the states of all neurons stablize. However, there is a technical problem because of the parasitic feedback; the requirement that $w_{ij} = w_{ji}$. This means that the next gate of the combinational circuit affects the current gate through the wire that connects the two gates, contrary to the properties of combinational circuits. To isolate the neurons from these undesired effects, we make the weights of the first stage far bigger than the second stage, and those in turn far bigger than the third stage, and so on. We also provide a buffered level of input neurons connected to the other input neurons by synapses of large positive weight. Figure 2 shows an example of such a network.

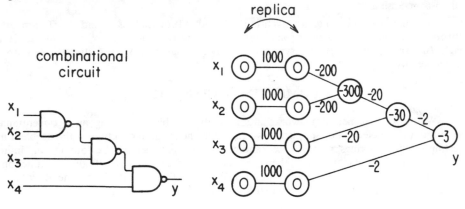

FIGURE 2

These networks have no effective feedback, and their size can be exponential in M, the number of inputs. They are only meant to prove that any computation can be done using some neural network. The efficiency and practicality considerations undermine the usefulness of this fact, as described later in this paper.

3. REPRESENTATION AND UNIFORMITY

We discuss two issues that often cause confusion in analysing computation and computation cost, namely representation and uniformity. Since the input of a computational problem, such as the Travelling Salesman Problem (TSP), contains all the *information* needed to determine the output, computation can be viewed merely as a change-of-representation process. Therefore, the boundary between input/output *represenation* and the actual computation is often fuzzy. For example, the input to the TSP is the array of distances $d_{ij}: 1 \leqslant i, j \leqslant M$ and the output is an optimal permutation $m_1 \cdots m_M$ of the cities $1 \cdots M$. If a computing system solves the TSP, one would expect to set some parameters of the system according to the input distances in a simple and straightforward manner and later on measure some parameters which will be directly related to the output permutation. The fuzziness occurs when one has to *compute* how to set the input parameters of the system in terms of the problem input, or how to interpret the problem output in terms the output parameters. As an extreme case, one may completely solve the TSP before hand and set the parameters of the system according to that solution, then read them off as the output! Even if only partial computation is done before using the system, the relative role of that 'preprocessing' has to be assessed in evaluating the actual computation done by the system.

Another critical point has to do with uniformity, which is the ability of the *same* system to handle many instances of the problem. In practice, this amounts to the system being reusable, which makes its initial design and building cost effectively divided by the large number of times it is used to solve the computational problem. For example, for a fixed-length input, there are exponentially many possible problem instances. Let us consider again solving the TSP using neural networks. Depending on the technology used to build the neural network, the synaptic strengths may or may not be considered hardware. If they are, then the solution must not require different synaptic strengths for different *instances* of the TSP with the same number of cities, since that would mean building a new network each time we have a new instance of the TSP. In general, what is considered hardware (e.g., number of neurons, synaptic connections, etc.) should be fixed for a large class of problem instances, and the nonpermanent parameters of the network (e.g., the initial states of the neurons) should be used to load a specific problem instance to the network.

4. COMPUTING LIMITATIONS

Although neural networks are capable of performing any computation, it is conceivable that this may come at a high cost. Do we save when we use neural networks instead of conventional computers to solve a hard computational problem? We shall focus on those computations which take an excessive amount of time when done on a sequential machine. By excessive time, we mean exponential, or faster than any polynomial, as a function of the length of the input. The TSP and other NP-complete problems are widely believed to belong to that category. The best known algorithm for these problems is essentially exhaustive and runs in exponential time. We shall show that, for problems that *require* exponential time on sequential machines, neural networks may provide speed only at the cost of an exponential size of the network.

In general, neural networks can be thought of as a special case of parallel computers, where the neurons play the role of the processors and the synapses play the role of the communication links. The particulars of neural networks, such as asynchronism and robustness, give them a special operating advantage. However, no parallel computer with N processors can speed up the computation by more than N times, which corresponds to 100% efficiency. To see this, let T_p be the running time of the parallel computer and T_s be that of the sequential computer. One can always resort in the sequential case to *simulating* the computation of all the processors of the parallel computer step by step. This will take at most T_p steps for each of the N processors, hence $T_s = O(N T_p)$.

In the special case of neural networks, the simulation amounts to computing the threshold functions of the neurons and implementing the state changes until a stable state is reached. Let N be the number of neurons and let $T(N)$ be the maximum number of neuron state changes that can take place before the network reaches a stable state (see appendix for conditions and proofs). If the neural network is simulated by a sequential machine, the machine will perform at most $N T(N)$ operations before a stable state is reached. Now suppose that this neural network solves (either exactly or approximately) a hard problem, and that any solution on a sequential machine would take a minimum of α^M steps, where M is the length of the input. Since the simulation of the network, which takes at most $N T(N)$ operations, is *a* solution, it follows that $\alpha^M = O(N T(N))$. The technology for building neural networks, e.g., optical technology, may allow a gigantic number of steps $T(N)$ in a short physical time. Therefore, it is still conceivable that we have a reasonable-size network (small N) and absorb the exponential in $T(N)$. However, from the appendix, the number of iterations $T(N)$ of a neural network can be at most a polynomial in N, specifically a square law. This means that $\alpha^M = O(N^3)$. Therefore, the number of neurons N is at least exponential in the size of the input. This means that the speed-up occurs at the cost of an impractically huge network. One may argue that the probabilistic nature of neural-network solutions may make the complexity polynomial in M. In this case, however, the probabilistic sequential solution will also be polynomial. In other words, the network may lead to a speed-up consistent with it being a parallel computer, but not beyond that.

The efficiency of a specific neural-network solution to some computational problem can be tested directly by checking how the degrees of freedom of the network are used. Let us take the solution of the TSP [4], for example. For M cities one needs a network with $N = M^2$ neurons. Such a network has the order of $N^3 = M^6$ degrees of freedom if the synaptic strengths are arbitrary real numbers [2], and $N^2 = M^4$ degrees of freedom if the synaptic strengths are quantized. The output solution is some $M \times M$ permutation matrix, with at most $\log(M!)$ degrees of freedom, which is the order of $M \log M$. Thus we are putting M^6 or M^4 in and getting $M \log M$ out.

A more general way of looking at the efficiency of embedding a problem in a neural network comes from a computational complexity point of view. Solving the problem on a sequential computer requires a certain number of steps (time complexity), a certain amount of scratch memory (space complexity) and a certain length of the algorithm (Kolmogorov complexity). When we have a neural network for solving the problem, the number of iterations accommodates the time complexity, the number of state variables or neurons accommodates the space complexity, and the degrees of freedom or information capacity of the synaptic connections, in which the algorithm is stored, accommodates the Kolmogorov complexity. However, when the number of neurons is N, the number of iterations is $O(N^2)$, and the capacity of the synapses is $O(N^3)$ bits. When we have a neural-network solution of a problem, we must meet the minimum requirement of time complexity *and* space complexity *and* Kolmogorov complexity of that problem. If the problem is very demanding in space complexity, for example, then the number of state variables or neurons N has to be large, and the capacity must follow suit, even if the Kolmogorov complexity of the problem is very modest. Having a large capacity without using it spells inefficiency. The above discussion of exponential-time problems is another example of this inefficiency, where time complexity is very demanding, while the other complexities may not be.

It is worth mentioning here that the role of *analog* computation by the networks is yet to be fully assessed. The lack of a uniform theory of analog computation leaves us with no parallels to the above simulation arguments in the analog case. It is vital to quantify the advantage of having a continuous spectrum of values (which are not necessarily exact), having soft decision capabilities, and having continuous-time processing.

The complexity analysis of neural networks leads us to identifying certain problems which best match the characteristics of the networks. The capacity of the network grows faster than the number of neurons. This means that problems which require a very long algorithm make the most use of the networks, whereas neural-network solutions of problems with short algorithms and high time complexity are very inefficient. Problems requiring a long algorithm are called random problems [6]. Some examples are pattern recognition problems in natural environments and artificial intelligence problems requiring huge data bases. These problems make the most use of the large capacity of neural networks. This fact is expected in biologically valid models of neural networks, since humans are very good at random problems, but not nearly as good in structured problems.

There is another property of random problems that makes them matched to the neural network methodology. In most cases, we do not have an explicit algorithm for solving a random problem, and we develop one by learning from training sets. Learning is quite natural in neural networks, where the accumulative adjustment of the synaptic strengths tunes the network automatically to the problem in question.

APPENDIX

We define the neural network model and prove a simple result about the number of iterations of a network. A neural network consists of N pairwise connected neurons. The ith neuron can be in one of two states: $u_i = -1$ (off) and $u_i = +1$ (on). The synaptic connections are undirected and have strengths which are fixed real numbers. The neurons repeatedly examine their inputs and decide to turn on or off by the following scheme. Let w_{ij} be the strength (could be negative) of the synaptic connection from neuron j to neuron i. ($w_{ij} = w_{ji}$ and $w_{ii} = 0$). Let t_i be the threshold voltage of the ith neuron. If the weighted sum over all of its inputs is greater than t_i, the ith neuron turns on and its state becomes $+1$. If the sum is less than t_i , the neuron turns off and its state becomes -1. If the sum equals t_i, we take the convention that the neuron maintains its previous state. The action of each neuron simulates a general threshold function of $N-1$ variables (the states of the rest of the neurons).

$$u_i = \text{sgn} \left(\sum_{j=1}^{N} w_{ij} u_j - t_i \right)$$

where sgn(x) is $+1$ for positive x, -1 for negative x, and undefined for $x=0$. The network starts in an initial state and runs with each neuron randomly and independently re-evaluating itself. Often, the network enters a stable point in the state space in which all neurons remain in their current state after evaluating their inputs. The basic operation of the network is to converge to a stable state if we initialize it with a nearby state vector (in the Hamming sense).

The internal thresholds t_i are often set to zero. In this case, the dynamics of the network is best described by a time-varying energy function which is defined in terms of the current states of the neurons as follows.

$$E = -\tfrac{1}{2}\sum_{i=1}^{N}\sum_{j=1}^{N} u_i w_{ij} u_j$$

It is easy to check that if neuron i changes its state u_i from -1 to $+1$ or vice versa, according to the above threshold rule, E will be decreased by $2 \times |\sum_{j=1}^{N} w_{ij} u_j|$, which must be positive. Since E is bounded below, the network must head towards stability.

The energy function also enables us to estimate an upper bound for the number of iterations a network may go through before it reaches a stable state. To prove the results of this paper, we only need the bound to have a polynomial form in terms of N. To simplify the analysis, we make the assumption that the weights w_{ij} have a finite resolution; they are quantized to L equispaced levels where L is a constant independent of N. Let Δ be the minimum resolution, i.e., the spacing between two consecutive quantization levels of w_{ij}. The value of E is now trapped in the range $-L\Delta N^2/2 \leqslant E \leqslant +L\Delta N^2/2$, for any state vector $u_1 \cdots u_N$. Any neuron that changes its state will decrease E by at least 2Δ (see the above expression for the change in E). Therefore, the maximum number of neuron state changes needed to reach a stable state, denoted by $T(N)$, is bounded above by

$$T(N) \leqslant \frac{(L\Delta N^2/2) - (-L\Delta N^2/2)}{2\Delta} = \frac{L}{2} \times N^2$$

which is polynomial (actually square-law) in the number of neurons N.

REFERENCES

[1] J. Hopfield, "Neural networks and physical systems with emergent collective computational abilities," *Proc. Nat. Academy Sci., USA*, vol.79, pp. 2554-2558, 1982.

[2] Y. Abu-Mostafa and J. St. Jacques, "Information capacity of the Hopfield model," *IEEE Trans. Inform. Theory*, vol. IT-31, pp. 461-464, 1985.

[3] R. McEliece, E. Posner, E. Rodemich, and S. Venkatesh, "The capacity of the Hopfield associative memory," submitted to *IEEE Trans. Inform. Theory*, 1986.

[4] J. Hopfield and D. Tank, "Neural computation of decisions in optimization problems," *Biol. Cybernetics*, vol. 52, p. 141, 1985.

[5] Z. Kohavi, *Switching and Finite Automata Theory*, McGraw-Hill, 1978.

[6] Y. Abu-Mostafa, "Complexity of random problems," in *Complexity in Information Theory*, Y. Abu-Mostafa (ed.), Springer-Verlag, 1986.

A BAYESIAN PROBABILITY NETWORK

C.H. Anderson
RCA Laboratories, Princeton, NJ   08540

E. Abrahams
Serin Physics Laboratory, Rutgers University, Piscataway, NJ   08854

ABSTRACT

A model of an associative neural network is developed in which the state of each node is described by a probability density. The realization of the network is based on the pairwise joint probabilities obtained from a training set of states. A positive definite "energy" functional of the probabilities may be constructed from Bayes' rule of statistical inference. When the states of some of the nodes are fixed as constraints, the minimization of the energy yields a set of probability functions which is consistent with the original pairwise correlations.

INTRODUCTION

Most neural net models start with the simplifying assumption that the output response of a neuron is a highly non-linear function of the input: the output is either on or off over much of the input range. However, real sensory and motor neurons are known to have a graded response which is linear over a significant range of inputs.

In this contribution, we suggest a model which has features that allow more general descriptions. The basic variables associated with the neurons are the probabilities of firing rates (or interspike intervals) and the correlation functions between them play the role of synapses. Thus we allow a distribution over an ensemble due to noise, for example, or uncertain knowledge about the inputs to the system. The model is analyzed by a statistical inference approach based on Bayes' rule. The purpose is to obtain, from a set of inputs, a distribution of outputs which is determined by the correlation functions built into the system from its previous exposure to training information.

DESCRIPTION OF MODEL

The system consists of a set of N units or nodes. A continuous variable x_i is associated with the i-th unit. For simplicity, we take all units to be identical with each x_i within $(0,1)$. The nodes have pairwise connections that are determined by the prior information by which the system is trained.

The prior information, or training set, is a set of data; the α-th datum consists of values $\{x_i^{\alpha}\}$ for all of the N nodes. All this information is represented by density distribution $\rho^o(\{x_i\})$ in N-dimensional space:

$$\rho^o(x_1 \ldots x_N) = \frac{1}{M} \sum_{\alpha=1}^{M} [\prod_i \delta(x_i - x_i^\alpha)] \tag{1}$$

where M is the number of data sets.

An input to the system is given by a density distribution for the x's of a certain set of input nodes {j}. This probability density $\rho^1(\{x_j\})$ is defined over an ensemble in which the input values x_j might vary due, for example, to noise or other uncertainties associated with the sources of input. Given an input ρ^1, we ask what is the probability distribution $\tilde{\rho}$ for the other, output, nodes {k} which is most consistent with the training set ρ^o, subject to the constraints imposed by ρ^1. An answer is given by the Bayes-Laplace rule:

$$\tilde{\rho}(\{x_k\}) = \int \prod dx_j \left[\frac{\rho^o(\{x_j, x_k\})\rho^1(\{x_j\})}{\int \prod dx_k \rho^o(\{x_j, x_k\})} \right] \tag{2}$$

In our realization of the system, we simplify the complex multinodal correlations implicit in Eq. (2). We intend to keep only two-node correlations so let us examine Eq. (2) when only two nodes are present

$$\tilde{\rho}(x_2) = \int dx_1 \rho^o(x_1 x_2)) P_1^1(x_1)/P_1^o(x_1) \tag{3}$$

where $P_i^o(x_1)$ is the margin of the training set

$$P_k^o(x_k) = \prod_{j \neq k} \int dx_j \rho^o(x_1 \ldots x_N) \ .$$

To build up a set of equations for the whole system, we first note that Eq. (3) is found as the minimum of an energy functional:

$$E[\tilde{P}_2(x_2)] = \frac{1}{2} \iint dx_1 dx_2 \rho^o(x_1 x_2) \left[\frac{\tilde{P}_2(x_2)}{P_2^o(x_2)} - \frac{P_1^1(x_1)}{P_1^o(x_1)} \right]^2 \tag{4}$$

We generalize Eq. (4) to many nodes with additive pairwise interactions. The energy is now given by

$$E\{P_m(x_m)\} = \frac{1}{2} \sum_{mm} \iint dx_m dx_n \rho_{mn}^o(x_m x_n) \left[\frac{P_m(x_m)}{P_m^o(x_m)} - \frac{P_n(x_n)}{P_n^o(x_n)} \right]^2 \tag{5}$$

where ρ_{mn}^o are the two-node margins derived from ρ^o. Eq. (5) can be shown to be equivalent to Eq. (2) if the pairwise correlations are sufficiently weak. This energy functional should be minimized with respect to the (output) P_m, subject to the constraint that the input P_j's (which are derived from ρ^1) are held fixed.

The pairwise correlations $C_{mn}(x_m, x_n)$ are defined by

$$\rho^o_{mn}(x_m, x_n) = P^o_m(x_m) P^o_n(x_n) [1 + C_{mn}(x_m, x_n)] . \qquad (6)$$

With the definition $P_m(x) = P^o_m(x)[1 + r_m(x_m)]$, Eq. (5) can be rewritten as

$$E\{(r_m(x_m)\} = (N-1)\sum_m \int dx_m P^o_m r^2_m - \sum_{mn} \int\int dx_m dx_n P^o_m r_m C_{mn} r_n P^o_n \qquad (7)$$

which shows explicitly how the deviations r_m from the P^o_m are driven by the correlation functions C_{mn}.

In practice, one could proceed from Eq. (7) by first fixing the constrained r_j from the inputs and then using a steepest descents method to minimize the energy with respect to the other r_k. We believe that the energy functional has only one minimum in $\{r_k\}$ space for a given set of fixed inputs $\{r_j\}$. The location of this minimum changes when the inputs change; this is how a variety of outputs arise as the inputs are varied.

EXAMPLES

For a simple example, assume that the probability functions can be described by

$$P_i(x < 1/2) = (1/2)(1 - r_i)$$

$$P_i(x > 1/2) = (1/2)(1 + r_i) , \qquad -1 \leqslant r_i \leqslant 1 .$$

This corresponds to a spin-1/2 model with all the margins $P^o_i(x) = 1/2$. The most general pairwise joint probability in this scheme is

$$\rho_{ij} = (1/4) \left[\begin{pmatrix} 1 & 1 \\ 1 & 1 \end{pmatrix} + C_{ij} \begin{pmatrix} 1 & -1 \\ -1 & 1 \end{pmatrix} \right] , \qquad -1 \leqslant C_{ij} \leqslant 1 \qquad (8)$$

The energy corresponding to Eq. (7) is

$$E = (N-1)\sum_i r_i^2 - \sum'_{ij} C_{ij} r_i r_j ,$$

which is equivalent to the Hopfield graded response model[1] of a neural network with a linear gain function. In addition, the rules for obtaining the correlation functions C_{ij} from Eqs. (1,6,8) are (apart from a constant factor) identical to those proposed in the Hopfield content addressible memory network.[2]

We now give an example which illustrates the general nature of the model. We ran a problem where the probabilities are described by four step functions of equal width (rather than 2 as in the previous example). The training values are taken with equal probability from two distinct sets:

First set

$$P_1^{(1)} = (.25, .25, .25, .25)$$

$$P_2^{(1)} = (0, .50, .50, 0)$$

$$P_3^{(1)} = (.50, 0, 0, .50)$$

$$P_4^{(1)} = (.125, .187, .313, .375)$$

Second set

$$P_1^{(2)} = (1, 0, 0, 0)$$

$$P_2^{(2)} = (0, 1, 0, 0)$$

$$P_3^{(2)} = (0, 0, 1, 0)$$

$$P_4^{(2)} = (0, 0, 0, 1)$$

(9)

The joint margins of ρ^o are obtained from

$$\rho_{ij}^o(x_i, x_j) = \frac{1}{2} [P_i^{(1)}(x_i)P_j^{(1)}(x_j) + P_i^{(2)}(x_i)P_j^{(2)}(x_j)]$$

which is appropriate when there are no correlations within each subset. If P_1 and P_3 are taken as inputs and fixed at the values $P_1^{(1)}$ and $P_3^{(1)}$ of Eqs. (9) respectively then the minimization of the energy gives

$$P_2 = (0, .59, .41, 0); \quad P_4 = (.1, .156, .246, .498).$$

If P_1 and P_3 are held at their second set values, we retrieve

$$P_2 = (0, .90, .10, 0); \quad P_4 = (.027, .039, .066, .868).$$

The results in the two cases are close to but not exactly the same as that in the training set. The question of precision is related to the storage capacity problem which we have not addressed.

CONCLUSION

We have only given a brief description of a model that has a number of tantalizing aspects. The idea of describing the state of the system in terms of probability functions came from the field of Labeled Relaxation,[3] which deals with abstract concepts of objects and their labels. Our model and Labeled Relaxation become essentially identical if the set of labels is made equivalent to a basis set of functions that provide a description of the $P_m(x)$. The connection to real neurons would be in an assignment of these basis functions to the independent degrees of freedom, or modes of activity, of the neurons. The neurons would broadcast their state by modulating their firing rate; they would demodulate the incoming signals as well. Thus the transfer of information could be more complex then simply excitation and inhibitory behavior, although the model does not presently handle coherent modulation.

Finally, deriving the coupling coefficient from statistical measures is, we believe, an important feature for pattern recognition tasks using neural networks. While the present model could be extended to include higher order statistics, we do not know how to do this efficiently using pairwise interactions and "hidden units".

REFERENCES

1. J.J. Hopfield, Proc. Natn. Acad. Sci. USA 81, 3088 (1984).
2. J.J. Hopfield, Proc. Natn. Acad. Sci. USA 79, 2554 (1982).
3. R.A. Hummel and S.W. Zucker, IEEE Trans. PAMI-5, 267 (1983); S. Peleg, IEEE Trans. PAMI-2, 326 (1980).

OPTICAL RESONATORS AND NEURAL NETWORKS

Dana Z. Anderson
Department of Physics, University of Colorado and
Joint Institute for Laboratory Astrophysics
University of Colorado and National Bureau of Standards
Campus Box 440, University of Colorado, Boulder, Co. 80309

ABSTRACT

It may be possible to implement neural network models using continuous field optical architectures. These devices offer the inherent parallelism of propagating waves and an information density in principle dictated by the wavelength of light and the quality of the bulk optical elements. Few components are needed to construct a relatively large equivalent network. Various associative memories based on optical resonators have been demonstrated in the literature, a ring resonator design is discussed in detail here. Information is stored in a holographic medium and recalled through a competitive processes in the gain medium supplying energy to the ring resonator. The resonator memory is the first realized example of a neural network function implemented with this kind of architecture.

INTRODUCTION

Optics is a natural context in which to contemplate the implementation of neural networks. The highly communicative nature of the processing elements in neural models begs for connection mechanisms that do not interact except at (or near) the processing element. This is particularly true in systems which are globally interconnected. Light waves do not interact except under the attendance of a material medium, and then only weakly. Electrical signals conducted by wires, on the other hand, love to talk to one another. Furthermore, optics lends itself very well to parallel computation. Lenses, holograms and other optical elements can easily manipulate large arrays of data in a parallel fashion.

What comes to mind most often with the word optics in association with neural networks is a system that implements optically a direct analogue of what one would like to implement electronically. Thus one pictures arrays of optical elements, such as bistable devices, interconnected by various other optical array elements through free-space propagation of light. A good example is the implementation of Hopfield's model for associative memory by Psaltis and Farhat using arrays of light emitting diodes and detectors connected through photographic masks.[1] Theirs is an electrooptic version. Somewhat closer to an all optical version is the adaptive optical network of Fisher and Giles.[2] I believe the efforts to develop entirely optical or nearly all optical neural processor is bound to become successful as optical device technology matures.

The drawback of these optical array architectures lies in the discreteness of the array elements. As with their electronic counterparts, high density manufacturing techniques are required to produce processors having acceptable power dissipation and reasonable computational power. The alternative I address is the use of optical devices that operate on the continuous function of the electric field distribution of an optical wave. Processors constructed using bulk optical elements to perform optical transforms, manipulate optical fields and provide nonlinearity will use temporally coherent light; this means that the signals will be inherently analog and bipolar: the field is everywhere describable by a complex amplitude. There are basically two classes of neural networks: One uses a layered architecture with feed-forward connections between network planes. An optical implementation of this type would use a series of bulk optical elements connected through free space propagation of light. The second type of network employs a feedback architecture. This type of network can be implemented in an optical resonator geometry. This paper deals with the latter. I shall discuss associative memory as an example of a neural network function that can be implemented with an optical resonator.

OPTICAL RESONATOR ASSOCIATIVE MEMORY

Several investigators have proposed and/or demonstrated associative memories based on optical resonators.[3,4,5,6,7] It should be possible to implement other kinds of networks as well. Figure 1 shows a generic ring resonator. The essential elements shown are: 1) a holographic recording medium (such as a photographic plate or a photorefractive medium), a gain medium (which in our case is an externally pumped photorefractive medium,[8]) a generalized connection operator T, which I discuss in more detail below, and a series of mirrors to provide feedback to the holographic medium. I should point out that this ring configuration is merely one possibility, other configurations using phase conjugating mirrors will be found in Refs. 3-7.

The operation of the device shown in the figure may be heuristically described in the following way: Ignore the gain medium for the moment. Initially the recording medium is unexposed. An information bearing coherent light beam is incident on the recording medium. (I'll call the information entity an object.) The object O beam becomes transformed to O' by the T operator as it travels around the ring. The O' beam then interferes with the original beam at the recording medium. The recording medium makes a record of this interference pattern. If the medium is a static one, such as a photographic plate, then it must be developed.

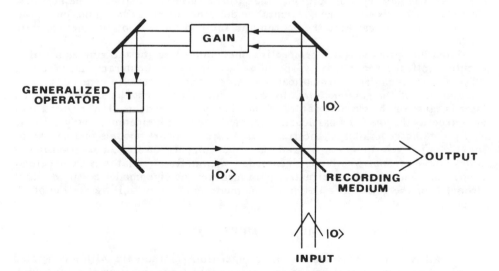

Fig. 1. An optical ring resonator consisting of a holographic recording medium, a gain medium and a generalized operator T that provides feedback to the recording medium.

The resulting hologram when replaced in its original position becomes a diffraction grating which will scatter the **O'** beam back into the path of **O**, that is, the hologram reconstructs **O** from **O'**. One can then say that **O** (or **O'**) is an eigenmode of this ring system; it propagates around the ring in a self-consistent manner. If the recording medium is dynamic, such as a photorefractive medium, then the hologram develops in real-time. Whereas initially the **O'** beam exits the ring, as the hologram forms, **O'** scatters back into the ring. Once again, **O** has become an eigenmode of the ring. In this case one may say the ring "learns" an eigenmode from an injected signal. In principle, several eigenmodes can be programmed by multiply exposing the hologram.

To some extent, an analogy can be drawn between the components of the ring resonator and a neural network: one cannot, however, take the analogy too seriously. The holographic recording medium plays the role of the neurons and synapses: during the exposure stage the medium adjusts in a particular way to the inputs. The recording medium may be "read only" so that the writing is permanent, as with a photographic emulsion, or it may be a real-time medium that can be programmed, read and erased in real time, as with some photorefractive media. The operator **T** serves as the wiring diagram, i.e., the dendrites that form connections among the neurons. **T** is represented schematically in the figure as a box; in fact it includes a description of the free-space propagation of the light around the ring and any number of optical elements. It determines which neurons are connected to which other neurons, but not the strength of the connection. Thus the nature of **T** determines the group properties, such as the rotational symmetries, of the stable states of the network. A conceptually straightforward literal connection technique consists of a fiber optic bundle that is braided in a manner to give the desired connection. Other examples are a lens which can be used to provide a fourier transform, a prism to provide image rotation and a (permanent) hologram, which could be designed to produce almost any physically realizable connection. One could even use the diffraction of the light itself to provide communication among the neurons. The gain medium provides the saturating nonlinearity for the network. Here the analogy is weak since, in neural models, the nonlinearity is inherent in the neuron itself, whereas here the neuron and the nonlinearity of this network are separate.

From an optical physics perspective one would describe the ring as a rather peculiar optical resonant cavity --peculiar because of the hologram that takes the place of a simple mirror. An optical resonator supports a set of eigenmodes. With the addition of the gain medium, this device acts very much like a laser; it isn't a laser because of the specific nature of the chosen gain medium. If the round-trip gain exceeds the loss for a particular eigenmode, then that eigenmode may oscillate (light will be generated in the ring). In neural network models one speaks of inhibitory or excitatory synapses and sometimes competition and cooperation among states of the system. Similarly, eigenmodes in an active resonator can compete or cooperate. With the photorefractive gain medium the eigenmodes compete rather strongly for the gain.[9] Oscillation of one mode tends to inhibit oscillation of all other modes.

MODE COMPETITION

Recall is achieved by biasing the competition for the gain with an injected signal. The injected signal may be decomposed into a superposition of eigenmodes. Normally one has a complete and orthogonal set of modes, in the present case the decomposition may be incomplete. The eigenmode with the largest coupling coefficient has the advantage in the competition and is most likely to win. Thus recognition in the resonator memory is done through a suppression of all but the eigenmode which most resembles the input.

An approach similar to that used to model a laser can be applied to the ring resonator having photorefractive gain.[9] When more the one mode has the potential to oscillate, an exact solution to the problem is a horrendous calculation. However, one can use a perturbative approach assuming that the resonator fields are weak relative to the external pump energy being supplied to the photorefractive medium. In this case the equations of motion for the mode intensities are similar to those of a laser: in a system having N modes, the equation of motion for the nth mode is given by

$$\dot{I}_n = \alpha_n I_n - \theta_{nn} I_n^2 - \sum_{\substack{i=1 \\ n \neq m}}^{\infty} \theta_{nm} I_n I_m \qquad (1)$$

where α_n is a linear net gain coefficient, θ_{nn} is a self-saturation coefficient and θ_{nm} is a cross saturation coefficient describing to what extent the presence of mode m lowers the gain for mode n. As an example, consider the competition between two modes. One can describe the degree of mode competition through a coefficient C:

$$C \equiv \frac{\theta_{12} \theta_{21}}{\theta_{11} \theta_{22}} . \qquad (2)$$

If C > 1, then the competition between the modes is strong and the presence of one mode will suppress the other. If C = 1 then the competition is described as neutral: both modes can oscillate simultaneously if they have the same net gain. If one mode has a slightly larger gain, it will suppress the other. If C < 1 then competition is considered weak and one mode will suppress the other only if the difference in net gains is sufficient. In the weak-field calculation, we have found that C is equal to or less than one for the photorefractive gain medium. The saturation coefficients are proportional to the overlap between intensity --not field amplitude-- distributions within the gain medium. Thus, the less two modes are alike in their intensity distribution, the less they compete.

CHARACTERISTICS OF A DEMONSTRATED MEMORY

Unfortunately there is no way to show the video tape that was a part of this talk at the conference, but I'll summarize the demonstration. In the first part of the video the ring resonator had been programmed with two very simple objects: two pairs of diagonal dots forming a square of dots. In a second demonstration the eigenmodes were again pairs of dots, but this time there were many eigenmodes rather than just two. With no injected signal the output of the resonator would wander, apparently randomly among its eigenmodes. Eventually it would settle into one state for quite sometime before making a transition to another mode. This wandering is expected; it is a characteristic feature of lasers having gain supplied by homogeneous medium. The transitions are induced by fluctuations: we are not certain in our case what the source of the fluctuations is, but thermal fluctuations in the gain medium is a reasonable suspect. The memory was addressed using a single injected dot. In all cases this caused the output of the oscillator to collapse to a mode having a pattern which included the injected dot; hence, the recall was associative. When the input was removed, mode wandering again commenced.

We have also demonstrated a ring resonator having lithium niobate as a real-time holographic medium. We have so far been able to store only one eigenmode in this resonator. Technical problems related to thermal drift of the resonator length have prevented us from storing more than one object. The objects we were able to store, however, were more complicated than the previous dot patterns. We have, for example, stored a picture of a bicyclist[4] and a picture of the word "jila".[10]

CONCLUSION

All of the associative memories based upon optical resonators so far demonstrated in the literature have been primitive: either only one or two objects have been stored, or the objects have been rather simple. Nevertheless, these demonstrations suggest that optical resonators are an interesting avenue to explore; the exploration will be a long hike. On the one hand there exist technical and materials problems with the present devices. On the other hand, we wish to go beyond memory and consider processor architectures. This requires a much deeper understanding of neural network field models that handle continuous distributed functions and of how they might best be mimicked optically.

This work was supported in part by a grant from the National Science Foundation (PHY82-00805).

REFERENCES

1. D. Psaltis and N. Farhat, Opt. Lett **10**, 98 (1985).

2. A.D. Fisher and C.L. Giles, in Proc. IEEE Compcon, (Feb. 1985) Available from the IEEE Computer Society Press, Silver Spring, Md.

3. D.Z. Anderson, Opt. Lett. **11**, 56 (1986).

4. D.Z. Anderson, in Proc. Conference on Optical Electronics and Laser Applications in Science and Engineering, Los Angeles Ca., (Jan. 1986, SPIE, to be published).

5. M. Cohen, in Proc. Conference on Optical Electronics and Laser Applications in Science and Engineering, Los Angeles Ca., (Jan. 1986, SPIE, to be published).

6. B.H. Soffer, G.J. Dunning, Y. Owechko and E. Marom, in Proc. Conference on Optical Electronics and Laser Applications in Science and Engineering, Los Angeles Ca., (Jan. 1986, SPIE, to be published) and Opt. Lett. **11**, 118 (1986).

7. A. Yariv and S. Kwong, in Proc. Conference on Optical Electronics and Laser Applications in Science and Engineering, Los Angeles Ca., (Jan. 1986, SPIE, to be published) and Opt. Lett. **11**, 186 (1986).

8. J. Feinberg, in Optical Phase Conjugation, R. Fisher ed., Academic Press, New York, pp. 431-434 (1983).

9. D.Z. Anderson and R. Saxena, "Theory of multimode operation of a unidirectional ring resonator having photorefractive gain: weak field limit," in preparation.

10. D.Z. Anderson and M.C. Erie, Opt. Lett. Submitted.

CONCEPTS IN CONNECTIONIST MODELS

James A. Anderson
Gregory L. Murphy

Department of Psychology, Brown University, Providence, RI 02912

A concept is a mental representation or rule that picks out a class of objects or events. We will discuss two aspects of human simple concept formation: first, representation of a category by a prototype, and, second, hierarchical structure.

There are two related but distinct ways of explaining simple concepts in connectionist models. First, there are prototype forming systems which involve taking a kind of average during the act of storage. Second, there are models which explain concepts as attractors in a dynamical system. There is a third way: the construction during the learning process of highly selective elements. Then the very selective elements stand for 'concepts'. Such 'grandmother cell' forming systems do not agree with the psychological or physiological data.

The first two concept models are related formally. We will briefly describe the two approaches and the relation between them and then present a computer simulation.

Prototypes. A prototype concept model is a representation of a simple concept by a prototype or best example (Rosch, 1978). Exemplars are judged to be category members by their closeness to the prototype. For example, sparrows or robins are close to a prototypical bird and are therefore 'good' birds, but vultures or penguins are not.

Hierarchical structure. Concepts often seem to be organized in a nested fashion: robin-bird-animal-living thing or rocking chair-chair-furniture-artifact, for example.

Stimulus Coding and Representation. Our fundamental represention assumption is that information is carried by the pattern or set of activities of many neurons in a group of neurons. This set of activities carries the meaning of whatever the nervous system is doing. We represent these sets of activities as state vectors.

The Linear Associator. We use an 'outer product' associator, also called the 'linear associator', as a starting point. (Kohonen, 1984; Anderson, 1970). The two primary quantitative assumptions are: First, the neuron acts as a linear summer of its inputs. That is, the ith neuron in the second set of neurons will display activity $g(i)$ when a pattern f is presented to the first set of neurons according to the rule,

$$g(i) = \sum_j A(i,j) f(j).$$

where $A(i,j)$ are the connections between the ith neuron in the second set of neurons and the jth neuron in the first set. We can then write g as the simple matrix multiplication, $g = A\ f$. Second, we assume that matrix elements are modifiable according to a generalized Hebb rule, that is, the change in an element of A, $\delta A(i,j)$, is given by

$$\delta A(i,j) \propto f(j) \, g(i).$$

We can then write the matrix A as a sum of outer products,

$$A = \eta \sum_{i=1}^{n} g \, f^T$$

where η is a learning constant.

Prototype Formation. The linear model forms prototypes as part of the storage process. Suppose a category contains many similar items associated with the same response. Consider a set of correlated vectors, $\{f_i\}$, with mean p.

$$f_i = p + d_i .$$

The final connectivity matrix will be (if $\eta = 1$)

$$= n \, g \, (p^T + \sum_{i=1}^{n} d_i^T)$$

If the sum of the d is small, the connectivity matrix is approximated by

$$A = n \, g \, p^T .$$

The system behaves as if it had repeatedly learned only one pattern, p. In this respect the distributed memory model behaves like a psychological prototype model.

If the sum of the d's is not relatively small, the response of the model will depend on the similarities between the novel input and each of the learned patterns, that is, the system behaves like a psychological exemplar model. Knapp and Anderson (1984) applied this model to the formation of simple 'concepts' composed of nine randomly placed dots.

Error Correction and the BSB Model. A related but more powerful model takes the basic linear associator and applies two additional steps: error correction and a simple limiting non-linearity. By using an error correcting technique related to the Widrow-Hoff procedure we can force the system to give us correct associations. Suppose information is represented by associated vectors $f_1 \rightarrow g_1$, $f_2 \rightarrow g_2$ ... A vector, f, is selected at random. Then the matrix, A, is incremented according to the rule

$$\Delta A = \eta \, (g - Af) \, f^T$$

where ΔA is the change in the matrix A and where the learning coefficient, η, is chosen so as to maintain stability.

If f and g are identical, the system is called 'autoassociative'. The virtue of autoassociation is that it reconstructs missing parts of stimuli. If we combine error correction with autoassociation, we want the system to behave after learning a set of stimuli $\{f\}$ as

$$A f_i = f_i$$

The autoassociative system combined with error correction is forcing the system to develop a particular set of eigenvectors.

The averaging prototype extractor is strongly affected by error correction. Consider the autoassociative system. Suppose we are learning a number of examples as before. If the d terms are small, then $A p \propto n p$, that is, an eigenvector of A is will be close to the prototype with eigenvalue proportional to n. There is a strong dependency of the eigenvalue (a measure of the strength of response) on the number of presentations. This is sensible for small n. However, if we consider the enormous relative frequencies of occurence of words, say, this result becomes unrealistic for large n. The more frequent items would completely dominate the system.

If error correction is perfect,

$$A f_i = f_i$$

In this case all the eigenvectors correspond to members of the stimulus set, and have eigenvalue 1. The system responds to examples.

A Prototype Region. Since all the eigenvalues associated with the stimulus set are 1, the set of eigenvectors in the stimulus set are degenerate. Any sum of exemplars (which includes the average, the prototype in the previous system) is an eigenvector, forming a subspace spanned by the exemplars. The exact geometry can be involved.

Internal to the subspace, the response to exemplars is not 'peaky' but 'flat' with no effect of frequency of presentation. The 'prototype' in a psychological sense perhaps can be interpreted as a region, not as a single vector. The actual average, the prototype in the simpler system, will be toward the center of this subspace.

Magnitude (Length) of Response to a Test Vector

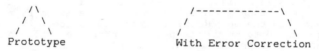

Some concept structures will be hard to learn. If a and b are regions related to particular concepts:

This geometry is related to the 'exclusive or' problem, which is known to be difficult to solve with distributed models of the type discussed here (see Minsky and Papert, 1969).

Retrieval Using a Non-linear Model. Assume we have an autoassociative matrix that has learned a set {f}. Starting information is represented by a vector constructed according to rules used to form the original vectors, except missing information is represented by zeros. The retrieval strategy will be to repeatedly pass the information through the matrix A and to reconstruct the missing information using the cross connections. The filled-in information will provide the required additional answers.

We will use the following nonlinear algorithm. Let x(i) be the current state vector of the system at quantized time (i). f(0) is the initial vector. Then, let x(i+1), the state vector at the next time quantum be given by

$$x(i+1) = LIMIT [\alpha A x(i) + \gamma x(i) + \delta f(0)].$$

The first term, (α A x(i)), passes the current state through the matrix and adds more information reconstructed from cross connections. The second term, γ x(i), causes the current state to decay slightly. The third term, δ f(0), can keep the initial information constantly present. Once the element values for x(i+1) are calculated, the element values are 'limited'. This process contains the state vector within a set of limits, and hence we have called this model the 'brain state in a box' or BSB model. After many iterations, the system becomes stable and will not change. The final state will be the output of the system, can be interpreted according to the rules used to generate the stimuli. This is a discrete approximation to an underlying continuous system. The dynamics of this system are similar to those discussed in Hopfield (1984) for continuous valued systems. It can be shown to be minimizing an energy function. In the more general autoassociative case, where the matrix is not symmetric because of limited connectivity, the system can also be shown be minimizing a quadratic energy function (Golden, in press). In this model, we are usually not concerned with the global energy minimum but with the nearest reachable local minimum. There are severe constraints on the state vector due to the geometry (the box).

General Description of Simulations. In the specific examples of state vector generation that we will use for the simulations to follow, letters, words and sets of words are coded as concatenations of the bytes in their ASCII representation. Zeros are replaced with minus ones. This is a highly arbitrary coding and is a 'distributed' coding because a single letter or word is determined by a pattern of many elements. In the outputs from the simulations the underline, '_', corresponds to to an uninterpretable character whose amplitude is below an interpretation threshold. All the BSB simulations described are 200 dimensional (25 characters). The autoassociative matrix is 50% connected, that is half the connections, randomly chosen, are set identically to zero. During learning, items to be associated are chosen randomly from the stimulus set.

Results of the Simulation. The set of stimuli used has a 'hierarchical' structure in that segments of the state vector to the right determine the segments to their left. Two separate hierarchies are stored in the matrix simultaneously.

Table 1: Stimuli in Learned Set

F[1]. VehiclesAirplaneJetPlane_   F[7]. FurniturChairs 0stuffed_
F[2]. VehiclesAirplaneX15Rockt_   F[8]. FurniturChairs Diningrm_
F[3]. VehiclesAirplanePro,pplne_   F[9]. FurniturChairs Swivelch_
F[5]. VehiclesNewcars Porsche _   F[10]. FurniturBgtablesWorkshop_
F[6]. VehiclesNewcars Mercedes_   F[11]. FurniturBgtablesKitchen _
F[6]. VehiclesNewcars Chevrolt_   F[12]. FurniturBgtablesClassrom_

This is an autoassociative system. The error correction
procedure applied to a set of exemplars, will, in the limit, have
eigenvalues with value 1. Without error correction, the eigenvalues
can be large. We would expect the eigenvalue spectrum of the largest
eigenvalues to contract toward 1 as error correction proceeds.

Table 2: Ratio of Largest to Tenth Largest Eigenvalue

Presentations	Largest e.v.	10th Largest	Ratio
12	3.830	0.331	11.57
50	1.000	0.374	2.67
100	1.000	0.555	1.80
200	1.000	0.765	1.31
400	1.000	0.928	1.08

Eigenvectors. The most frequently occuring patterns (Vehicles,
Furniture) appear in the eigenvectors. More eigenvectors become
meaningful as learning progesses. The system is gradually structuring
its eigenvectors for most accurate recall or, one might say,
developing a more useful set of vector valued 'features'
(macrofeatures). The eigenvectors are not quite the same as the
stimuli.

Table 3: Interpreted Largest Eigenvector

Number Pres.	Interpretation	Rank	Eigenvalue
12	VuxmidesBetabpg!Cmtscdod_	1	3.830
50	VehiclesAirplanePro,pplne_	1	1.000
100	FurniturBfeir~!2GM`kossm_	1	1.000
200	FurniturChairs 0stuffdd_	1	1.000

Using the Hierarchy. Often, concepts are arranged
hierarchically, and we have structured the stimulus set to show this.
Retrieval respects the hierarachy. If we put in one field,
information to the left is usually reconstructed correctly, in whole
or part, whereas information to the right (which is ambiguous) is
extremely slow to be reconstructed and sometimes incorrect. This can
be interpreted as superordinate general information (to the left) can
be retrieved, but it is often impossible to give a specific example
(to the right). If the eigenvectors representing the stimuli are
degenerate, then sums of stimuli should give about the same response
as the learned stimuli.

If 'JetPlane' is input, the recall process reconstructs
'Vehicles' after 16 iterations and 'Airplane' after 15 iterations.
The sum of 'JetPlane' and 'X15Rockt' gives 'Vehicles' after 17
iterations and 'Airplane' after 16 iterations. Suppose we want to
know what 'JetPlane' and 'Chevrolt' have in common. They are neither

'Airplanes' or 'Newcars' but both are 'Vehicles'. If we put in the average of 'JetPlane' and 'Chevrolt', the system reconstructs 'Vehicle' after 20 iterations. Apropriate use of commonalities can automatically raise or lower the level of computation in a concept hierarchy.

Table 4: Computation with a Hierarchy

Input	Result		
JetPlane	Vehicles (16)	Airplanes	(15)
JetPlane + X15Rockt	Vehicles (17)	Airplanes	(16)
JetPlane + X15Rockt + Propplne	Vehicles (16)	Airplanes	(15)
JetPlane + Chevrolt	Vehicles (20)	--	(50)
Mercedes + X15Rockt	Vehicles (23)	--	(50)
JetPlane + X15Rockt + Propplne + Porsche + Mercedes + Chevrolt	Vehicles (15)	Airplane	(33)

The number in parenthesis is the number of iterations required to completely reconstruct the characters shown. '--' means nothing successfully reconstructed.

This work is sponsored by the National Science Foundation under grants BNS-82-14728 and BNS-83-15145 administered by the Memory and Cognitive Processes section.

References

Anderson, J.A. (1970). Mathematical Biosciences., 8, 137-160.

Anderson, J.A. (1986a). In E. Bienenstock, F. Fogelmann, and G. Weisbuch (Eds.), Disordered Systems and Biological Organization. Berlin: Springer.

Anderson, J.A., Silverstein, J.W., Ritz, S.A. & Jones, R.S. (1977). Psychological Review. 84, 413-451.

Golden, R. M. (in press). Journal of Mathematical Psychology.

Hopfield, J. J. (1984). Proceedings of the National Academy of Sciences USA, 81, 3088-3092.

Knapp, A.G. & Anderson, J.A. (1984). Journal of Experimental Psychology: Learning, Memory, and Cognition., 10, 616-637.

Kohonen, T. (1984). Self Organization and Associative Memory. Berlin: Springer.

Minsky, M., & Papert, S. (1969). Perceptrons: An introduction to computational geometry. Cambridge, MA: MIT Press.

Rosch, E. (1978). In E. Rosch & B.B. Lloyd (Eds.), Cognition and Categorization., Hillsdale, NJ: Erlbaum.

COMPLEX DYNAMICS IN SIMPLE NEURAL CIRCUITS

K.L. Babcock and R.M. Westervelt
Harvard University, Cambridge, MA 02138

ABSTRACT

Simple circuits incorporating one or two electronic 'neurons' are shown to be capable of complex nonlinear behavior. Underdamped transients and instability leading to oscillation are possible when the neurons possess a finite frequency response or a delay in their transfer characteristic. Inertial terms in the dynamics allow intertwined basins of attraction and chaotic response to an external drive. These phenomena contrast with the steady state asymptotic behavior and overdamped transients exhibited by Hopfield's deterministic model with symmetric RC couplings, and indicate that care should be exercised when adding inertia to enhance performance of optimizing networks or when implementing networks electronically.

INTRODUCTION

A deterministic neural network model introduced by Hopfield[1] describes a set of electronic 'neurons' coupled through a network, T_{ij}, made up of inverters and resistors. It is well known that when the T_{ij} are symmetric the network dynamics are relatively simple and show useful associative memory and optimizing capabilities[2,3]. In that special case, the dynamics are governed by a Liapunov function E which monotonically decreases in time, and the neuron voltages approach steady states which correspond to memories, patterns, or near-optimal solutions. Furthermore, there is no oscillation in the approach to these steady states, because the equations of motion are determined as a pseudo-gradient of this function: $dU_i/dt = -\partial E/\partial V_i$, where U_i and V_i are the input and output voltages of the i_{th} neuron, respectively. However, little is known about the dynamics when the network assumes characteristics which deviate from this model.

Much of the current research on this and similar models is concerned with just such deviations. For example, alterations of the usual Hebbian 'learning rule', which determines the symmetric T_{ij} required to store a specified set of memories[2], may result in an increased memory capacity at the cost of non-symmetric or non-local T_{ij}[4]. Inertial terms in the network dynamics, associated with either the neurons or their couplings, may be useful in accelerating the global minimum-seeking of an optimizing network, or in allowing the dynamics to escape local minima in the energy function. The electronic neurons used in hardware implementations may be constructed from operational amplifiers or MOS buffers and inverters which have finite frequency response that introduces phase shifts into the dynamics. Also, effective delays in the response of an electronic neuron to changes at its input can arise due to a finite slew rate. These features, whether useful or unavoidable, prevent a dynamical description in terms of a Liapunov function[5], and the behavior may become more complex than the steady-state, overdamped transient behavior found with symmetric RC couplings and pseudo-gradient dynamics.

To determine the effects such features can have on the dynamics, we have analyzed and performed numerical simulations on simple one and two neuron systems incorporating symmetric inertial couplings, neurons with finite frequency response, or neurons with delays in their transfer characteristic. The behavior exhibited by these systems is briefly described in this paper; the reader is referred to references 6 and 7 for a more complete and detailed account. As discussed below, the nonlinear differential equations governing the input voltages of the neurons showed underdamped (ringing) transients, instability and spontaneous oscillation, intertwined basins of attraction, and chaotic response[8] to an external drive.

NEURON WITH INERTIAL FEEDBACK

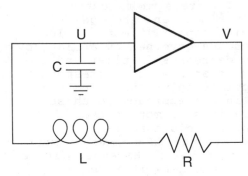

The simple neural circuit shown in Fig. 1 exhibits a variety of complex nonlinear behavior. The triangular neuron is assumed to have a transfer characteristic of the form: $V = \tanh(\beta U)$. The input voltage U is then governed by the following nonlinear differential equation:

Fig.1. Neuron with inductive feedback.

$$LC\, d^2U/dt^2 + RC\, dU/dt + U = \tanh(\beta U) \qquad (1)$$

For gains $\beta > 1$, this system is a bi-stable oscillator with two stable steady states $\pm U_s$ which satisfy $U_s = \tanh(\beta U_s)$. To each steady state there corresponds a basin of attraction consisting of the set of initial voltages and their time derivatives which evolve to that state. When the inverse quality factor $Q^{-1} = (R/2)(C/L)^{1/2}$ is sufficiently small, these basins become intertwined, as shown in Fig. 2a; note that an initial condition need not

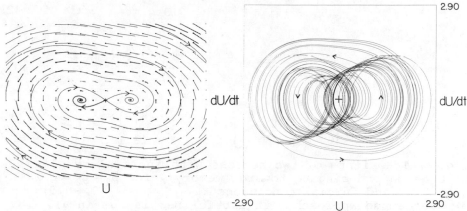

Fig. 2a. Intertwined basins
of attraction in Eq. 1, for
$Q^{-1} = 0.125$ and $\beta = 1.6$. The
grey region contains initial
conditions which evolve to
the steady state on the left.

Fig. 2b. Chaotic evolution
of Eq. 1 with added drive,
for $Q^{-1} = 0.15$ and $\beta = 1.6$.

evolve to the nearest steady state. Adding a harmonic
driving current f cos(ωt) to the neuron input results in a
chaotic response; a trajectory is shown in Fig. 2b. A very
rich set of nonlinear phenomena, including period doubling
sequences, horseshoes, and fractal basin boundaries[9,10],
underlies this behavior; see references 6 and 7.

TWO NEURONS WITH INERTIAL COUPLINGS

The circuit diagram in Fig. 3 shows a two neuron
'network' with inductors added to symmetric RC connections. The gains β_i may
take either sign to allow
for 'excitory' and 'inhibitory' couplings (the T_{ij}
are of the form $\pm 1/R$).
Two coupled differential
equations of the same form
as Eq. 1 describe the
dynamical evolution. When
the gains satisfy $\beta_1\beta_2 > 1$,
this system has two stable
steady states of the form
$\pm(U_{1s}, U_{2s})$. In the high
gain limit, these approach
$\pm(1, 1)$ and correspond to
'memories' or 'patterns'.
These steady states may

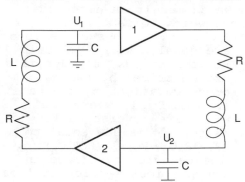

Fig. 3. Neurons with sym-
metric inertial connections.

be characterized as a function of the gains $\beta_1\beta_2$ and
the inverse of the quality factor $Q^{-1} = (R/2)(C/L)^{1/2}$.

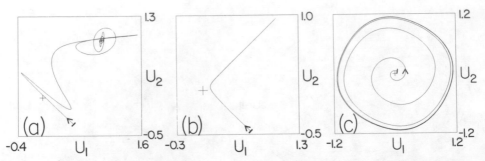

Fig. 4. Transients of two neurons with inertial couplings.
a) Underdamped ringing toward 'memory', for $\beta_1 = \beta_2 = 2.0$,
$Q^{-1} = 0.2$, and initial conditions $(U_{10}, U_{20}) = (1.0, -0.9)$.
b) Overdamped approach with little inertia, as in a),
except $Q^{-1} = 10.0$. c) Instability and oscillation for
$\beta_1 = 2.0$, $\beta_2 = -2.0$, $Q^{-1} = 0.5$, and $(U_{10}, U_{20}) = (0.1, 0.1)$.

In Ref. 7, a bifurcation diagram is constructed showing the
position in the complex plane of the eigenvalues of the
steady states versus these parameters. For low values of
Q^{-1}, the transients ring toward the memories, as shown
in Fig. 4a. This is contrasted in Fig. 4b with the
overdamped transients which result in the absence of
inertia. Another regime of interest is $\beta_1\beta_2 < 0$, where the
origin $(U_{1s}, U_{2s}) = (0,0)$ is the only fixed point. For small
values of Q^{-1}, the origin is unstable and the neuron
voltages oscillate spontaneously, as in Fig. 4c. This
system also shows intertwined basins and chaotic response
to an external drive for values of Q^{-1} less than $\cong 0.5$.

TWO NEURONS WITH FINITE FREQUENCY RESPONSE

Figure 5 shows a neuron which models an electronic
neuron with a frequency response which rolls off at
frequencies greater than $\omega \cong 1/R_nC_n$. When two such neurons
are coupled with symmetric
RC connections, the dynamics
are described by two second-
order equations of the same
form as Eq. 1, where the
effective inverse quality
factor Q^{-1} is now equal to
$(RC + R_nC_n + RC_n)/2(RCR_nC_n)^{1/2}$.
The behavior described in
the previous section applies
to this system, but with Q^{-1}
restricted to be greater

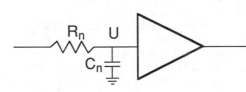

Fig. 5. Model neuron with
response that rolls off at
high frequencies.

than unity. Underdamped ringing onto the fixed points is
seen, even though the RLC couplings are separately
overdamped, and instability and oscillation are found for
$\beta_1\beta_2 < -4$. We have found no intertwined basins or driven

chaotic response, but with larger numbers of neurons such behavior might exist.

TWO NEURONS WITH DELAYED RESPONSE

Two neurons with delays in the response of the output to changes at the input are coupled with symmetric RC connections. The input voltages obey:

$$RC \, dU_1/dt + U_1 = \tanh[\beta_2 U_2(t-\tau_2)]$$

$$RC \, dU_2/dt + U_2 = \tanh[\beta_1 U_1(t-\tau_1)]$$

(2)

where the τ_i are the delays and the gains β_i may again take either sign. Equations similar to Eq. 1 also arise in models of small biological systems of interacting neurons, where the delays correspond to finite signal propagation times[11]. Initial conditions in such delay differential equations[12] are determined by specifying forms for $U_1(t)$ and $U_2(t)$ for $-\tau_i \leq t \leq 0$; we chose constant initial conditions, i.e., $U_i(t) = U_{i0}$ over these intervals, and the network was allowed to evolve freely for $t \geq 0$. Two stable steady states again exist for $\beta_1\beta_2 > 1$, and there is an oscillatory form of transient behavior for non-zero delays, sometimes coexisting with an unstable limit cycle, as shown in Fig. 6a. When $\beta_1\beta_2 < -1$ and the total delay $\tau_1 + \tau_2$ is sufficiently small, the origin $(U_1, U_2) = (0,0)$ is stable and transients collapse toward it. As the total delay passes through a critical value, the origin becomes unstable and the voltages oscillate. Near this critical value, the transients can be extremely long, as shown in

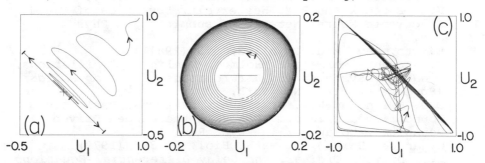

Fig. 6. Transients of Eq. 2. a) Oscillatory transient for $\tau_1 = \tau_2 = 1.5$, $\beta_1 = \beta_2 = 2.0$, $U_{10}(t) = 0.2$ and $U_{20}(t) = -0.18$ for $-\tau_i \leq t \leq 0$. An unstable limit cycle lies along $U_1 = -U_2$. b) Long approach to a limit cycle for $\tau_1 = 0.3$, $\tau_2 = 0.4$, $\beta_1 = 2.0$, $\beta_2 = -2.0$, and $U_{10}(t) = U_{20}(t) = 0.1$ for $-\tau_i \leq t \leq 0$. c) Wandering transient settling onto a limit cycle for $\tau_1 = 5.0$, $\tau_2 = 10.0$, $\beta_1 = \beta_2 = 2.0$, and initial conditions $U_{10}(t) = 2\sin[2\pi(7.5)t]$ and $U_{20}(t) = 2\cos[2\pi(7.5)t]$ for $-\tau_i \leq t \leq 0$.

Fig. 6b.   Initial conditions other than the constant forms described above can lead to more complicated transients. For example, a form for the $U_i(t)$ which varies sinusoidally over $-\tau_i \leq t \leq 0$ can lead to such extremes as the trajectory in Fig. 6c, which shows a long, wandering transient that settles onto a limit cycle.   This occurs for values of β_1 and β_2 for which stable steady states away from the origin exist in the zero-delay limit or with constant initial conditions, and shows that even the asymptotic behavior depends strongly on the magnitude of the τ_i and the form of the initial conditions.

REFERENCES

1.   J.J. Hopfield, P.N.A.S. **81**, 3088 (1984).
2.   J.J. Hopfield, P.N.A.S. **79**, 2554 (1982).
3.   D.W. Tank and J.J. Hopfield, Biol. Cyb. **52**, #3, 141 (1985).
4.   See, for example, A.S. Lapedes and R.M. Farmer, "A Self-Optimizing, Nonsymmetrical Neural Net For Content Addressable Memory and Pattern Recognition", to appear in Physica D, and articles in this volume.
5.   For the case of inertial terms with symmetric RC connections, a Liapunov function may be constructed using the neuron voltages, their time derivatives, and the coupling strengths (John Denker, private communication).   However, this does not preclude oscillatory transients.
6.   K.L. Babcock and R.M. Westervelt, "Stability and Dynamics of Simple Electronic Neural Networks With Added Inertia", to be published in Physica D.
7.   K.L. Babcock and R.M. Westervelt, "Dynamics of Simple Electronic Neural Networks", submitted to Phys. Rev. A.
8.   E. Ott, Rev. Mod. Phys. **53**, 643 (1981).
9.   J. Guckenheimer and P. Holmes, *Nonlinear Oscillations, Dynamical Systems, and Bifurcations of Vector Fields* (Springer-Verlag, New York, 1983).
10. E.G. Gwinn and R.M. Westervelt, "Fractal Basin Boundaries and Intermittency in the Damped Driven Pendulum", to appear in Phys. Rev. A.
11. U. an der Heiden, J. Math. Biol. **8**, 345 (1979).
12. R.D. Driver, *Ordinary and Delay Differential Equations* (Springer-Verlag, New York, 1977).

BIFURCATION ANALYSIS OF OSCILLATING NETWORK MODEL OF PATTERN RECOGNITION IN THE RABBIT OLFACTORY BULB

Bill Baird
University of California, Berkeley, Berkeley, CA 94720

ABSTRACT

A neural network model describing pattern recognition in the rabbit olfactory bulb is analysed to explain the changes in neural activity observed experimentally during classical Pavlovian conditioning. EEG activity recorded from an 8x8 array of 64 electrodes directly on the surface of the bulb shows distinct spatial patterns of oscillation that correspond to the animal's recognition of different conditioned odors and change with conditioning to new odors. The model may be considered a variant of Hopfield's model of continuous analog neural dynamics.[1] Excitatory and inhibitory cell types in the bulb and the anatomical architecture of their connection requires a nonsymmetric coupling matrix. As the mean input level rises during each breath of the animal, the system bifurcates from homogeneous equilibrium to a spatially patterned oscillation. The theory of multiple Hopf bifurcations is employed to find coupled equations for the amplitudes of these unstable oscillatory modes independent of frequency. This allows a view of stored periodic attractors as fixed points of a gradient vector field and thereby recovers the more familiar dynamical systems picture of associative memory.

INTRODUCTION

The olfactory bulb is the first sensory cortex of the odor recognition system and receives input directly from chemo-receptors on the nasal surface. It is generally thought that the spatial distribution of activity on these input lines is the code that distinguishes odor type. The system must learn to recognize any particular spatial pattern of receptor stimulation given by random convection of an odor in the nose as an instance of the class of spatial patterns corresponding to that odor. On each inspiration of the animal (at 3-8 Hz), the EEG of the bulb shows a large DC shift and goes from a background state of low level spatially phase locked but temporally irregular activity into a high amplitude bulk oscillation at a single frequency in the range of 40-80 Hz. Like a standing wave, this oscillation exhibits a particular spatial pattern of rms amplitude.[2]

Before any conditioning is done, a single cluster of such patterns is observed which does not serve to distinguish any odors presented. After a particular odor is paired with water in a classical conditioning paradigm, statistically distinct clusters of patterns emerge that correlate with the animal's recognition behavior. When a conditioning odor is changed, new clusters appear that again suffice for statistically significant prediction of the animal's responses.[3] Such observations contradict the notion that the bulb is

a passive contrast enhancing preprocessor. On the hypothesis that
the neural activity generating these EEG patterns serves to classify
stimuli for the animal as the patterns themselves do for us, we have
undertaken a theoretical characterization of the system as an associative memory. This is the first stage of processing in the rabbit,
as the variance of patterns within clusters may indicate, and later
stages such as prepyriform cortex and even motor cortex may be expected to further refine the categorization. It is the mapping from
stimulus environment to response that the animal learns.

Both the bifurcation to form an ordered pattern from the homogeneous background state and the use of periodic attractors are
unusual features for a pattern recognition system that are forced on
us by this natural design. Stephen Grossberg[4] has also discussed the
olfactory system and suggested that standing waves of oscillation in
the bulb and prepryiform cortex could encode the spatial patterns
required in his theory of pattern recognition by "adaptive resonance"
between cortical levels. We consider here how a single level of
cortex with its own crosscorrelating feedback might use oscillation
for pattern recognition. Using the tools of multiple bifurcation
theory, I will present a candidate mathematical portrait of the
olfactory bulb in which its observed behavior can be clearly visual-
ized. The impact on the function of pattern recognition of dynamical possibilities peculiar to the periodic dynamics, such as chaos
and resonance, can be examined as well. Oscillation is ubiquitous in
biology and may underly new principles of neural computation (for
full details, see [5]).

MATHEMATICAL MODEL

The EEG is generated largely by the average voltage level in
the inhibitory granule cells whose geometry produces a vertical
dipole field of extracellular current visible at the surface. We
employ spatial and temporal filtering and deconvolution of the
effects of volume conduction to optimally resolve this component of
the EEG. This gives us a readout of the activity of one element of
the neural medium from which the behavior of the rest is inferred
via the model.[2] The EEG itself is only an "epiphenomenon" or indi-
cator of underlying activity. However, since an individual neuron
in the bulb gives at most a few spikes output to the rest of the
brain during a sniff that is sufficient for the rabbit to respond,
and since a seemingly irreducible stochasticity is introduced by
synaptic events, it can be argued that reliable computation must
occur as an average over some ensemble of neurons. The macroscopic
patterns revealed by the EEG in the densities of neural activity
may in fact be the information bearing degrees of freedom in the
system. There are many parallels here to the formation of ordered
macrostates of concentration in reaction diffusion systems, convection in fluid mechanics, or populations in ecology, which we can
exploit for mathematical inspiration.

Receptor input appears to be spatially coarse grained by its
distribution in clumps (called glomeruli) to a population of excitatory neurons connected to inhibitory neurons that forms the olfac-

tory version of a cortical column. This is the natural ensemble or local unit of tissue to model since we find it to be the minimum element of space over which the average density of activity is constant and between which it varies. There are estimated to be over 2000 of these in the blub. The standard equation of neural dynamics can be used to describe this system. Here we consider ensembles instead of single neurons and take the average cell voltage v of a local excitatory or inhibitory subpopulation to be the state variable. As Grossberg[4] emphasizes, a simple Gaussian distribution of thresholds in the population gives the required sigmoid function of mean cell voltage input v to pulse density output $g(v)$. In vector notation (with average membrane time constant = 1),

$$v = -v + T \underline{g}(\underline{v}) + \underline{b} , \quad \text{where } T = \begin{bmatrix} A & & (+) \\ & A & \\ & & \ddots \\ (+) & & A \end{bmatrix} \quad (1)$$

$$\text{and } A = \begin{bmatrix} a_{11} & -a_{12} \\ a_{21} & -a_{22} \end{bmatrix}$$

The particular structure of our system appears in the coupling matrix T. The constant local coupling within columns is indicated by the block matrix A along the diagonal. The signs of the coupling are always positive for excitatory and negative for inhibitory populations. This leads to the essential assymmetry of the matrix and sets the stage for oscillation. A single column by itself is assumed to have internal coupling such that it will exhibit a Hopf bifurcation to oscillation as the mean input level b, considered to be a bifurcation parameter, rises to a critical value.[5] This models the input driven DC shifts in the EEG during respiration and the creation and termination of the burst. The bifurcation in and out of oscillation accomplishes memory reset so that the animal can resample its environment (since the tissue must inverse bifurcate to escape from the last basin of attraction). Respiration thus supplies the natural sampling operation for olfaction and controls the bifurcation as well by the mechanism of input bias variation. Most of the bias may in fact be generated by inspiration sensitive mechanoreceptors designed for this purpose.

Evoked potential studies show long range excitatory coupling in the bulb, indicated here by the off diagonal (+) coupling in the T matrix. Freeman[2] has evidence that axon collaterals of the output neurons, which accomplish this feedback cross coupling of columns, modify with learning as would be expected in an associative memory. Our conditioning results suggest that modification occurs only during motivationally significant sensory events. To see how this system might operate as an associative memory, we consider the system linearized at the bifurcation point. The spatially uniform bias is subtracted out and the new perturbation variables of the linearized equations are variations in space about this activity level. The spatial pattern or "information content" of input is now an initial condition for this system and it evokes a linear combination of eigenvectors (modes) from the feedback matrix T as in Anderson's view of dynamical associative memory.[6] In our case the eigenvectors are complex and correspond to competing unstable oscillatory modes.

The activity level in the network rises as eigenvectors grow from positive feedback until nonlinear terms become significant. Components then begin to saturate and the system goes to a final attracting state at some nearby maximum of a "saturation landscape." For T matrix couplings that give rise to a relatively flat eigenvalue spectrum, input gives a lead to related modes such that the most similar eigenvector wins the growth competition. The state space is thus partitioned into many basins of attraction that classify input.

MULTIPLE BIFURCATIONS

Alternatively we can look at pattern recognition from the perspective of the bifurcation which is so prominent in our system. We need to visualize a multiple bifurcation and a simple mechanical analogy may be of assistance. Consider a thin symmetric column subject to a compression load that represents our spatially uniform input bias. At some pressure corresponding to a bifurcation point the column will buckle to the side. It can do so in any direction in 360° around the original vertical axis. Some slight asymmetry in the load, representing the spatial pattern of input above, will determine this direction. Thus the column bifurcation can be viewed as an input asymmetry detector that amplifies small deviations into large macroscopic ones. Learning might be represented as a serrated collar around the column which quantizes the possible buckling directions and acts as the saturation landscape that guides the buckling to some final position or "attractor." All input asymmetries in a range of directions are now classified to the same final state.

The symmetry of an originally circular collar is broken arbitrarily by successive dents of learning experiences that form the serrations. The collar here is only a visual aid for what could be broken symmetries in the equations for the structure of the column itself. In either the structure of the column or the coupling matrix of the network above, learning establishes the pre-determined modes in which the symmetry of the system can be broken by input. We are currently investigating the application of symmetry groups and singularity theory to characterize the properties of coupling matrices and the attractors and pattern classes they create in multiple bifurcations. It may be that neural network models can usefully be viewed as operating as a bias level beyond an implicit bifurcation point where multiple basins of attraction are already established.

It is specifically the spatial pattern of amplitude and not the patterns of phase or the detailed temporal evolution of the EEG oscillation that we have found to be correlated with pattern recognition in the bulb. It may be that amplitude is the only degree of freedom that an oscillating system can use for pattern recognition. Necessarily or fortuitously, the theory of multiple Hopf bifurcations may be employed to find equations for the amplitudes of oscillatory modes independent of their phase when the Jacobian at the bifurcation point has all purely imaginary eigenvalues with nonresonant frequencies (not integral multiples <5).[7] This allows us to recover a view of our periodic attractors as fixed points of these new equations in a new vector field. This is the vector field in our system that must be examined for its pattern recognition character. A four dimensional system, to illustrate, is first expanded in a Taylor series about the bifurcation point. The Jacobian is then diagonalized to separate modes and find the center manifold. A further transformation yields the "normal form" of amplitude equations shown here and two equations for frequency which we neglect.[5]

$$\dot{r}_1 = u_1 r_1 + a_{11} r_1^3 + a_{12} r_1 r_2^2$$

$$\dot{r}_2 = u_2 r_2 + a_{22} r_2^3 + a_{21} r_2 r_1^2$$

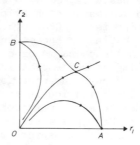

Fig. 1. Amplitude vector field

Because this analysis is "local" (i.e., restricted to the neighborhood of the bifurcation and the low order terms of the Taylor expansion of the sibmoid nonlinearity), it gives us a mechanism for pattern selection that is consistent with the near sinusoidal oscillations we observe. For pattern recognition, we need T matrix couplings that not only create the multiple Hopf bifurcation, but also give negative coefficients a_{ij} in the nonlinear terms. In ecology these are then called "competitive" equations and our modes could be rabbits and foxes. The amplitude variables are now, after diagonalization, "modes" or linear combinations of the original space variables which correspond to the "spatial patterns" of oscillation we see in the animal. The u_i are unfolding parameters representing the real parts of the eigenvalues of the Jacobian in the vicinity of the bifurcation. They give the exponential growth rates of mode amplitudes, and the largest dominates in a purely linear picture.

The nonlinear terms modify this story--the cubic terms describe "selfdamping" and the cross terms "intermode suppression." This terminology from nonlinear stability theory[8] is useful to describe the saturation landscape for natural selection of these patterns of activity in competition for the scarce resources of the neural tissue. When intermode suppression is strong relative to self damping, we get the kind of balanced competition which leads to multiple basins of attraction. This type of gradient vectorfield is shown in figure 1, where A and B are stable, interior point C is a saddle, and the orbit from the origin to C forms a separatrix.

PORTRAIT OF EXPERIMENTAL OBSERVATIONS

The recognition behavior of the animal is pictured here by assuming that the basin of A represents all spatial variations of receptor stimulation in the nose due to the background odors of the experiment box. On each inspiration, the symmetry of the bifurcation is broken by one of these patterns and the system state lies in the basin of A after the bifurcation. The system then relaxes to attractor A which we observe as one of the patterns of the "waiting" cluster and the animal gives no response. Since the other modes are of zero amplitude at A, we see only a single frequency. When the odor previously paired with water is introduced, the system is placed in the basin of B by the new input and we observe a shift to this pattern in the EEG. This new output of the bulb falls within different basins in succeeding associative structures (also modified by learning) such as motor cortex, and the rabbit gives an anticipatory licking response that indicates recognition of the odor. Given the range of patterns that form the clusters we observe, it is likely that classification in the bulb is preliminary and a single odor corresponds to many attractors which are further classified by later structures. We are investigating clusters which reorganize when seemingly unrelated aspects of the conditioning are changed (such as the introduction of a new unpaired control odor) to see if some subset of patterns remains invariant.

When odors are released, we often observe "disorderly" bursts with broad spectral content whose spatial patterns vary too much to form a distinct cluster.[3] Such bursts could occur in this picture (viewed as a submanifold of a larger space) when self damping exceeds intermode suppression. Then only the interior point C is stable, which implies coexisting spatial patterns that beat against each other in a quasiperiodic oscillation. The system has failed to decide between modes and in fact the animal most often sniffs to get a better sample.[3] Should A and B be stable, with the interior containing a chaotic flow, a fractal separatrix could arise as it does in the Lorenz equations. Here the boundary between basins A and B is so convoluted that initial conditions in the region go to A or B with an uncertainty given by the fractal dimension of the separatrix. In this fasion the system could choose to make probabilistic decisions in some areas of the state space.

Supported by grant MH06686 from the National Institute of Mental Health. Support of Walter Freeman is gratefully acknowledged.

REFERENCES

[1] J. J. Hopfield, Proc. Natl. Acad. Sci. USA 81, 3088 (1984).
[2] W. J. Freeman, Mass Action in the Nervous System (New York: Academic Press, 1975).
[3] K. A. Grajski, this volume (1986).
[4] S. Grossberg, Biol. Cyber. 23, 187 (1976).
[5] B. Baird, Physica D, Conference on Evolution, Games, and Learning, to be published (1986).
[6] J. Anderson, Hinton (eds.), Parallel Models of Associative Memory (New Jersey, LEA, 1981).
[7] J. Guckenheimer, D. Holmes, Nonlinear Oscillations, Dynamical Systems, and Bifurcations of Vector Fields (New York: Springer, 1983).
[8] L. A. Segel, J. Fluid Mech., 21, 359 (1965).

Caging and Exhibiting Ultrametric Structures

Pierre Baldi
Eric B Baum
Caltech,Pasadena Ca 91125

Abstract: We review ultrametricity and state several theorems restricting and exhibiting ultrametric structures. Applications to spin glasses, optimization problems, and neural nets are discussed.

Ultrametricity is a simple topological concept[1]. It appeared first in number theory in the study of p-adic fields and has since been used in taxonomy. Recent advances in theoretical physics seem to indicate that ultrametric structures are of more general interest and occur in a variety of contexts including spin glasses, optimization problems, and neural networks. We will review ultrametricity and state bounds on and constructions of ultrametric structures, as well as applications of these results.

Definition:A *metric space* is a set X with a distance function d such that, for any x,y,z in X:

(1) $d(x,x)=0$
(2) $d(x,y)=d(y,x)$
(3) $d(x,z) \leq d(x,y)+d(y,z)$

An *ultrametric space* is a metric space (X,d) which in addition satisfies:

(4) $d(x,z) \leq \max(d(x,y),d(y,z))$.

This is equivalent to the statement that any triangle in X is *isosceles with the third side shorter than or equal to the other two*. This concept is useful because ultrametric spaces correspond to hierarchic structures. Any ultrametric space can be represented by a tree of constant height where the elements of the space are the nodes at the bottom of the tree and the distance between any two points is the height one must ascend up the tree in order to reach a common predecessor.

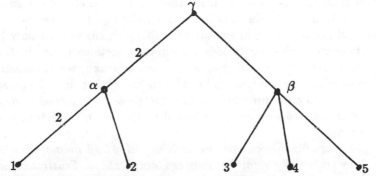

Figure 1: A tree of constant height $= 4, d(1,2)=2, d(1,3)=4, d(3,5)=2$

To see that every ultrametric space corresponds to a tree we can apply the triangle relationship twice[1]. We first find the unusual property that any two points inside a ball of radius r are themselves no further apart than r. Let a and b be any two points in the ball of radius r about c, $B_r(c)$; we have

$$d(a, b) \leq max(d(a, c), d(b, c)) \leq r \tag{1}$$

We then see that two balls of radius r are either disjoint or identical. Indeed let $B_r(c)$ and $B_r(e)$ have point b in common. Then any point $a \in B_r(e)$ is also contained in $B_r(c)$ as:

$$d(c, a) \leq max(d(c, b), d(b, a)) \leq r \tag{2}$$

Therefore any finite ultrametric space is contained in one ball of radius r_1 (the longest distance), which is comprised of two or more disjoint balls of radius r_2, which themselves split into balls of radius r_3,.... This gives the tree structure.

Some Examples of Ultrametric Structures

(1) Words of fixed length on an alphabet give an example of a set with a natural tree structure.

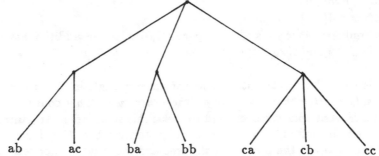

Figure 2: A language of 2 letter words on the alphabet $\{a, b, c\}$

(2) We know from the theory of evolution that the natural classification of living species is a family tree. The distance between two living species can be measured by the "age" of the closest common ancestor. Note this is not the only possible distance between "species" or "words". Hamming distances, e.g. in the space of genes, can be used. These will lead only to an approximate ultrametricity.

(3) More generally, any neutral evolution process acting on a large space of features gives rise to an approximately ultrametric set. Conversely every ultrametric structure can be seen as an unbiased realization of a certain stochastic branching process.[1,2]

(4) Spin glasses are disordered systems such as a Hopfield memory pushed past capacity. They are usually studied at nonzero temperature. *Frustration* occurs in that there is no configuration of spins (neurons) which satisfies all the bonds, i.e.

puts each bond in its lowest energy state. Frustration creates a huge multiplicity of metastable states and many *thermodynamic equilibrium states* - deep valleys in the free energy landscape which are stable in the limit $N \to \infty$, for N the number of sites. These thermodynamic equilibrium states have been shown to be ultrametric under euclidean distance in configuration space[3].

(5) Combinatorial optimization problems such as the travelling salesman problem (TSP)[4], graph partitioning[5], and graph coloring[6] have been studied as spin glasses. These problems are hard, at least in part, because frustration creates a multiplicity of local minima. Simulated annealing, an algorithm where one adds a temperature to the problem and slowly lowers it, has been suggested by the analogy to spin glasses. Recent evidence indicates that many of these problems may have an underlying ultrametricity.

Theorems on Ultrametric Structures

Let (E, d_E) be a given metric space and let (X, d) be an ultrametric structure. Two questions naturally arise.

(1) Can we find a 'copy' of X in E?

(2) What is the largest possible X that we can fit in E?

Question (1) is relevant to our ability to implement a given tree in a given system. This may perhaps be important for hierarchic memory models. Question (2) will give constraints on systems known to be ultrametric. For example, it is easy to find 3 points in the plane which are ultrametric, but given 4 points the constraint that each of the 4 possible triangles must be ultrametric implies they can not lie in a plane. Interesting cases for E are:

(a) subsets (A_i) $i = 1, ... k$ of an N-element set with the metric:

$$d(A_i, A_j) = max(|A_i|, |A_j|) - |A_i \cap A_j| \tag{3}$$

(b) N-dimensional hypercube, ie (0,1) or (1,−1) vectors of length N with Hamming distance.

(c) $\Re^N$ with euclidean distance.

Theorem: Let (E, d_E) be one of the metric spaces (a),(b), or (c). If (X, d) is an ultrametric structure that can be embedded in E then:

$$|X| \leq N + 1 \tag{4}$$

Moreover this bound can be attained.

We will sketch a proof in case (b). The other cases are similar[7].

Consider a maximal set $X_1, ..., X_k$ of (1,-1) vectors in $\Re^N$ such that Hamming distance d_H is ultrametric. Form the $k \times N$ matrix having X_j for its j-th row. Let B be the $k \times k$ matrix $B = AA^t = (b_{ij})$. Obviously $\text{rank}(B) \leq N$. Moreover $b_{ij} = N - 2d_H(X_i, X_j)$. A careful study of B shows that $\text{rank}(B) \geq k - 1$. Therefore $k \leq N + 1$.

The upper bound is attained by the N vectors having one (-1) and $N - 1$ (1) entries together with the vector having all (-1) entries.

This theorem completely answers the second question, at least for (a), (b), and (c). In some of the applications one will find that most but not all of the triangles are of ultrametric form. We define a structure to be *q-almost ultrametric* if the ratio of triangles violating ultrametricity to triangles obeying it is less than N^{-q}. Then we have the

Theorem: A q-almost ultrametric set in cases (a),(b), or (c) has fewer than $(\frac{3\sqrt{3}}{2}N)^{\frac{2}{q}}$ elements.

This is obtained[7] by first applying a result of Spencer[8] to prove the existence of an exactly ultrametric subset Y of size at least $(2/(3\sqrt{3}))k^{q/2}+1$. The previous theorem then yields the result.

For the first question, it can be shown that under "reasonable" assumptions every ultrametric space (X,d) can be realised in (E,d), for E as in (a),(b), or (c) and N large enough[7]. In the euclidean case the embedding is *always* possible and if we assume $|X| = m$ then we can take $N = m - 1$. We will give here an example of how to map an ultrametric structure(fig 1) onto a hypercube with Hamming distance. We start by associating to each leaf of the tree the 1-vector (1). As we work our way up, at each branching, we make the vectors longer in such a way as to preserve ultrametricity. At intersection (α) we associate 2-vectors (1,0), and (0,1) to leaves 1 and 2 respectively. At intersection (β), for leaves 3,4,5, the 3-vectors (1,0,0),(0,1,0),(0,0,1) are equidistant in Hamming distance. Notice that at each level of the tree we require all of our vectors to have the same weight, ie. number of 1's. This allows us to keep extending the process. At intersection (γ) we must amalgamate these two sets of vectors. We add three 0's to the end of vectors 1, and 2 and two 0's to the beginning of vectors 3,4,5. This gives us the set of 5-vectors: (1,0,0,0,0),(0,1,0,0,0);(0,0,1,0,0),(0,0,0,1,0), (0,0,0,0,1) which are equidistant. To increase the distance of the first two from the last three, while keeping equidistant the first set and the last set, and keeping all vectors of equal weight, we add to the end of the first two 1,0; and to the end of the last three 0,1, to get:(1,0,0,0,0,1,0),(0,1,0,0,0,1,0);(0,0,1,0,0,0,1) (0,0,0,1,0,0,1),(0,0,0,0,1,0,1). This embeds the tree in the 7 dimensional hypercube. One can show that this construction can be continued for all trees, except certain classes which can not be embedded in a hypercube[7].

Applications

(1) *Spin Glasses:*

Mezard et al[3] showed that if we choose three equilibrium states in a spin glass, in the $N \to \infty$ limit, and compute the euclidean distances between them the probability is unity that the triangle is ultrametric. This motivated our definition of 'q-almost ultrametric'. If the ratio of the number of triangles violating ultrametricity to those satisfying it is bounded by N^{-q}, we have shown a polynomial bound on the number of equilibrium states[9]. This implies that the zero temperature entropy is zero.

One should note, however that finite N effects may render all triangles non-ultrametric. It may be instead that almost all the triangles are approximately ultrametric. It is clear that the metastable states, known to be exponentially numerous[10], are at best quasi ultrametric in this sense. They may be approximately ultrametric because they are clustered near the bottoms of the deep valleys (equilibrium states). Their deviations from ultrametricity then reflect both the width of the deep valleys and any deviations of the equilibrium states from ultrametricity.[9]

(2) *Optimization problems:*

The travelling salesman problem (TSP) is to find the shortest tour visiting each of n cities.[11] A tour is said to be λ-opt if it can not be improved by replacing λ links.[12] There is evidence that the 2-opt and 3-opt tours are approximately ultrametric under metric (a), ie:

$$D(tour_1, tour_2) = n - (\#links\ in\ common) \tag{5}$$

Our theorems bound the size of any ultrametric or q-almost ultrametric set of tours. Polynomial bounds on the number of 2-opt tours would be surprising[11] and could lead to effective heuristics[11,13]. More probably 2-opt and 3-opt tours are analogous to metastable states in the spin glass, and are only quasi-ultrametric, this quasi-ultrametricity perhaps due to their clustering around underlying ultrametric "equilibrium states".

(3) *Hierarchical memories:*

Parga and Virasoro[2] have noted that (i) a hierarchy of errors is important for biological memories, and (ii) we seem to categorize before memorizing. Since the Hopfield memory is in some respects like a spin glass, they suggested using the natural ultrametricity of spin glasses to obtain hierarchical storage. We have shown how to embed any tree on the hypercube. This could possibly be useful for hierarchic encodings. Our constructions show there is no strong limitation on capacity arising strictly from the constraint of ultrametricity, since even richly branching trees with N leaves can be embedded in hypercubes of reasonable dimension, ie dimension of order $O(N)$[7].

EBB was supported in part by DARPA through arrangement with NASA

References

(1)R.Rammal,G.Toulouse,M.A.Virasoro, "Ultrametricity for Physicists", to appear Rev. Mod. Phys.(1986)

(2)N.Parga,M.A.Virasoro,"The Ultrametric Organization of Memories in a Neural Network", Trieste preprint (1985) and M.A.Virasoro, in "Disordered Systems and Biological Organization", ed. E. Bienenstock,F. Fogelman, G. Weisbuch Springer,Berlin

(3)M.Mezard,G.Parisi,N.Sourlas,G.Toulouse,M.A.Virasoro, Phys. Rev. Lett. 52(1984)p1156 and J. de Physique 45(1984)p843

(4)S.Kirkpatrick,G.Toulouse,J. de Physique 46(1985)1277

(5)Y.Fu,P.W.Anderson,"Application of Statistical Mechanics to NP Complete Problems in Combinatorial Optimization", Princeton preprint (1985)

(6)J.P.Bouchal,P.le Doussal,"Ultrametricity Transition in the Graph Coloring Problem",1985 preprint

(7)M.Aschbacher,P.Baldi,E.B.Baum,R.M.Wilson, "Embeddings of Ultrametric Spaces in Finite Dimensional Structures", submitted to SIAM J. Disc. Alg. Meth. (1986)

(8)J. Spencer, Disc. Math. 2(1972)183

(9)P.Baldi,E.B.Baum,Phys. Rev. Lett. April 1986

(10)R.J.McEliece,E.C.Posner,"The Number of Stable Points of an Infinite Range Spin Glass",Caltech Preprint (1985)

(11)E.B.Baum,"Towards Practical Neural Computation for Combinatorial Optimization Problems",this volume

(12)S.Lin,Bell Sys. Tech. J. 44(1965)p2245

(13)E.B.Baum,"Iterated Descent: A Better Algorithm for Local Search in Combinatorial Optimization Problems", submitted to Oper. Res.(1986)

GAME-THEORETIC COOPERATIVITY
IN NETWORKS OF SELF-INTERESTED UNITS

Andrew. G. Barto[1]
Department of Computer and Information Science
University of Massachusetts, Amherst MA 01003

ABSTRACT

The behavior of theoretical neural networks is often described in terms of competition and cooperation. I present an approach to network learning that is related to game and team problems in which competition and cooperation have more technical meanings. I briefly describe the application of stochastic learning automata to game and team problems and then present an adaptive element that is a synthesis of aspects of stochastic learning automata and typical neuron-like adaptive elements. These elements act as self-interested agents that work toward improving their performance with respect to their individual preference orderings. Networks of these elements can solve a variety of team decision problems, some of which take the form of layered networks in which the "hidden units" become appropriate functional components as they attempt to improve their own payoffs.

INTRODUCTION

The behavior of neural networks is often described in terms of competition, cooperation, and coalition formation [1,2,3,4]. Because these terms belong to the technical lexicon of game theory and economics, one is prompted to ask if they mean the same thing when used by network theorists? Certainly, as applied to networks, their ordinary meanings are preserved, but what of their more technical intent? Much of game theory and economics is based on the ideas of individual utility functions and utility maximization by self-interested agents (the assumption of individual rationality). These fields consider questions such as what social rationality should mean and how socially "optimal" behavior or resource allocation might be produced by inter-acting self-interested agents, where optimality is a far more complicated concept than it is in many other contexts. Competition arises due to conflicts of agent self-interest as embodied in the payoff structure of a game, and cooperation, that is, the formation of coalitions within which agents coordinate their activity, provides a means for agents to take advantage of overlapping interests.

To some extent, these basic ideas of game theory and economics have been faithfully interpreted in context of theoretical neural networks. For example, learning methods in which total synaptic strength is conserved (e.g., as in "competitive learning" [4]) might be regarded as including a resource allocation mechanism for a non-production economy, where the resource is the sum of the connection weights. Similarly, it may not be misleading to regard a unit in a Boltzmann machine [3] or a Hopfield network [5] as preferring to participate in low energy configurations with its neighbors.

Missing from these cases, however, are versatile agents that will attempt to make the best (according to their own interests) of whatever situation they may find themselves in. This requires agents whose decision strategies are not task-specific. In this paper I outline some aspects of our study of the collective behavior of self-interested neuron-like elements, an approach first suggested to us by the "hedonistic neuron" hypothesis of Klopf [6]. I begin by describing a type of adaptive agent that has been extensively studied in game and team decision problems.

---

[1]This research was supported by the Air Force Office of Scientific Research and the Avionics Laboratory (Air Force Wright Aeronautical Laboratories) through contract F33615-83-C-1078.

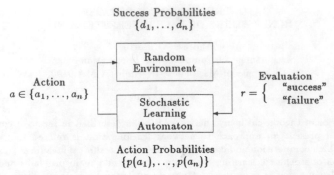

Success Probabilities
$\{d_1, \ldots, d_n\}$

Random Environment

Action
$a \in \{a_1, \ldots, a_n\}$

Stochastic Learning Automaton

Evaluation
$r = \left\{ \begin{array}{l} \text{"success"} \\ \text{"failure"} \end{array} \right.$

Action Probabilities
$\{p(a_1), \ldots, p(a_n)\}$

Figure 1. Stochastic learning automaton interacting with a random environment.

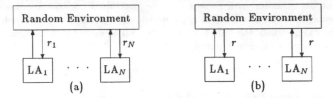

Figure 2. (a) The game problem. (b) The team problem.

STOCHASTIC LEARNING AUTOMATA

The theory of learning automata originated with the independent work of the Soviet cybernetician Tsetlin [7], mathematical psychologists studying learning, and statisticians studying sequential decision problems (e.g., the "n-armed bandit problem" [8]). Although this theory has an extensive modern literature in engineering (reviewed in ref. 9), it is unfortunate that there has been very little cross-fertilization between this theory and neural-network research.

Fig. 1 shows a learning automaton interacting with an environment. At each step in the processing cycle, the automaton randomly picks an action from a set of possible actions, $A = \{a_1, \ldots, a_n\}$, according to a vector of action probabilities, $P = \{p(a_1), \ldots, p(a_n)\}$. The environment then evaluates that action by selecting an evaluation signal that it transmits back to the automaton. Fig. 1 shows the case in which the evaluation, r, is either "success" or "failure" and is selected according to probabilities $\{d_1, \ldots, d_n\}$, where $d_i = \text{prob}\{\text{success}|a_i\}$ (other formulations allow a countable number or a bounded continuum of evaluations). Upon receiving the evaluation, the automaton updates its action probabilities as a function of its current action probabilities, the action chosen, and the environment's evaluation of that action. Beginning with no knowledge of the environmental success probabilities, the objective of the automaton is to improve its expectation of success over time. Ideally, it should eventually choose action a_l with probability 1, where $d_l = \max\{d_1, \ldots, d_n\}$. Many different algorithms have been studied under a number of different performance measures, and many convergence results have been proven [9].

Theorists have become increasingly interested in the collective behavior of learning automata. Fig. 2 shows collections of N learning automata interacting with an environment. In Fig. 2a, each automaton recieves a different evaluation signal that depends, in general, on the actions of all N automata. This models the situation in which the automata have differing, and possibly conflicting, interests. This a game decision problem. In contrast to the problems studied in classical game theory, the automata operate in total ignorance of the payoff structure of the game and the presence of the other automata. In the case of zero-sum games (games of pure conflict), theoretical results show that when employing certain algorithms, the learning automata converge to the game's solution (by finding the appropriate mixed, or probabilistic, strategies if necessary) [9].

Fig. 2b shows a collection of learning automata in the team situation, which is the special case of the game situation in which the automata receive the same evaluation signal. In this case, the automata have a common goal but each automaton only has partial control over the evaluation. As in the case of games, the learning process in this case is incompletely understood, but a number of mathematical results have been proven, the strongest of which shows that certain stochastic learning automaton algorithms lead to monotonic increases in performance [10].

COMPARISON OF LEARNING AUTOMATA AND NETWORK ELEMENTS

Comparing stochastic learning automata and the typical adaptive elements used in theoretical neural-network research reveals several important differences. First, a typical neuron-like adaptive element has multiple input pathways that carry patterned stimulus information. Such an element might also have a pathway specialized for training, such as the pathway for the desired response of a Widrow/Hoff Adaline [11] or a Perceptron element [12]. The learning process causes the element to implement or approximate a desired mapping from stimulus patterns to responses. A learning automaton, on the other hand, only has a single input pathway for the evaluation signal. Learning either results in the selection of a single optimal action or a suitable action probability vector—no (nontrivial) mapping is produced. On this dimension of comparison, then, the usual adaptive elements are doing something more sophisticated than are learning automata.

However, the usual adaptive element requires an environment that directly provides either a desired response or a signed error that directly tells the element what response it should have produced. In contrast, a learning automaton has to discover, in a stochastic environment, which action is best by sequentially producing actions and observing the results. Since there are no constraints on the success probabilities, information gained from performing one action provides no information about the consequences of the other actions. This can be a nontrivial problem even in the case of two possible actions and is fundamentally different from the supervised learning problem [13,14,15]. Therefore, in terms of the amount of information required for successful learning, a stochastic learning automaton implements a form of learning more powerful than the supervised learning performed by most neuron-like adaptive elements.

Because typical network adaptive elements and learning automata excel on different dimensions, it has been fruitful to study learning elements that combine the capabilities of these two types of systems. The resulting elements are able to learn mappings in the absence of explicit instructional information. In the next section, I describe one algorithm that has resulted from our study of this class of learning elements. Then I discuss the implications of this class of hybrid algorithms for game and team decision problems.

THE ASSOCIATIVE REWARD-PENALTY ELEMENT

The Associative Reward-Penalty element [14], or A_{R-P} element, combines aspects of familiar neuron-like adaptive elements with properties of stochastic learning automata. It is a refinement of similar elements that my colleagues and I have studied [13,16,17,18]. An A_{R-P} element has input pathways $x_1, \ldots, x_n$, which carry pattern input to the element. Each input pathway x_i has an associated weight w_i. An additional input pathway is specialized for delivering environmental evaluation to the element. We call this pathway the reinforcement pathway, r. There is a single output pathway for the element's action, a. Let $\vec{x}(t)$, $\vec{w}(t)$, $r(t)$, and $a(t)$ respectively denote inout pattern vector, weight vector, reinforcement signal, and action at time t. The action is determined by comparing the inner product of the pattern and weight vectors with a randomly varying threshold:

$$a(t) = \begin{cases} 1, & \text{if } \vec{w}(t) \cdot \vec{x}(t) + \eta(t) > 0; \\ 0, & \text{otherwise;} \end{cases} \tag{1}$$

where the $\eta(t)$ are independent identically distributed random variables, each having distribution function Ψ, which is a known and fixed characteristic of the element. Let $p(t)$ denote prob$\{a(t) = 1|\vec{x}(t)\}$. Then

$$p(t) = \text{prob}\{\vec{w}(t) \cdot \vec{x}(t) + \eta(t) > 0\} = 1 - \Psi(-\vec{w}(t) \cdot \vec{x}(t)).$$

Thus, the action probabilities depend on the element's input in a manner parameterized by the weights and the distribution function Ψ. This input-to-probability mapping is adjusted by updating the weight vector according to the following equation:

$$\vec{w}(t+1) - \vec{w}(t) = \begin{cases} \rho[a(t) - p(t)]\vec{x}(t), & \text{if } r(t) = \text{reward}; \\ \lambda\rho[1 - a(t) - p(t)]\vec{x}(t), & \text{if } r(t) = \text{penalty}; \end{cases} \tag{2}$$

where $0 \leq \lambda \leq 1$ and $\rho > 0$.

As $|\vec{w}(t) \cdot \vec{x}(t)|$ increases for all $\vec{x}(t)$, the mapping (1) approaches a deterministic linear discriminate function. According to (2), $\vec{w}(t)$ changes in such a way that in the case of reward, the element is more likely to produce the same action, $a(t)$, when patterns similar to $\vec{x}(t)$ occur in the future; in the case of penalty, $\vec{w}(t)$ changes in such a way that the element is more likely to produce the other action, $1 - a(t)$, when patterns similar to $\vec{x}(t)$ occur in the future. Therefore, the A_{R-P} element is an associative learning automaton capable of learning a mapping rather than just a single optimal action. It reduces to a (nonassociative) stochastic learning automaton algorithm when the input pattern is constant and nonzero over time. When it is deterministic ($\eta(t) = 0$ for all t), it becomes the Perceptron algorithm if one treats the terms $a(t)$ and $1 - a(t)$ as the training input giving the desired response. It is most closely related to the "selective bootstrap adaptation" algorithm of Widrow, Gupta, and Maitra [10].

If the input vectors are linearly independent and the distribution function Ψ is continuous and strictly monotonic, one can choose parameters ρ and λ so that the A_{R-P} element converges as closely as desired to the optimal mapping. This holds for arbitrary environmental success probabilities specified by a function $d : X \times A \rightarrow [0, 1]$, where $d(\vec{x}, a) = \text{prob}\{r(t) = \text{success}|\vec{x}(t) = \vec{x}, a(t) = a\}$ (X is the set of input vectors and $A = \{0, 1\}$ is the set of actions). This means that the element will eventually respond to each input vector $\vec{x} \in X$ with the action $a_{\vec{x}}$ with probability as close to 1 as desired, where $a_{\vec{x}}$ is such that $d(\vec{x}, a_{\vec{x}}) = \max\{d(\vec{x}, 0), d(\vec{x}, 1)\}$. Details are provided in ref. 14. Note that if there is a single input pattern, this task reduces to the learning automaton task described above (Fig. 1). When there are many input patterns, the task reduces to a conventional supervised learning pattern classification task with a noisy teacher only if for all $\vec{x} \in X$, $d(\vec{x}, 0) + d(\vec{x}, 1) = 1$. This restriction implies that the evaluation signal provides as much information as a noisy teacher in the supervised learning task [14,15].

TEAM DECISION PROBLEMS WITH ASSOCIATIVE LEARNING AUTOMATA

The simplest team of associative learning automata, such as A_{R-P} elements, consists of N automata sharing the same pattern input and evaluation signal. We have called this system an "Associative Search Network" [16] and have produced several illustrations of its capabilities [13,17,20]. Using the terminology of team decision theory [21], the "information structure" is one in which the team members share the same environmental information but have different points of control over the evaluation. The associative mappings this kind of network can form have exactly the same properties as the mappings formed by the more familiar non-recurrent associative memory networks [22], but they are formed in the absence of explicit instructional information.

Fig. 3 shows a team of associative learning automata with a more complex information structure (the evaluation pathways are not shown). The elements in the leftmost layer have access to environmental state information but have no direct control over the evaluation; the rightmost elements are in the opposite situation; and the interior, or "hidden", elements neither directly sense nor control the team's external environment. However, A_{R-P} elements are able to take advantage of signals provided by other elements in order to improve performance. For example, we set the environmental evaluation criteria so that in order to maximize success

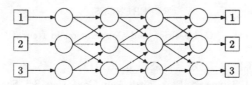

Figure 3. Layered network as a team.

probability the network in Fig. 3 had to implement transmission pathways from input 1 to output 3, input 2 to output 2, and input 3 to output 1. Thus, for a team member to be successful it had to learn to participate in a network that logically formed crossed tranmission pathways through the four layers. A_{R-P} elements solved this task by setting appropriate interconnection weights. This is a form of cooperation by which the elements coordinate their actions in order to increase success probability. As illustrated elsewhere [15], A_{R-P} elements cooperate in a similar manner to form nonlinear mappings if doing so maximizes success probability.

Williams [23] has shown that if all the A_{R-P} elements in an arbitrary acyclic network use the logistic distribution $\Psi(s) = 1/(1 + e^{-s})$ for the random variable in (1) and $\lambda = 0$, then the expected change of any weight in the network is proportional to the partial derivative of the overall network's reward probability with respect to that weight. This means that in a probabilistic sense the collective activity of A_{R-P} elements is gradient-following in the appropriate way (although λ has to be nonzero for networks to avoid absorbing at nonoptimal states). In fact, it may not be misleading to regard an A_{R-P} network as a kind of stochastic approximation to a network using the error-backpropagation algorithm recently developed by Rumelhart, Hinton, and Williams [24] in which weights change according to exact gradient information.

Because the weights in A_{R-P} networks only have available an unbiased estimate of the gradient, many more presentations of the input patterns are required than for comparable backpropagation networks but less computation per step is needed. For example, the transmission task just described requires approximately 15,000 to 20,000 steps for solution by an A_{R-P} network, which is probably considerably more than a backpropagation network would require. This assumes, however, that only one action for each element is evaluated for each presentation of an input pattern. Letting the elements perform many actions for each presentation can improve the accuracy of the gradient estimate and reduce the number of presentations required. We do not yet have precise data on the relative amounts of computer time required by these learning methods, but it appears that backpropagation has a substantial edge for conventional sequential simulations [25]. Note, however, that networks of associative learning automata, such as A_{R-P} elements, are applicable to tasks in which error-backpropagation does not apply. These are tasks in which the environment can evaluate the consequences of the network's actions but cannot provide individual error signals for the network's output elements. This type of task arises in learning control problems and is discussed elsewhere [13,26].

CONCLUSION

Cooperativity in neural networks probably takes many forms, some of which are surely represented in the mathematical models and computer simulations to which the label cooperative computation has been applied. The research described in this article is an attempt to add another level of meaning to computational cooperativity by starting with self-interested elements possessing robust adaptive strategies for furthering those interests in relatively unconstrained environments. I have discussed how teams of such elements learn to coordinate their collective behavior in order to solve problems that individual elements cannot solve due to lack of information, control, or representational power. We have not yet investigated more general game decision problems in which the elements have conflicting interests, but the elements' ability to act conditionally upon the decisions of other elements should lead to collective behavior that is more complex than that obtainable from nonassociative learning automata in similar situations.

REFERENCES

1. S. Amari, M. A. Arbib, Competition and Cooperation in Neural Nets (Springer-Verlag, NY, 1982).

2. Feldman, J. A., Ballard, D. H., Cog. Sci. 6, 205 (1982).

3. G. E. Hinton, T. J. Sejnowski, Proc. Fifth Ann. Conf. Cog. Sci. Soc. (1983).

4. D. E. Rumelhart, D. Zipser, In D. E. Rumelhart, J. L. McClelland, eds., Parallel Distributed Processing: Explorations in the Microstructure of Cognition (Bradford Books/MIT Press, Cambridge, MA, to appear).

5. J. J. Hopfield, Proc. Nat. Acad. Sci. 79, 2554 (1982).

6. A. H. Klopf, The Hedonistic Neuron: A Theory of Memory, Learining, and Intelligence (Hemishere, Washington, D.C., 1982).

7. M. L. Tsetlin, Automaton Theory and Modeling of Biological Systems (Academic Press, NY, 1973).

8. H. Robbins, Bull. Amer. Math. Soc. 58, 527 (1952).

9. K. S. Narendra, M. A. L. Thathachar, IEEE Trans. Sys., Man, Cybern. 4, 323 (1974).

10. K. S. Narendra, R. M. Wheeler, IEEE Trans. Sys., Man, Cybern. 13, 1154 (1983).

11. B. Widrow, M. E. Hoff, 1960 WESCON Convention Record Part IV, 96 (1960).

12. F. Rosenblatt, Principles of Neurodynamics: Perceptrons and the Theory of Brain Mechanisms (Spartan Books, Wash., D.C., 1961).

13. A. G. Barto, R. S. Sutton, and C. W. Anderson, IEEE Trans. Sys., Man, Cybern. 13, 846 (1983).

14. A. G. Barto, P. Anandan, IEEE Trans. Sys., Man, Cybern. 15, 360 (1985).

15. A. G. Barto, Human Neurobiology 4, 229 (1985).

16. A. G. Barto, R. S. Sutton, and P. S. Brouwer, IEEE Trans. Sys., Man, Cybern. 40, 201 (1981).

17. A. G. Barto, R. S. Sutton, Biol. Cybern. 42, 1 (1981).

18. R. S. Sutton, PhD Dissertation, University of Massachusetts (1984).

19. B. Widrow, N. K. Gupta, and S. Maitra, IEEE Trans. Sys., Man, Cybern. 5, 455 (1973).

20. O. Selfridge, R. S. Sutton, A. G. Barto, Proc. Ninth IJCAI (1985).

21. Y. C. Ho, Proc. IEEE 68, 644 (1980).

22. T. Kohonen, Associative Memory: A System Theoretic Approach (Springer, Berlin, 1977).

23. R. J. Williams, Technical Report, Institute for Cognitive Science, University of California at San Diego (to appear).

24. D. E. Rumelhart, G. E. Hinton, and R. J. Williams, In D. E. Rumelhart, J. L. McClelland, eds., Parallel Distributed Processing: Explorations in the Microstructure of Cognition (Bradford Books/MIT Press, Cambridge, MA, to appear).

25. C. W. Anderson, PhD Dissertation, University of Massachusetts (to appear).

26. A. G. Barto, Proc. IEEE Workshop on Intelligent Control, Rensselaer Polytechnic Institute, Troy, NY, 1985 (to appear).

Generalizing Back Propagation to Computation

Eric B. Baum
Caltech
Pasadena Ca 91125

Abstract: We generalize back propagation from an algorithm for supervised learning to an algorithm for discovering computational algorithms by a constrained gradient descent in algorithm space. Our generalization is analogous to the Hopfield,Tank generalization of the Hopfield Memory to computation. We discuss extensions and applications.

Recently Rumelhart et al.[1] have proposed an interesting algorithm for supervised learning called 'back propagation'. In this paper we generalize back propagation to computation. This is analogous to the Hopfield,Tank[2] generalization of Hopfield's content addressable memory[3] to computation. We give an algorithm for performing gradient descent in algorithm space. The locally optimal solution found by such gradient descent will be a fast parallel heuristic for whatever problem we are considering. For hard problems gradient descent directly on configuration space is plagued by local minima and ineffective. Nonetheless abstract reasoning frequently leads humans to heuristics which work well. Our gradient descent is a form of abstract reasoning. It is worth investigating whether the locally optimal algorithms we generate will be effective. It is also interesting that the approach allows machines to find algorithms without supervision.

The back propagation algorithm adjusts the weights in a feedforward neural network[*1] consisting of several layers of neurons including an input layer, some intermediate layers, and an output layer. The goal is to teach the network to associate specific output states, called target states, to each of several input states.The output of each neuron is assumed to be a differentiable semilinear function of its inputs. That is, if v_j denotes the output of the j-th neuron, and $w_{ji}, j < i$ denotes the connection of the output of the j-th neuron to the input of the i-th neuron, then the i-th neuron takes value $v_i = f(\sum_j w_{ji} v_j)$. For simplicity we for the moment take the differentiable function f to be the same for all neurons. f is called the activation function. Now the value of the output neurons is an easily computable differentiable function of the values of the input neurons and the weights w_{ji}. Rumelhart et al. give a set of input values i_{pi} together with associated target values t_{pi}, and ask to vary the weights so as to

---

[*1]   As Rumelhart et al note their algorithm can be extended easily to systems with feedback by considering the state of the system at time i to be the i-th level in a feedforward net. We have applied our methods to systems with limited feedback using this approach.

minimize the error E in achieving these target values, where

$$E = \sum_p E_p = \sum_{p,j} (t_{pj} - o_{pj})^2 \qquad (1)$$

Here o_{pj} represents the value of the j-th output neuron when the input neurons are held in the p-th pattern. To do this they compute the gradient of E. They note first

$$\frac{\partial E_p}{\partial w_{ji}} = 2 \sum_k (t_{pk} - o_{pk}) \frac{\partial o_{pk}}{\partial w_{ji}} \qquad (2)$$

and then compute $\partial o_{pk}/\partial w_{ji}$ by the chain rule. This chain rule calculation yields a path integral which sums over all the paths by which w_{ji} affects o_k. Rumelhart et al. give a fast iterative method for computing this path integral.[*2] They then adjust the weights by cycling through the input patterns and for each in turn moving a small distance ϵ in the direction of the gradient of E_p. If ϵ is sufficiently small, this is an accurate approximation to gradient descent for E. Eventually this descent converges to a local minimum.

Because this algorithm yields only a locally and not a globally optimal choice of weights the question arises whether gradient descent is sufficient for practical applications.[*3] Rumelhart et al. consider several learning problems and conclude that frequently it is. Sejnowski[5] used this algorithm to teach a network to translate text into phonemes. Whether these applications represent hard computational tasks is unclear. Back propagation, and our generalization, may prove useful for practical tasks which in principle are easy, and yet are difficult for humans. Another question we hope to answer is whether back propagation can generate effective heuristics for fundamentally intractable problems.

We now give a generalization of the back propagation algorithm to computation. Instead of presenting the network with input and target values, and asking to minimize the error, we ask the network to minimize a differentiable function of the input and output values. Such a function can be chosen for example to represent any optimization problem. In general we consider the function to specify an optimization problem and the values of a given input to specify an instance. We present the system with a variety of inputs, corresponding to a large sample of possible instances, and apply gradient descent. The network converges to a heuristic. If we then present an instance and wait time equal only to the number of levels in the feedforward net, we can read the output values to solve our problem.

[*2] The algorithm for fast computation of this path integral has been previously discovered[4] in a different context with intriguing formal similarities.

[*3] Another question is: what size network is necessary to solve a given problem, if the network is globally optimal?

The computation of the gradient direction is possible by a trivial extension of the Rumelhart calculation. We wish now to minimize

$$E = \sum_p F(\{o_{pl}\}, \{i_{pk}\}) \tag{3}$$

where F is the function specifying the problem and $\{i_{pk}\}$ is the set of input values specifying instance p. But

$$\frac{\partial E}{\partial w_{ij}} = \sum_p \frac{\partial F}{\partial o_l}|_p \frac{\partial o_l}{\partial w_{ij}}|_p \tag{4}$$

We easily compute $\partial F/\partial o_l|_p$, and back propagation computes $\partial o_l/\partial w_{ij}$.

For example we might consider the Travelling Salesman Problem(TSP) [*4] using the energy function of Hopfield and Tank[2].

$$F = \sum_i (\sum_j o_{ij} - 1)^2 + \sum_j (\sum_i o_{ij} - 1)^2 + \sum_{ijk} i_{jk} o_{ij} o_{i+1,k} \tag{5}$$

Here we have written our (previously single) subscripts as double subscripts in accord with Hopfield,Tank since a tour is now represented by an N×N matrix. The distances between city j and city k are specified by the (analog) value of input neuron i_{jk}. When F is minimized the matrix o_{ij} is a permutation matrix specifying the order in which the cities are visited in the shortest tour.

If we hold the inputs fixed to represent a given instance, and vary the weights in the gradient direction, the system converges, in our experiments, to a locally stable configuration where the output matrix represents a fairly short tour. Now instead of doing this, we present many instances randomly chosen from some distribution. We hope that the algorithm will abstract statistical features of our distribution of instances and arrive at a heuristic.

This is a hard task. It is generally believed that no fast algorithm to exactly solve TSP exists. Many workers have sought heuristics for TSP and similar problems which work well with high probability when presented with a randomly chosen instance[6]. Human generated heuristics generally do not take advantage of the particular distribution of instances to be considered. Humans would for instance apply the same heuristic to TSP instances generated by choosing each distance randomly and uniformly between 0 and 1 as they would to instances generated by distributing points randomly within a euclidean unit cube. On the other hand, our gradient descent algorithm will generate a network which cannot be applied to an arbitrary number of cities. If the feedforward network we adjust has input neurons to describe a distance matrix between N cities, there

---

[*4] For statement of the TSP, remarks on its difficulty of solution, and examples of heuristics, see Baum[6] and references therein.

is no obvious way that the algorithm generated could be applied to $N+1$ cities. Finding a way to do this is a major unresolved question for practical applications and for models of thought, generalization, learning, or abstract reasoning based on back propagation.

We have been unable to generate satisfactory heuristics for TSP using this approach. The Hopfield Tank energy is unsuitable for our back propagation in two regards. Firstly, they code an N city tour by an N by N matrix. This means that each level of our network must have $O(N^2)$ neurons, so that the number of weights we consider grows as $O(N^4)$. Even small instances of TSP require an impractical amount of computation. Secondly, the Hopfield,Tank energy function requires the system to learn the syntax, that is to learn to produce tours. Two trivial local minima therefore plague us. One is where the output is set to the same valid tour,independent of the input. This minimizes the first two sums in the energy. The other is where the system turns off the output neurons, failing to give a tour but minimizing the last term in the energy. We are currently implementing a different energy function requiring less computation and having implicit syntax. We hope this energy, based on an analog system we proposed at the Santa Barbara conference[7], will solve our difficulties.

We have been able to solve problems where we ask the output values to be elementary functions of the input neurons. For example the system has no trouble in setting output neurons to be threshold functions of input neuron values. If the network is composed of several layers, each of which connects only to the next layer, the input information is scrambled by the processing and it would be very difficult for a human to construct weights to calculate even simple functions. The paper of Pearlmutter and Hinton[8] at this conference seems to have applied a special case of this algorithm to the problem of feature extraction with considerable success. They seem, however, to have considered only two level systems, without hidden neurons.

We have found, in these implementations, that the choice of neural activation function is important to the performance of the network. The logistic activation function of Rumelhart et al[1]:

$$o_{pj} = \frac{1}{1 + e^{-(\sum_i w_{ij} o_{pi} + \theta_j)}} \tag{6}$$

has not been as effective in some applications as:

$$o_{pj} = sin(\sum_i w_{ij} o_{pi} + \theta_j) \tag{7}$$

where θ_j are bias functions varied by the gradient descent algorithm.*5 It is not hard to extend the gradient descent by considering neural activation functions

---

*5 In practice the argument of sin ranges over several periods. This means there are many equivalent sets of weights for a given input. This redundancy perhaps allows more freedom in adapting to the range of inputs we present.

depending on several parameters and optimizing over these as well. The learning logic algorithm incorporates such parameters[9]. We are studying in this manner the question of which activation functions are most generally useful. We have also been informed that Parker has results along these lines[10].

The procedure of presenting many input values chosen from some distribution of instances for which we desire a heuristic is essentially a Monte Carlo integration over the space of instances. For simple choices of activation function we will be able to compute this integration analytically. Symbolic manipulation programs such as SMP[11] may be valuable in integrating. Once we have integrated out the input instances, we will be able to perform the back propagation algorithm rapidly.[*6] This will correspond directly and exactly to a gradient descent in the space of heuristics for the given problem and distribution of instances. Once the gradient descent has converged to a local optimum, we present any specific instance to the input of the network we have constructed and the output will hopefully be a good solution. What is missing from this analysis is a rapid algorithm, like back propagation, for computing the integrals over input distributions. We are searching for such an algorithm. Without it this procedure could only be practically useful for networks with one hidden layer, which however are the main sort so far utilized[1,5].

As we have noted, the outputs of these feedforward nets can be written, in principal at least, as explicit differentiable functions of the inputs and the weights. As such they can be expanded, for instance in Taylor series. Instead of moving in a gradient direction in weight space for a given network, we might simply write outputs as expansions in inputs and in parameters which we vary in a gradient direction.[*7] If we choose appropriate expansions it is easy, at least in principle, to integrate over appropriate distributions of inputs and perform gradient descent in heuristic space. For realistic problems it appears that we need a huge number of parameters in our expansion to achieve useful results, so the remarks of this paragraph are probably of more theoretical than practical

---

[*6] This method also avoids a putative pitfall. By moving a distance ϵ in the gradient direction for each pattern p, the back propagation algorithm moves approximately in the gradient direction for $E = \sum_p E_p$. In our case, where we have a continuous distribution of patterns, we might be forced to choose ϵ very small to get convergence. This problem has not arisen in our simulations.

A simple trick for speeding up gradient descent is to increase ϵ, perhaps by a factor of 1.1, at each iteration where the energy does not overshoot, and decrease it, by perhaps a factor of 2, when the energy overshoots. This eliminates guesswork of the best value and can save orders of magnitude of computing time in practice. To apply this method to back propagation, when the distribution of inputs has not been integrated out, one must average the energy over many inputs for purposes of updating ϵ.

[*7] Analysis like this may shed light on the results of Maxwell et al[12] that nonlinear neurons improve the computational potential of nets.

interest. We are curious, however, as to why this method is less effective, if indeed it is, then back propagation on neural nets involving the same number of parameters. Perhaps the importance of nets lies solely in that they let us compute an exponential number of paths in polynomial time. If so we should devote more study to multilevel nets, where this distinction emerges. As often in physics, the key may be the choice of parameters we expand in.

In conclusion it is straightforward to generalize back propagation to computation. We hope that this will prove valuable for generating heuristics. It is unclear whether it will prove a good model for human cognition, however it may lead to models for unsupervised learning. There is much work to be done in developing applications of these ideas. It is also crucial to discover how many interneurons organized in how many levels are necessary for particular amounts of computing power. These questions are under active study and I hope to report on some of them next year.

Acknowledgement: I would like to thank J.J. Hopfield for conversations, and a critical reading of the manuscript. This work was supported in part by DARPA through arrangement with NASA.

References
(1)D.E.Rumelhart, G.E.Hinton, R.J.Williams,"Learning Internal Representations by Error Propagation", ICS Report 8506(UCSD)(1985), to appear in D.E.Rumelhart,J.L.McClelland(Eds),"Parallel Distributed Processing:Explorations in the Microstructure of Cognition. Vol. 1:Foundations",Cambridge MA Bradford Books/MIT Press

(2)J.J.Hopfield,D.W.Tank,BioCyber 52,p 141(1985)

(3)J.J.Hopfield,Proc.Natl.Acad.Sci USA 79, p2554(1982) and 81, p3088(1984)

(4)L.E.Baum,"An Inequality and Associated Maximization Technique in Statistical Estimation for Probabilistic Functions of Markov Processes",in O. Shisha (ed) "Inequalities 3",Acad. Press NY (1972) p1, and L.E. Baum,T. Petrie,G. Soules,N.Weiss, Ann. Math. Stat. 41 p164 (1970)

(5)T.Sejnowski,"NETtalk: A Parallel Network That Learns to Read Aloud",this volume.

(6)E.B.Baum,"Towards Practical Neural Computation", this volume.

(7)E.B.Baum,"Graph Orthogonalization", submitted to Discr. Math. (1985)

(8)B.Pearlmutter,G.Hinton, "G-Maximization: a Deterministic Unsupervised Learning", this volume.

(9)D.Parker,"Learning-Logic", MIT TR-47,(1985)

(10)D.Parker, personal communication

(11)S.Wolfram,"SMP Reference Manual",Computer Math Group, Inference Corp, Los Angeles(1983)

(12)T.Maxwell, C.L.Giles,Y.C.Lee,"Nonlinear Dynamics of Artifical Neural Systems", this volume.

Towards Practical 'Neural' Computation
for Combinatorial Optimization Problems

Eric B. Baum[†]
Caltech, Pasadena CA 91125

Abstract: We propose a hill climbing attachment called Iterated Descent useful in conjunction with any local search algorithm, including neural net algorithms.

In this paper we introduce a hill climbing attachment we call 'iterated descent' useful in conjunction with local search algorithms for optimization problems. In particular we hope iterated descent will improve neural net algorithms.

The Travelling Salesman Problem (TSP) is a widely studied combinatorial optimization problem. We study this as an example. The *Travelling Salesman Problem* is: given a matrix d_{ij} describing the distance between any two of a set of N cities, find the shortest tour visiting each city exactly once and returning home. Note these cities need not lie on a plane, nor need d_{ij} obey the triangle inequality. We assume d_{ij} is symmetric.

This problem is finite and thus in principle soluble by exhaustive search. This requires time of order $O(N!)$ and is impractical. We want an algorithm to find the shortest tour in time bounded by some polynomial in N. The TSP is known to be NP-complete[1], meaning a polynomial time algorithm for TSP could be modified to give a p-time algorithm for any other problem in the class NP, a broad class of problems. Many workers have searched without success for a p-time algorithm for NP-complete problems. It is widely believed that for deep, but little understood reasons no such algorithm exists.[1]

Since we have no p-time algorithm for many optimization problems but nonetheless need practical answers, we search for heuristics. We are interested particularly in heuristics which when applied to randomly generated instances frequently give a good solution. Such heuristics are often for practical purposes more useful than even a true algorithm and would likely satisfy biology.

Local search algorithms are an intuitive and useful class of heuristics. We perform a discrete version of gradient descent. For each allowed configuration we choose a *neighborhood* of nearby allowed configurations, eg. for each tour we choose a set of nearby tours. Typically this neighborhood is specified by a set of transformations which act on the candidate tour to generate small perturbations. The local search algorithm starts with a pseudorandomly generated tour. We then search in the neighborhood of this candidate tour until we find a shorter tour. If we find one we replace the candidate by this more optimal tour and search the new candidate's neighborhood. Eventually we generate a locally optimal tour, that is a tour with no shorter tours in its neighborhood. For example, a tour is λ-*optimal*[1,2] if no tour found by breaking λ links and reforging

---

[†] Supported in part by DARPA through arrangement with NASA

a tour is shorter. A λ-*opt transformation* breaks λ links and reforges a tour. The set of 3-opt transformations gives a useful neighborhood for TSP.

Since there will usually be many locally optimal tours, one repeats the local search starting from many different pseudorandomly chosen initial tours. By choosing the different initial candidates uniformly, one hopes to explore the gross features of configuration space. Further, it is not clear how to construct a heuristic for generating good starting states which is better or faster than local search applied to an initially random tour[1,3]. Iterated descent is a method for using information contained in previously found local minima to find better starting states. In simple terms, the idea of iterated descent is to take the best local minimum so far found and perturb it.

Simulated annealing(SA)[4] is another approach to hill climbing. One chooses random neighbors of the candidate tour, but now accepts them not only when they have shorter length, but also with probability $P = e^{-\beta\Delta l}$ when they have length longer by Δl. Here $T = 1/\beta$ is an effective temperature; this algorithm puts the system in a Boltzman distribution. T is now slowly lowered. Simulated annealing is effective for some problems[5]. Physicists, however, seem more excited by SA than workers in combinatorial optimization. SA makes manifest the analogy of combinatorial optimization problems to spin glasses. An interesting feature is that the system converges, as the temperature is lowered, by first roughing out the large scale, global properties of the solution and then fixing the local details. This is characteristic of neural net algorithms as well.

I first became interested in optimization problems when I realized one could smooth away many local minima by embedding discrete systems to be optimized in continuous analog systems with the same global minima.[6,7*1] We can view the discrete system as a constrained continuous system. We expect to avoid local minima for two main reasons. First, removing the constraints allows more degrees of freedom, and thus more directions to descend. Thus some of the local minima in the discrete system are no longer local minima in the continuous system. Perhaps more importantly, the constraints on the discrete system prevent free propagation of forces. In the continuous system, global collective action keeps one part of the system from becoming stuck independent of the rest. False minima exist, but have a more global nature. The system first blocks out the gross features of the solution and then works on the details.

This field took off when Hopfield and Tank embedded TSP[8] into a continuous neural net which can be built from transistors, wires, and resistors, and which relaxes extremely rapidly, in parallel. Their algorithm allows the em-

---

*1 I have embedded an NP-complete problem so there are no local minima.[6] What happens is that I removed all the local hills, but left a flat energy landscape like a putting green. The system does not know how to roll toward the minimum. This embedding is useless for computation, but interesting as a manifestation of intractability. Perhaps it exhibits the limitations of sculpting energy surfaces to remove local minima.

bedding of any optimization problem which can be expressed with a quadratic energy. If one allows some amplifiers in the net to be faster than others[9], which is easy to realize in practice, then we can extend their construction to higher order energies as well[6]. These proposals allow one to consider hardwiring computers to perform optimization calculations.

Their net works much better than a discrete version would work. However it works poorly compared to good digital algorithms. For example, on a 30 city example (in the plane) for which the Lin-Kernighan algorithm produced a tour of length 4.26, Hopfield and Tank found "less than 7 commonly, less than 6 occasionally."[8] As I said at last year's Santa Barbara conference, I know of no NP-complete problem where a neural net algorithm is competitive with the best known discrete algorithm. One possibility is that neural net algorithms have so far been outperformed by digital algorithms only because so much effort has gone into designing good digital algorithms. We hope that, given the advantages in minima avoidance we expect neural nets to have, we may soon discover neural net algorithms which outperform digital algorithms. One message of my computational results, however, will be that better algorithms will beat more repetitions of worse algorithms. A good algorithm on a PC will usually out-optimize a poor algorithm on a Cray.

Two points summarize this brief discussion of neural net optimization. Firstly, hard wired neural nets compute extremely rapidly. Secondly we have good motivation and some experimental results indicating well designed neural net ciruits may avoid most local minima and give good optima. I created iterated descent as a hill climbing attachment to be useful in conjunction with a hard wired neural net computer. Iterated descent spends its time in runs of a local search algorithm, so that if we hard wired a neural net, iterated descent would run extremely rapidly, taking maximal advantage of the minima avoidance and the speedups generated by hardwiring.

To understand why iterated descent is useful, we ask why local search is. Consider a silly algorithm for TSP: Monte Carlo search. Choose a large number of tours, say N^3 for an instance of N cities, and keep the shortest one. Now this is a horrible algorithm since, for N large, the distribution of tours is Gaussian and sharply clustered about the mean. If we choose N^3 tours, they will all have about the same length. Thus we are driven to algorithms like 2-opt or 3-opt which allow us to move purposefully away from the mean. We soon find a tour shorter than almost all other tours. However, as N gets large, we *conjecture* the central limit theorem will cluster the distribution of 3-opt tours so that we gain little from repeated applications of 3-opt starting from random starting states.

Thus we propose: since local search is superior to random search, why not iterate and use local search on the starting states for local search, rather than random search on the starting states. Iterated descent implements this idea. We now state iterated descent. We start by listing 'Subroutine Optimize' which begins at a given candidate solution, S, applies local search, and returns the local optimum whose basin of attraction S is in. In this listing, subroutine optimize uses 3-opt. It could use any local optimization such as a neural net.$d(S)$ is the length of tour S. T_3 is the set of 3-opt transforms.

Subroutine Optimize

1) Do 3 for $j = 1, |T_3|$
2) $t = $ j-th 3-opt transform
3) If $d(tS) < d(S)$ then $S = tS$ and go to 1
4) Return
5) End

Standard local search starts from many, pseudorandomly chosen initial states, calls subroutine optimize, and keeps the best local optimum found, S_o.

Local Search

1) Do 4 until exhausted
2) Pseudorandomly choose S
3) Call Subroutine Optimize
4) If $d(S) < d(S_o)$ then $S_o = S$
5) End

Iterated Descent

1) Intialize Pseudorandomly
2) Call Subroutine Optimize
3) If $d(S) < d(S_o)$ then $S_o = S$, $ntim1 = 0$, else $ntim1 = ntim1 + 1$
4) If $ntim1 = 50$ go to 7
5) Pseudorandomly choose t_2 and set $S = t_2 S_o$
6) Go to 2
7) If $d(S_o) < d(S_{oo})$ then $S_{oo} = S_o$ and $ntim2 = 0$, else $ntim2 = ntim2 + 1$
8) If $ntim2 > 10$ go to 12
9) Pseudorandomly choose t_2, t_2'. Take $S_o = S = t_2 t_2' S_{oo}$
10) $ntim1 = 0$
11) Go to 2
12) Print "Optimal distance =", $d(S_{oo})$
13) End

Iterated descent starts pseudorandomly (line 1). Now compare the loop of lines 2-6 to local search. Iterated descent keeps the best local optimum it has found as S_o. Further candidates presented to subroutine optimize are found by applying a small perturbation to S_o. In general, for this small perturbation, we used a single randomly chosen 2-opt transformation (t_2).[*2] This loop finds a local optimum among local optima. That is it finds a 3-opt tour which no t_2

[*2] The small perturbation should be chosen so that there is small probability of falling back to the same local minimum, else we will waste computation time finding the same local minimum many times. This probability should be large enough that we are assured of searching a local region well. A single 2-opt transform is a small easily generated perturbation. It was possibly too large.

transformation perturbs into the basin of attraction of a shorter 3-opt tour.[*3]

We can add outer loops, such as that specified by lines 7-11, which search for local minima among local minima among local minima. We store the best local minimum among local minima so far found, the best S_o as S_{oo}, and generate new starting states for the inner loop by a larger perturbation of S_{oo}, for instance generated by two successive pseudorandomly chosen 2-opt transformations. We can extend this with further nested loops.

When there is no correllation between the length of nearby locally optima, we expect iterated descent will do as well as standard local search, since generating starting states by small perturbation of known good states is equivelent to choosing a random starting state. We expect that almost always there will be positive correlation, and iterated descent will improve performance.

We expect iterated descent to work particularly well when we are minimizing a function which is approximately a superposition of continuous functions varying on different length scales. For example, iterated descent will work rapidly to find the minimum of $f=A\sin(\omega_1 t)+B\sin(\omega_2 t)$. Reiterated descent, with several nested loops, is tailored to the scaling limit where we have fluctuations on all length scales. Scaling is known to occur in physics in many aspects from the distribution of matter in the universe to critical point phenomena. It is interesting to ask whether large instances of NP-hard problems scale.

Ultrametricity, now associated with a variety of optimization problems and spin glasses, is another motivation for iterated descent. See Baldi,Baum[10] in this volume for a general discussion. For example, approximate ultrametricity has been found for the 2-optima and 3-optima in TSP.[11] We believe this approximate ultrametricity is evidence the 3-optima are clustered near the bottoms of hierarchically organized deep valleys in the energy landscape of TSP[10]. The ultrametric hierarchy is a hierarchy not in energy, necessarily, but in configuration space, ie. in the distance between one optima and another, measured in terms of how many links must be broken and reforged to transform one to the other. Iterated descent is tailored to exploring such a space. It first looks for the bottom of one valley. A series of nested loops using larger perturbations can then search out the tree structure of the ultrametric space. In contrast, simulated annealing, which moves up in energy rather than laterally in configuration space, seems less tailored to ultrametricity.

We give the results of experimental evaluations. We considered 100 city instances generated by randomly choosing each distance uniformly between 0 and 1. We compared iterated descent(ID), local search starting from random tours(Rand) and simulated annealing(SA). We used 3 different neighborhoods, T_i, 2-city interchange where we generate neighbors by interchanging the po-

---

[*3] In fact, in our listing, we have a cutoff on ntim1, the number of attempts we made to search the neighborhood. Thus we are not assured of finding a local optimum among local optima. This is necessary in practice since there are $\binom{N}{2}$ possible 2-opt transformations.

sitions of 2 cities, 2-opt, and 3-opt. We considered approximately the same number of transformations in comparing ID and SA, so we used slightly less computer time in the ID runs than the SA runs. We considered the same number of local optima in ID and Rand, so we used rather less time in ID, by a factor of about 1.4, as ID was always faster to converge since its initial candidates are always close to an optimum. We present our results in Table 1.

Table 1

T	I.D.	Rand.	S.A.	Runs	Ntim1	Ntim2
T_i	5.11	6.76	5.11	5	20	It=3000
	4.82		5.20	3	100	It=3000
2-opt	2.69		2.62	8	1000	0
	2.49	2.51		6	500	0
	2.57		2.52	3	100	0
	2.23		2.22	1	3000	0
	2.51	2.53		2	200	8
3-opt	2.17	2.20		9	20	4

Note that never once, even in a single run, did random starting states beat iterated descent. Also iterated descent was substantially faster. Note however that the most important factor is choice of neighborhood. Rand. using 3-opt is better by more than a factor of 2 than any algorithm using 2-city interchange. It is not intuitive to me that 2-city interchange should be so inferior to 2 link interchange. One should be warned that neither iterated descent nor simulated annealing is a substitute for finding a good neighborhood.

The comparison between ID and SA is close. ID is substantially better on the worst of the neighborhoods, T_i, when enough neighbors were tried to allow ID to find accurately a local minimum among local minima. For 2-opt,however, even with substantial local searching, simulated annealing seems slightly better.

In conclusion, iterated descent appears clearly to both improve and speed up local search. It should prove useful whenever local search is necessary. It should improve performance of neural algorithms. We hope it will be possible to implement iterated descent in conjunction with hard wired neural nets.

References

(1)C.H.Papadimitriou,K.Steiglitz,"Combinatorial Optimization,Algorithms,and Complexity",Prentice-Hall Englewood Cliffs NJ(1982) and references therein
(2)S. Lin,Bell Sys. Tech. J. 44 p2245(1965)
(3)S. Lin,B.W. Kernighan,Oper. Res. 21 p494(1973)
(4)S.Kirkpatrick,C.D.Gelatt,Jr,M.P.Vecchi,Science 220p671(1983)
(5)C.R.Aragon,D.S.Johnson,L.A.McGeoch,C.Shevon,unpublished notes(1985)
(6)E.B.Baum,unpublished
(7)E.B.Baum, "Graph Orthogonalization",submitted to Disc.Math.(1985)
(8)J.J.Hopfield,D.W.Tank,Bio Cyber 52 p141 (1985)
(9)J.J.Hopfield,D.W.Tank,IEEE Circ. Sys.,in press
(10)P. Baldi,E.B. Baum,this volume
(11)S.Kirkpatrick,G.Toulouse,J.de Physique 46p1277(1985)

NUMERICAL SIMULATIONS OF BOLTZMANN MACHINES

D. G. Bounds

Royal Signals and Radar Establishment, St Andrew's Road, Malvern, Worcs WR14 3PS, England.

ABSTRACT

Statistical mechanics methods have been used to investigate the Boltzmann Machine algorithm proposed by Hinton and Sejnowski[1]. Exact calculations of the partition function for a ten-unit Boltzmann Machine show that there is a "window" of annealing temperatures at which learning is possible. For the 4-2-4 encoder problem it is found that the optimum learning rate is obtained when the number of energy states thermally accessible is approximately twice the number needed to store the hidden unit codes.

INTRODUCTION

The computational abilities of networks of simple processing units is of great interest. Networks have been proposed which show fascinating emergent properties including the ability to store patterns, and to recall perfectly stored patterns given noisy or incomplete cues. Apart from the intrinsic interest of this novel behaviour, these models are promising candidates for pattern recognition devices and for fault-tolerant, content-addressable memories (CAM's) with some capacity for error correction. Such networks loosely resemble neural networks in that they have high connectivity, a highly non-linear response at the "neuron", and each neurons output is determined by whether or not a weighted sum of the inputs from other neurons exceeds some threshold.

Networks consist of a set of N two-state units $S = \{\sigma_i; i=1,\ldots,N\}$, ($\sigma = 0,1$) joined by a set of scalar links $W = \{w_{ij}; i,j=1,\ldots,N\}$. If $w_{ji} = w_{ij}$, this is sufficient to define a Hamiltonian

$$E(W,S) = -\sum_{i=1}^{N}\sum_{j=i}^{N} w_{ij}\,\sigma_i\,\sigma_j \qquad (1)$$

A pattern to be stored defines the states of V visible units, $V \subseteq S$, and a learning algorithm subsequently adjusts the connections W so that stored patterns correspond to low-energy states of the network. In the simplest networks V=S, i.e. a pattern fixes the states of all units. There is then no internal coding of patterns (beyond the fact that they correspond to energy minima) and similar input patterns are mapped onto similar output patterns. Networks that contain a set of "hidden" units, H=S−V, whose states are not fixed directly by the input patterns are capable of richer behaviour: because the hidden units may generate an internal coding of the input patterns, they may act as feature detectors. Networks with hidden units include Boltzmann Machines[1,2] and the error propagation networks of Rumelhart et al[3].

Here we report some preliminary results of a statistical mechanical study of Boltzmann Machines. Direct calculations of the partition function for small Boltzmann Machines give information about the number of states which must be thermally accessible if the machine is to learn successfully, and about the sensitivity of the learning rate to temperature.

THE 4-2-4 ENCODER PROBLEM

The encoder problem[2] is an abstraction of the task of communicating information between components of a parallel network[2]. It may also be viewed as a pattern recognition problem. As shown in figure 1, the visible set is split into two groups, V(1) and V(2), with 4 units each. All units in V(1) are connected to each other, as are all units in V(2), but there are no direct connections between V(1) and V(2). Instead, the visible groups communicate via a pair of hidden units. The hidden units are not connected to each other, but each is connected to all the visible units. The problem is to evolve a set of link strengths which allows V(1) and V(2) to communicate their states to each other.

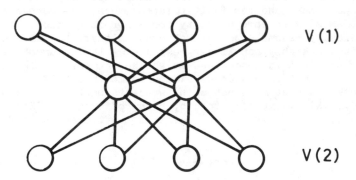

V (1)

V (2)

Fig. 1   The 4-2-4- encoder network (connections within V(1) and V(2) are not shown).

There are 2^4 possible states for each visible group. However, in the version studied here and in[2], although all possible states have some probability of occuring in the training set, the statistics are dominated by patterns where only one unit in V(1) and the corresponding unit in V(2) are in the $\sigma = 1$ state. Because there are only two hidden units, the system can only communicate 2^2 states. The machine must not only recognise which four vector pairs occur most often, but also develop a set of weights which encode these patterns so that they correspond to the four possible hidden unit patterns (0,0), (0,1), (1,0), (1,1). This is a non-trivial problem.

SIMULATION METHODS AND RESULTS

Boltzmann Machines are so called because they are stochastic systems in which the relative probability of two global states α and β depends on their energy difference through the Boltzmann relation:

$$P_\alpha /P_\beta = \exp[-(E_\alpha - E_\beta)/T] \qquad (2)$$

where the temperature T is in inverse energy units. For a description of Boltzmann Machines and a derivation of the mathematics, readers should consult the elegant paper of Ackley, Hinton and Sejnowski[2]. Details of the original 4-2-4 encoder experiments are also given there. Except where stated, the same methods and parameters have been used here. We have verified that our program behaves in a similar way to that of Ackley, Hinton and Sejnowski.

Thermal equilibrium, for which eqn. (2) holds, is attained by repeated iteration of Kirkpatrick's simulated annealing procedure[4]. The annealing temperatures are key factors which influence the learning rate[5]. Temperature affects the learning rate by altering the number of energy states which are thermally accessible at any stage of the learning algorithm. The partition function, Z

$$Z = \sum_{i=1}^{2N} \exp[-(E_i- E_{min}) /T] \qquad (3)$$

is one measure of the number of states readily accessible at temperature T. In the low T limit Z takes the value of the ground state degeneracy of the system. At high T, Z approaches the total number of states in the system.

The partition function may be used to investigate several aspects of Boltzmann Machine behavior. A Boltzmann Machine with all connections $w_{ij}=0$ has clearly not learned to solve the encoder problem, and, because all states have the same energy, $Z=2^N$ for all temperatures. However, a solution to the 4-2-4 problem must require at least 4 low energy states at the lowest temperature of the annealing schedule, T_{min}. Given that the number of states which must be thermally accessible at T_{min} when a Boltzmann Machine has evolved connections which solve the encoder problem is somewhere between 4 and 1024, what number is typically observed?

Table I shows the mean values (over the number of runs at each annealing schedule - see table II) of the partition function at T_{min}, and at the highest annealing temperature, T_{max}. In each case Z was calculated from the weights after the final iteration. First, consider results for the original annealing schedule[2], shown in

Table I. Summary of partition function data. Z_{min} is the mean
value of Z at T_{min} ; Z_{max} is the mean value at T_{max} .

	low T	normal T	high T
Z_{min}			
converged	3.6 +/- 1.30	7.75 +/- 1.85	15.75 +/- 1.39
unconverged	5.04 +/- 1.99	7.61 +/- 1.73	11.99 +/- 1.79
all machines	4.24 +/- 1.69	7.72 +/- 1.82	12.37 +/- 1.89
Z_{max}			
converged	12.54 +/- 4.84	35.98 +/- 6.00	61.75 +/- 0.62
unconverged	15.05 +/- 5.51	32.51 +/- 4.79	48.73 +/- 4.99
all machines	13.64 +/- 5.16	35.24 +/- 5.82	50.03 +/- 5.43

Table II. The effect of annealing temperatures on convergence.

	low T	normal T	high T
annealing schedule (number of timesteps at temperature T)	2 @ 4 4 @ 3 4 @ 2	2 @ 20 2 @ 15 2 @ 12 4 @ 10	2 @ 100 2 @ 75 2 @ 60 4 @ 50
Number of runs	50	150	20
Fraction of runs that converged in 400 cycles	0.56	0.79	0.10
median learning time	350	149	>400

the column headed "normal T". There are no significant differences
between machines which have converged to a stable solution and
those which did not converge in the cut-off time of 400 learning
cycles[5]. Differences in the connection strengths, which decide
whether or not the network has formed a satisfactory model of its
environment, are not evident in the underlying energy surface.

Approximately twice the minimum number of states necessary to solve the problem are available at T_{min}, and approximately 35 states make a significant contribution to Z at T_{max}.

At sufficiently low temperatures one might expect that learning would be difficult because not enough states are easily accessible to store a set of four codes. Fifty low temperature Boltzmann Machines were run with the annealing schedule (2@T=4, 4@T=3, 4@T=2). The results are summarised in tables I and II. Z_{min} shows that only just enough low energy states are available, and convergence of the learning algorithm is much worse. Only 28 machines converged in the cut-off time of 400 learning cycles, and the median learning time was 350 cycles. Thus some redundancy in the number of low energy states appears to be necessary for convergence.

At high temperatures, convergence should also be poor because too many states are available resulting in "confusion" - the machine skips between a large number of states most of which are not good solutions. Only 2 out of 20 high temperature Boltzmann Machines converged in 400 cycles.

There is therefore some window of temperature at which learning in Boltzmann Machines is possible. At higher temperatures too many states are available; at low T, too few. In much larger systems it might be profitable to look for phase transitions separating these three regions.

Finally, it is interesting to ask how Z_{min} changes during the learning process. We have followed Z_{min} as a function of time (= number of learning cycles). Because of the small system size and the relatively large step size in the weight update, there are quite large fluctuations in Z_{min} from cycle to cycle; especially when the weights are small early in the learning process. However, some clear trends do emerge. Z_{min} drops rapidly over the first few cycles. For the normal T runs, the latest cycle on which $Z_{min}>30$ is approximately 30 cycles. $Z_{min}(t)$ then wanders about at values of 15- 25 for, typically, another 30 cycles and then drops to final values consistent with the means in table I. This behaviour is reminiscent of Ackley, Hinton and Sejnowski's description of the three stages of learning[2] and it is tempting to relate these stages to particular Z regimes. If, however, there are any quantitative correlations, they are hidden by the large fluctuations in small systems.

ACKNOWLEDGEMENTS

I thank Roger Moore, John Bridle, Geoffrey Hinton and David Wallace for many enlightening discussions.

REFERENCES

1. G. E. Hinton and T. J. Sejnowski, Proc. 5th Ann. Conf. Cog. Sci. Soc., Rochester, N.Y., May 1983, p 1.
2. D. H. Ackley, G. E. Hinton and T. J. Sejnowski, Cog. Sci. 9, 147 (1985).

64

3. D. E. Rumelhart, G. E. Hinton and R. J. Williams, Inst. Cog.
    Sci. UCSD Report #8506 (1985).
4. S. Kirkpatrick, C. D. Gelatt and M. P. Vecchi, Science, 220,
    671 (1983).
5. D. G. Bounds, "Exact results for a learning algorithm",
    submitted.

LEARNING AND MEMORY PROPERTIES IN FULLY CONNECTED NETWORKS

A.D. Bruce, A. Canning, B. Forrest, E. Gardner and D.J. Wallace
Physics Department, University of Edinburgh, Scotland U.K.

ABSTRACT

This paper summarises recent results of theoretical analysis and numerical simulation, in fully connected networks of the Little-Hopfield class. The theoretical analysis is based on methods of statistical mechanics as applied to spin-glass problems, and the numerical work involves massively parallel simulations on the ICL Distributed Array Processor (DAP). Specific applications include: (i) exact results for the fraction of nominal vectors which are perfectly stored by the usual Hebbian rule; (ii) a numerical estimate of the position of the second phase transition in the Hopfield model, at which there is effectively total loss of memory capacity; (iii) a numerical study of the nature of the spurious states in the model; (iv) an exploration of the performance of a learning algorithm, including the exact storage of up to 512 (random) nominal vectors in a 512 node model; (v) a theoretical study of the phase transitions in generalisations where the energy function is a monomial in the state vectors.

INTRODUCTION

The Little[1]-Hopfield[2] model has its origins in the early work of McCulloch and Pitts[3] and Hebb[4]. The state of a network of N nodes is described by a vector with binary components S_i, $i = 1,2,...,N$ i.e. S_i takes the values 1,0 or 1,-1 say. Given a state vector $\{S\}$, the new state at site i is determined by

$$S_i' = \text{sign} \left[\sum_{j=1}^{N} T_{ij} S_j \right], \qquad (1,-1) \text{ case} . \qquad (1)$$

(In the 1,0 model the sign function is replaced by the step-function θ). If one wishes to store p "nominal" vectors $\{S^{(r)}\}$, $r = 1,...p$, a natural choice for the T_{ij} component of the connection strength matrix is the Hebbian rule which adds +1 (-1) if the ith component of a nominal vector is the same as (differs from) the jth component, so that e.g. in the (1,-1) case,

$$T_{ij} = 0 \ (i=j), \qquad \sum_{r=1}^{p} S_i^{(r)} S_j^{(r)} \ (i \neq j) . \qquad (2)$$

It is clear that the stability of a <u>nominal</u> vector under the iteration (1) is favoured by a signal of $O(N)$ from the contribution to T_{ij} from <u>that</u> nominal vector in the sum in (2), while there is an interference of $O((pN)^{1/2})$ from the remaining nominal vectors in (2), if they are statistically random. Provided $T_{ij} = T_{ji}$ and $T_{ii} = 0$ (as in (2)), there exists an energy function[2]

$$E = - \frac{1}{2} \sum_{i,j=1}^{N} S_i \, T_{ij} \, S_j \tag{3}$$

which is a monotonically decreasing function for serial updating of the nodes as in (1). The intuitive picture which emerges is therefore of stable vectors corresponding to local minima of this energy function, and well defined "memory states" for each nominal vector provided p << N.

The ICL DAP on which the simulations are done is a 64×64 array of bit-serial processing elements whose SIMD parallelism can readily be exploited for this work: the update rate for entire vectors on 4096 fully connected nodes is greater than 1 per second. Further information on other scientific applications is given in Ref.5.

We turn now to summarising some results.

PERFECT MEMORY FRACTION

Given the storage prescription (2), how does the fraction of (randomly chosen) nominal vectors which are perfectly stored depend upon the number of nodes N and nominal vectors p? This fraction can be calculated exactly in the limit $N \to \infty$, with p/N fixed[6]. The result has the asymptotic form $F = [f(p/N)]^N$, and is obtained by writing an integral representation for the condition $S_i \sum T_{ij} S_j > 0$, for each i, and evaluating this integral by steepest descent. Figure 1 shows the excellent agreement between this exact calculation for the $S_i = 1,0$ model and the results of numerical simulation on the DAP, up to "finite size effects".

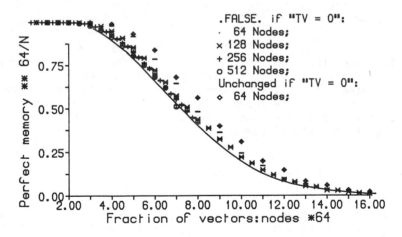

Fig. 1 Numerical and asymptotic results for perfect storage.

SECOND PHASE TRANSITION

In the absence of external noise, phase transitions in the sense of statistical mechanics correspond to changes in the structure of minima of the energy function (3); they are clearly very important for characterising the memory properties of the network. Their existence has been identified and their properties explored by Amit et al[7], using ideas developed in the study of spin-glasses. For example, in the absence of external noise, Amit et al predict two phase transitions as the ratio $\alpha \equiv p/N$ is increased. At the first, spurious states become of lower energy than the memory states associated with the nominal vectors; at the second the memory states disappear and only spurious states remain. For the (1,-1) model they estimate these two critical values to be $\alpha_1{}^c \approx 0.051$; $\alpha_2{}^c \approx 0.138$.

We have extended these results to the (1,0) model[6]. It turns out that within the same theoretical approximation (technically, no replica symmetry breaking), the critical values are close to half those for the (1,-1) model i.e.

$$\alpha_1{}^c \approx 0.025; \quad \alpha_2{}^c \approx 0.069 \quad . \qquad (4)$$

Figure 2 shows estimates obtained for $\alpha_2{}^c$ versus $1/N$ for networks of size N between 64 and 4096. The prescription for $\alpha_2{}^c(N)$ is such that 50% of nominal vectors are well stored; for large N there is an unambiguous identification because the distribution of Hamming distances between initial nominal and final vectors has two distinct peaks. We note the strong size dependence: in the (1,0) model the capacity decreases from around 0.16N at N = 64 (cf Ref.2) to 0.07N for N large. A precise numerical estimate for the critical value is difficult to obtain because of ambiguities in the extrapolation N → ∞. A fit of the form $A + B/N + C/N^2$ to the six data points yields[8] A = 0.0731±0.005, suggesting the need for small corrections to the replica symmetric mean-field results (4).

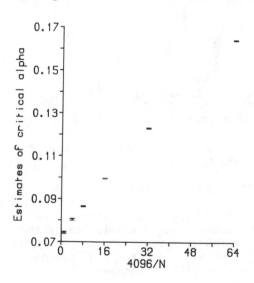

Fig. 2   Estimate of critical value of p/N for loss of memory states

SPURIOUS STATES

Amit et al[7] discuss some properties of the spurious minima states which are generated by the storage prescription (2). A competition between energy and entropy favours spurious states which are compositions of nominal states. In Fig. 3 we show the distribution for the Hamming distance between all pairs of vectors obtained by iteration from a random start, for N = 4096 nodes, in the (1,-1) model. For p as small as 15, the spurious states which are simple mixtures of nominal vectors are clearly visible. For larger N the width of the distribution is still much larger than $O(N^{1/2})$, showing clear evidence for the continuing clustering of spurious states in the subspace of nominal vectors. We obtain no evidence for an ultrametric structure for the minima obtained by iteration of (1) from random starts; any ultrametric structure may exist only for the relatively few[9] deep minima which would be identified only by an annealing dynamics.

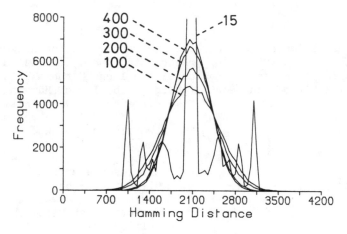

Fig. 3 Distribution of Hamming distances between pairs of minima for p = 15,100,200,300 and 400, and N = 4096.

PERFORMANCE OF A LEARNING ALGORITHM

We have studied the performance of an algorithm of the perceptron type[10] for storing exactly p nominal vectors while maintaining an energy function E with a T_{ij} which is symmetric and zero on the diagonal. Given a starting T_{ij}, define a mask for each nominal vector $\varepsilon_i^{(r)} = 1$ (if $S_i^{(r)}$ flips), 0 (if $S_i^{(r)}$ is stable). Then T_{ij} is modified by

$$\Delta T_{ij} = \sum S_i^{(r)} S_j^{(r)} [\varepsilon_i^{(r)} + \varepsilon_j^{(r)}] \quad (i \neq j), \quad O(i=j) . \qquad (5)$$

It can be shown that this algorithm will converge (to $\varepsilon^{(r)} = 0$) in a finite number of steps provided a solution exists.

We have studied on the DAP a modification of this algorithm, with a mask factor $[\varepsilon_i^{(r)} + \varepsilon_j^{(r)} - \varepsilon_i^{(r)}\varepsilon_j^{(r)}]$. Figure 4 shows the number of learning cycles needed to store exactly up to 512 vectors on 512 nodes. The approximately scaling form (with logarithmic corrections) can be understood in terms of the exponential decrease of the fraction of the bits wrong. In the context of phase transitions, the exact learning has raised α_2^c to the limit of the existence of a solution. It is likely that there is less impact on the value of α_1^c. However, we already have examples where learning is crucial for disentangling <u>correlated</u> nominal vectors and related spurious states; it will surely play an important role in assessing whether spurious states can indeed play a significant beneficial role.

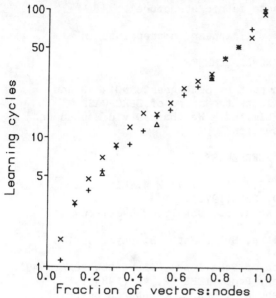

Fig. 4 Learning cycles required for N = 128(Δ), 256(+) and 512($\times$).

HIGHER ORDER CORRELATIONS

We have considered also the memory capacity and phase diagrams for models with higher order correlations:

$$E = \frac{m!}{N^{m-1}} \sum_{i_1 < i_2 \ldots < i_m}^{N} T_{i_1 i_2 \ldots i_m} S_{i_1} S_{i_2} \cdots S_{i_m} \,,$$

$$T_{i_1 i_2 \ldots i_m} = \sum_{r=1}^{p} S_{i_1}^{(r)} S_{i_2}^{(r)} \cdots S_{i_m}^{(r)} \tag{6}$$

It can be shown that any particular nominal vector is likely to be perfectly stored provided that $p < N^{m-1}/(2(m-1)!\ell nN)$. Such models therefore enable many more pictures to be stored on a given number of nodes, with the penalty of the increase in the number components of T, so that the efficiency of information storage is effectively unaltered. The increased capacity is also reflected in the positions of the phase transitions (in the sense of a previous section). For example spurious states become lower in energy than the nominal states at a critical value of $\alpha \equiv pm!/2N^{m-1}$. Amusingly, even the replica symmetry breaking can be calculated exactly in the limit $m \to \infty$: $\alpha_1{}^c = 1/(4\ln 2)$. (Technically, the Parisi order parameter[9] is a step function, and the ground states have a discrete metric structure).

Details of the above work will appear in preprints in preparation.

ACKNOWLEDGEMENTS

This work is supported in part by SERC grants NG14840 and NG15908. A.C. and B.F. acknowledge the award of SERC CASE studentships with ICL and GEC Limited. We thank David Bounds and David Willshaw for useful discussions.

REFERENCES

1. W.A. Little, Math. Biosc. 19, 101 (1974); W.A. Little and G.L. Shaw, Math. Biosc. 39, 281 (1978).
2. J.J. Hopfield, Proc. Nat. Acad. Sc. USA 79, 2554 (1982) and 81, 3088 (1984).
3. W.S. McCulloch and W.A. Pitts, Bull. Math. Biophys. 5, 115 (1943).
4. D.O. Hebb, The Organisation of Behaviour (Wiley, New York,1949).
5. K.C. Bowler and G.S. Pawley, Proc. IEEE 72, 42 (1984).
6. E. Gardner, D.J. Wallace and A.D. Bruce, Edinburgh preprint, in preparation.
7. D.J. Amit, H. Gutfreund and H. Sompolinsky, Phys. Rev. Lett. 55, 1530 (1985) and Phys. Rev. A32, 1007 (1985).
8. D.J. Wallace, in Lattice Gauge Theory - A Challenge in Large Scale Computing, eds. B. Bunk and K.H. Mutter (Plenum, 1986), to be published.
9. M. Mezard, G. Parisi, N. Sourlas, G. Toulouse and M. Virasoro, Phys. Rev. Lett. 52, 1156 (1984).
10. M.L. Minsky and S. Papert, Perceptrons (MIT Press, Cambridge, MA, 1969).

Influence of Noise on the Behaviour of an Autoassociative Neural Network

J. Buhmann and K. Schulten
Physik–Department
Technische Universität München, D-8046 Garching

Abstract

Recently, we simulated the activity and function of neural networks with neuronal units modelled after their physiological counterparts[1,2]. Neuronal potentials, single neural spikes and their effect on postsynaptic neurons were taken into account. The neural network studied was endowed with plastic synapses. The synaptic modifications were assumed to follow Hebbian rules, i.e. the synaptic strengths increase if the pre- and postsynaptic cells fire a spike synchronously and decrease if there exists no synchronicity between pre- and postsynaptic spikes. The time scale of the synaptic plasticity was that of mental processes, i.e. a tenth of a second as proposed by v.d. Malsburg[3]. In this contribution we extend our previous study and include random fluctuations of the neural potentials as observed in electrophysiological recordings[4]. We will demonstrate that random fluctuations of the membrane potentials raise the sensitivity and performance of the neural network. The fluctuations enable the network to react to weak external stimuli which do not affect networks following deterministic dynamics. We argue that fluctuations and noise in the membrane potential are of functional importance in that they trigger the neural firing if a weak receptor input is presented. The noise regulates the level of arousal. It might be an essential feature of the information processing abilities of neuronal networks and not a mere source of disturbance better to be suppressed. We will demonstrate that the neural network investigated here reproduce the computational abilities of formal associative networks[5,6,7].

Introduction

The neural system investigated is composed of a set of interconnected neurons the membrane potentials of which evolve according to deterministic rules and according to stochastic fluctuations. The connections to sensory organs or to other neural networks are taken into account by a primary set of receptors which send input to the neurons. The receptor–neuron connections form a local projection of the activity pattern presented by the receptors as modelled by a one–to–one or a center–surround connectivity. The system is schematically presented in Fig 1.

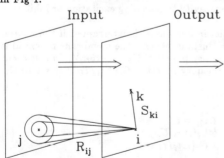

Input   Output

Receptors   Neurons

Figure 1: Schematic presentation of the neural model investigated: Receptors send spikes to a network of neurons. The resulting activity of the neural network is affected by an activity-dependent alteration of the synapses $S_{ik}(t)$, i.e. the network experiences a feed–back as indicated.

Dynamics of the Membrane Potential

The dynamics of the membrane potentials involves two processes, the relaxation of the membrane potential and the neural interaction as determined by the somatic integration rule. Axonal spikes are generated whenever the membrane potential reaches a threshold value. The postsynaptic excitation by presynaptic spikes is described by an exponential activity function with decay time $T_U = 1ms$

$$G_k(\Delta t_k/\tau) = exp\left(-\frac{\Delta t_k}{\tau}\right), \text{ with } \Delta t_k = t - t_{0k}. \tag{1}$$

$\Delta t_k = t - t_{0k}$ measures the time that has elapsed since the last spike of neuron k at t_{0k}.

The kinetic equations of the membrane potentials $U_i(t)$ which also include the stochastic fluctuations are given by a system of non–linear coupled Langevin equations

$$\frac{dU_i(t)}{dt} = -\frac{U_i(t)}{T_R} + \rho[\Delta t_i]\left(\omega\sigma[A_i(t)] + \frac{\eta}{\sqrt{T_R/2}}\xi(t)\right). \tag{2}$$

The first term in (2) approximates the relaxation of the membrane potential $U_i(t)$ to its resting value $U_0 = 0mV$ within a time interval $T_R = 2.5ms$. The second term in (2) describes the communication of the postsynaptic cell i with the connected neurons and receptors and adds a Gaussian white noise $\xi(t)$ with the strength $\eta/\sqrt{T_R/2}$. The noise produces a Gaussian distribution of the membrane potential $U_i(t)$ with mean value $U_0 = 0mV$ and variance $\eta = 10mV$. Afferent impinging activities in addition to the noise are integrated to the total postsynaptic excitation $A_i(t)$. The activity of the presynaptic neurons k or receptors j are weighted by the time-dependent synaptic strengths $S_{ik}(t)$ or the static receptor connection strengths R_{ij}, respectively

$$A_i(t) = \sum_k S_{ik}(t)G_k(\Delta t_k/T_U) + \sum_j R_{ij}G_k^R(\Delta t_k^R/T_U). \tag{3}$$

The sigmoidal function $\sigma[A_i(t)]$ with a linear behaviour for small $A_i(t)$ and a saturation value for strong activity prevents potential changes which are unphysiologically large. The total and relative refractory periods are taken into account by the function $\rho[\Delta t_i]$ which suppresses the sensitivity of neuron i to afferent excitation during a total refractory period $T_F = 5ms$. The function also lets the neuron gradually regain its sensitivity to incoming excitation or inhibition during a relative refractory period of $5ms$.

The continuous time evolution of the potential in our model is interrupted if the neuron reaches the threshold $U_T = 30mV$ and fires a spike. Instantaneously the membrane potential is set to a value normally distributed around the refractory potential $U_F = -15mV$. In this event the time of the last spike t_{0i} is updated and the memory function $G_i(\Delta t_i/T_U)$ is set to the value 1. This behaviour is represented as follows:

$$\text{if } U_i(t) \geq U_T \text{ then } \begin{cases} t_{0i} = t, \\ U_i(t) \approx U_F, \\ G_i(\Delta t_i/\tau) = 1. \end{cases} \tag{4}$$

The reaction of a neuron to a receptor input depends on the coupling constant ω and the connection strength R_{ik}. In the case of strong coupling the excited neuron will always reach the threshold whereas weak coupling causes only small postsynaptic potentials which never reach the threshold. Figure 2 shows the probability that a neuron which received a receptor spike at $t = 0ms$ will fire within $5ms$. This probability is presented as a function of the coupling strength ωR_{ik} for three different noise levels $(\eta = 0, 6, 10mV)$. Due to the synapse dynamics the mean spike probability of the neuron $\omega \overline{A_i(t)}$ is time-dependent and can be shifted by learning.

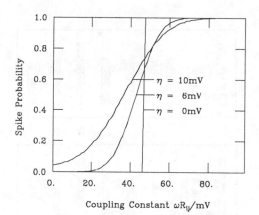

Coupling Constant $\omega R_{ij}/mV$

Figure 2: The probability to reach the threshold within $5ms$ after a receptor spike depends on the coupling between receptors and neurons. The gain of the curve strongly depends on the noise level η. In our computer simulations we have employed in most cases the value $\eta = 10mV$ and $\omega R_{ij} = 45mV$.

SYNAPTIC PLASTICITY IN THE STOCHASTIC NEURAL NETWORK

In our neural network with stochastic firing we introduced a plasticity of the synapses on a time scale of $0.2 - 0.5s$ [2]. According to the Hebbian rules the synaptic dynamics was assumed to depend on the synchronicity or asynchronicity of the pre- and postsynaptic spikes. In addition to the Hebbian rules we require for synaptic modifications in the present study that the mean spike frequencies $\overline{\nu_i}, \overline{\nu_k}$ of both neurons exceed considerably the spontaneous spike rate $\nu_s \approx 5s^{-1}$. If both neurons satisfy this condition in the case of synchronous firing the synapse can be strengthened. If only the presynaptic neuron fires with a high spike rate the synapse $S_{ik}(t)$ is weakened after each presynaptic spike. Details are described in Ref. 2.

The plasticity of the synapse with the strength $S_{ik}(t)$ connecting neuron k to neuron i is governed by the equation

$$\frac{dS_{ik}}{dt} = \begin{cases} -\dfrac{S_{ik}(t)-S_{ik}(0)}{T_S} + \Omega\, G_k\left(\dfrac{\Delta t_k}{T_M}\right)\, \kappa(G_i, G_k), & \text{if} \quad S_u \geq |S_{ik}| \geq S_l; \\ -\dfrac{S_{ik}(t)-S_{ik}(0)}{T_S}, & \text{else} \end{cases} \tag{5a}$$

with

$$\kappa(G_i, G_k) = \begin{cases} 1, & \text{if } G_i > G_k > e^{-1} \;\wedge\; \overline{\nu_i} \gg \nu_s \;\wedge\; \overline{\nu_k} \gg \nu_s; \\ -1, & \text{if } G_k > e^{-1} > G_i \;\wedge\; \overline{\nu_i} \ll \nu_s \;\wedge\; \overline{\nu_k} \gg \nu_s; \\ 0, & \text{else.} \end{cases} \tag{5b}$$

Equation (5a) holds both for excitatory and inhibitory synapses. The first term describes a relaxation process which leads to the gradual loss of stored information. The second term effects a change of the synaptic strength. The influence of this term decays exponentially with the presynaptic activity $G_k(\frac{\Delta t_k}{T_M})$. The short decay time $T_M = 2.5ms$ guarantees the Hebbian synchronicity condition for synaptic changes. The function $\kappa(G_i, G_k)$ switches between increase of the synaptic strength $(\kappa = 1)$, decrease $(\kappa = -1)$ and passive relaxation $(\kappa = 0)$ of the synapses to the initial value $S_{ik}(0)$. The characteristic time Ω^{-1} determines the time scale for synaptic modifications. The values assumed for Ω^{-1} were in the range $0.2 - 0.5s$.

LEARNING AND ASSOCIATION OF A PATTERN

The neural network presented showed remarkable associative properties in spite of the stochastic fluctuations of the membrane potentials. Starting from a homogeneous structure of synaptic connections with equal numbers of excitatory and inhibitory neurons the network learned a pattern presented by the receptors and associatively reconstructed the original pattern when only incomplete or disturbed patterns were presented.

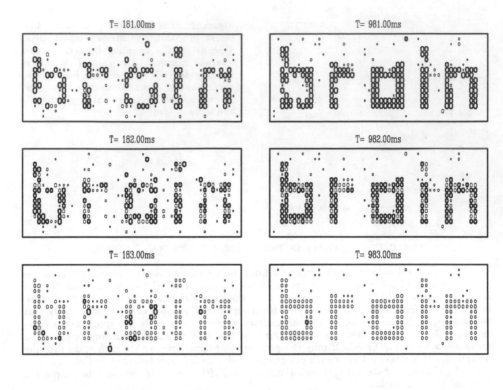

T= 181.00ms

T= 981.00ms

T= 182.00ms

T= 982.00ms

T= 183.00ms

T= 983.00ms

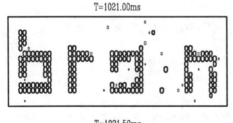

T=1021.00ms

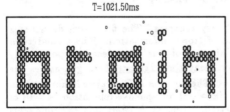

T=1021.50ms

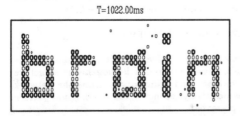

T=1022.00ms

Figure 3 (above):
The activity function $G_i(\Delta t_i/T_M)$ is shown for an untrained (left) and an instructed (right) network. At the beginning of the learning session (t=181ms,182ms,183ms) 75 percent of the excited neurons fire after a receptor spike. At the end of the learning stage (t=981ms,982ms,983ms) nearly all excited neurons have synchronized their firing behaviour and reach the threshold.

Figure 4 (left):
Network activity during the association task: The network associates the missing letter i by excitatory interaction within 2 milliseconds (t=1021ms,1021.5ms,1022ms).

The simulations of the network were carried out in three different stages. During a first stage which lasted $0.3 - 1.5s$ the neural network had to learn the pattern **brain** synchronously presented by the receptors with a frequency of $50s^{-1}$. A homogeneous background noise with a spike rate of $10s^{-1}$ was superimposed on the pattern. The coupling constant ωR_{ii} was set to $45mV$ which effected the firing of about 75 percent of excited neurons. In a second stage lasting $50ms$ the receptors rested quiescent and the electrical activity of the network relaxed to the spontaneous spike rate. During a third stage the receptors presented the test pattern **bra n** which differed from the originally learned pattern by the letter i being left out.

Figure 3a shows the activity of the network at the beginning of the learning phase. At $t = 180ms$ the receptors corresponding to the pattern **brain** had just fired. Within $3ms$, 75 percent of the excited neurons reach the threshold and fire. The other neurons are only gradually excited and fail to fire. The network reaction to a receptor input at the end of the learning stage is shown in Fig 3b. Due to the acquired excitatory synaptic connections between neurons receiving input directly from the pattern **brain** (pattern neurons) the assembly reacts more synchronously and the fault level, given by the number of pattern neurons which fail to fire, nearly vanishes.

The success of the learning session is documented in Fig 4. The incomplete test pattern **bra n** is associatively restored by the network. The neurons representing the missing letter i react with a delay time of $1 - 3ms$, i.e. they fire nearly synchronously with the neurons excited by the test pattern.

The synchronization of the neural activity and the associative abilities of the network can be understood on account of the synaptic structure acquired during the learning session. Figures 5a,b show the afferent synapses of neuron $(37,4)$ (presented by a star) after the training. All the neurons representing the pattern **brain** have developed saturated excitatory or inhibitory synapses to the reference neuron. During the association task the excitatory synapses saturated at a strength value S_u support the firing of the reference cell, whereas the inhibitory synapses saturated at $-S_l$ do not prevent the reference cell from firing. Afferent synapses of the reference cell $(37,4)$ coming from a background neuron rest at the initial synaptic strength.

Afferent Synapses from Neuron (37,4)

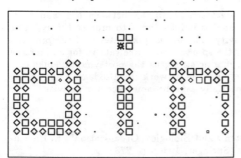

Figure 5: The size of the squares and the diamonds encodes the changes $S_{ik}(t) - S_{ik}(0)$ which the excitatory and inhibitory synapses acquired during the learning session, respectively.

Due to fluctuations of the membrane potential which raise the sensitivity of the neurons the network can also learn a pattern which at any given time is only partially presented by the receptors. At each time interval the invisible fraction of the pattern (50 percent of the receptors) is chosen randomly. The uninstructed network has to learn the total pattern from the detected spike coincidences. The evolution of the synapses is demonstrated for the case of the afferent synapses of neuron $(37,4)$ which represents the dot on the letter i. During the learning stage which lasts $3.7s$ the network has build up a synaptic structure which contains the information of the whole pattern (Fig 6). This simulation demonstrates that the synchronization of all pattern receptors at any given time is not a necessary condition for learning.

76

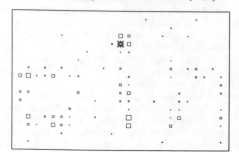

Figure 6: Evolution of the afferent synapses of neuron (37,4) for the times $t = 1s$ (left) and $t = 3.6s$ (right) during the learning stage. 50 percent of the pattern brain is invisible.

CONCLUSION

We have presented a model neural network with a high level of endogeneous noise acting on the cellular potentials. This noise which is inherent in all biological neurons does not destroy the abilities of the network to learn and associatively reconstruct patterns. On the contrary, the noise controls the level of arousal and makes the network capable to react to a weak receptor input otherwise neglected. We argue that noise has a functional importance in neural systems. The explicit simulation of single spikes allows to test the influence of single neural events which are averaged over by mean spike rate models[8]. In addition the nonspecific influence of large neural nets (neural activity bath) on small neural assemblies can also be studied by the stochastic dynamics.

On the basis of the Hebbian rules which detect synchronicities between pre- and postsynaptic spikes a second condition for synaptic changes is introduced to protect the synaptic structure against destruction by spontaneous activity. The mean spike rates $\bar{\nu}$ of the pre- and postsynaptic neurons have to exceed considerably the spontaneous spike rate ν_s for an increase of the synaptic strengths. For a decrease of the synaptic strengths the postsynaptic spike rate must be considerably below ν_s. With this modified rules the network can also learn highly noisy patterns and patterns which are presented by a partially asynchronous receptor activity.

REFERENCES

1. J. Buhmann, K. Schulten, in "Disordered Systems and Biological Organization", E. Bienenstock, F. Fogelman Soulié, G. Weisbuch, Eds. (Springer, 1986), p. 273.
2. J. Buhmann, K. Schulten, submitted to Biol. Cybern. (1986).
3. C.v.d. Malsburg, Int. Rep. 81/2, Dept. Neurobiologie, MPI f. Biophysikalische Chemie, Göttingen (1981).
4. M. Abeles, Local Cortical Circuits (Springer, 1982).
5. L.N. Cooper, Proceedings of the Nobel Symposium on Collective Properties of Physical Systems, eds. Lundqvist (Academic, 1973), p. 252.
6. J.J. Hopfield, Proc. Natl. Acad. Sci. USA 79, 2554 (1982).
7. T. Kohonen, Selforganization and Associative Memory (Springer, 1984).
8. J.J. Hopfield, Proc. Natl. Acad. Sci. USA 81, 3088 (1984).

ABSOLUTELY STABLE LEARNING
OF RECOGNITION CODES
BY A SELF-ORGANIZING NEURAL NETWORK

Gail A. Carpenter†

Department of Mathematics, Northeastern University, Boston, MA 02115,
and Center for Adaptive Systems, Boston University, Boston, MA 02215

Stephen Grossberg‡

Center for Adaptive Systems, Boston University, Boston, MA 02215

ABSTRACT

A neural network which self-organizes and self-stabilizes its recognition codes in response to arbitrary orderings of arbitrarily many and arbitrarily complex binary input patterns is here outlined. Top-down attentional and matching mechanisms are critical in self-stabilizing the code learning process. The architecture embodies a parallel search scheme which updates itself adaptively as the learning process unfolds. After learning self-stabilizes, the search process is automatically disengaged. Thereafter input patterns directly access their recognition codes, or categories, without any search. Thus recognition time does not grow as a function of code complexity. A novel input pattern can directly access a category if it shares invariant properties with the set of familiar exemplars of that category. These invariant properties emerge in the form of learned critical feature patterns, or prototypes. The architecture possesses a context-sensitive self-scaling property which enables its emergent critical feature patterns to form. They detect and remember statistically predictive configurations of featural elements which are derived from the set of all input patterns that are ever experienced. Four types of attentional process—priming, gain control, vigilance, and intermodal competition—are mechanistically characterized. Top-down priming and gain control are needed for code matching and self-stabilization. Attentional vigilance determines how fine the learned categories will be. If vigilance increases due to an environmental disconfirmation, then the system automatically searches for and learns finer recognition categories. A new nonlinear matching law (the 2/3 Rule) and new nonlinear associative laws (the Weber Law Rule, the Associative Decay Rule, and the Template Learning Rule) are needed to achieve these properties. All the rules describe emergent properties of parallel network interactions. The architecture circumvents the saturation, capacity, orthogonality, and linear predictability constraints that limit the codes which can be stably learned by alternative recognition models.

† Supported in part by the Air Force Office of Scientific Research (AFOSR 85-0149 and AFOSR F49620-86-C-0037) and the National Science Foundation (NSF DMS-84-13119).

‡ Supported in part by the Air Force Office of Scientific Research (AFOSR 85-0149 and AFOSR F49020-86-C-0037), the Army Research Office (ARO DAAG-29-85-K0095), and the National Science Foundation (NSF IST-84-17756).

Acknowledgement: We wish to thank Carol Yanakakis for her valuable assistance in the preparation of the manuscript.

SEARCH CYCLE:
INTERACTIONS BETWEEN ATTENTIONAL
AND ORIENTING SUBSYSTEMS

The neural network outlined herein is called an ART system, after the adaptive resonance theory introduced by Grossberg[1]. More recently, ART networks have been further characterized, and their dynamic properties have been derived in a series of theorems[2-4]. A single cycle of the search process carried out by this ART network is depicted in Figure 1. In Figure 1a, an input pattern I generates a short term memory (STM) activity pattern X across a field of feature detectors F_1. The input I also excites an *orienting subsystem A*, but pattern X at F_1 inhibits A before it can generate an output signal. Activity pattern X also elicits an output pattern S which, via the bottom-up adaptive filter, instates an STM activity pattern Y across a category representation field, F_2. In Figure 1b, pattern Y reads a top-down template pattern V into F_1. Template V mismatches input I, thereby significantly inhibiting STM activity across F_1. The amount by which activity in X is attenuated to generate X^* depends upon how much of the input pattern I is encoded within the template pattern V.

When a mismatch attenuates STM activity across F_1, the total size of the inhibitory signal from F_1 to A is also attenuated. If the attenuation is sufficiently great, inhibition from F_1 to A can no longer prevent the arousal source A from firing. Figure 1c depicts how disinhibition of A releases an arousal burst to F_2 which equally, or nonspecifically, excites all the F_2 cells. The cell populations of F_2 react to such an arousal signal in a state-dependent fashion. In the special case that F_2 chooses a single population for STM storage, the arousal burst selectively inhibits, or resets, the active population in F_2. This inhibition is long-lasting. One physiological design for F_2 processing which has these properties is a *gated dipole field*[5,6]. A gated dipole field consists of opponent processing channels which are gated, or multiplied, by habituating chemical transmitters. A nonspecific arousal burst induces selective and enduring inhibition of active populations within a gated dipole field.

In Figure 1c, inhibition of Y leads to removal of the top-down template V, and thereby terminates the mismatch between I and V. Input pattern I can thus reinstate the original activity pattern X across F_1, which again generates the output pattern S from F_1 and the input pattern T to F_2. Due to the enduring inhibition at F_2, the input pattern T can no longer activate the original pattern Y at F_2. A new pattern Y^* is thus generated at F_2 by I (Figure 1d).

The new activity pattern Y^* reads-out a new top-down template pattern V^*. If a mismatch again occurs at F_1, the orienting subsystem is again engaged, thereby leading to another arousal-mediated reset of STM at F_2. In this way, a rapid series of STM matching and reset events may occur. Such an STM matching and reset series controls the system's search of long term memory (LTM) by sequentially engaging the novelty-sensitive orienting subsystem. Although STM is reset sequentially in time via this mismatch-mediated, self-terminating LTM search process, the mechanisms which control the LTM search are all parallel network interactions, rather than serial algorithms. Such a parallel search scheme continuously adjusts itself to the system's evolving LTM codes. In general, the spatial configuration of LTM codes depends upon both the system's initial configuration and its unique learning history, and hence cannot be predicted *a priori* by a pre-wired search algorithm. Instead, the mismatch-mediated engagement of the orienting subsystem realizes a self-adjusting search.

The mismatch-mediated search of LTM ends when an STM pattern across

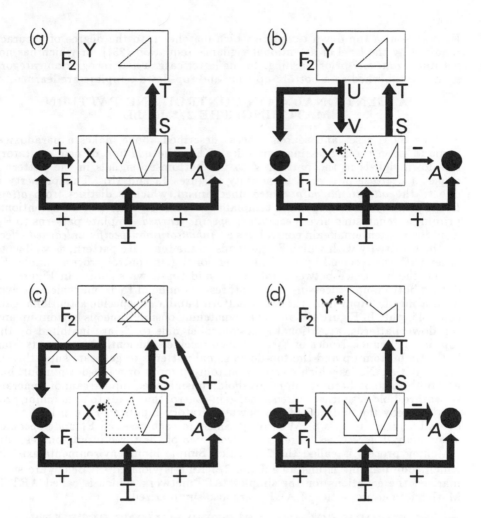

1. Search for a correct F_2 code: (a) The input pattern I generates the specific STM activity pattern X at F_1 as it nonspecifically activates A. Pattern X both inhibits A and generates the output signal pattern S. Signal pattern S is transformed into the input pattern T, which activates the STM pattern Y across F_2. (b) Pattern Y generates the top-down signal pattern U which is transformed into the template pattern V. If V mismatches I at F_1, then a new STM activity pattern X^* is generated at F_1. The reduction in total STM activity which occurs when X is transformed into X^* causes a decrease in the total inhibition from F_1 to A. (c) Then the input-driven activation of A can release a nonspecific arousal wave to F_2, which resets the STM pattern Y at F_2. (d) After Y is inhibited, its top-down template is eliminated, and X can be reinstated at F_1. Now X once again generates input pattern T to F_2, but since Y remains inhibited T can activate a different STM pattern Y^* at F_2. If the top-down template due to Y^* also mismatches I at F_1, then the rapid search for an appropriate F_2 code continues.

F_2 reads-out a top-down template which matches I, to the degree of accuracy required by the level of attentional vigilance (equation (23)), or which has not yet undergone any prior learning. In the latter case, a new recognition category is then established as a bottom-up code and top-down template are learned.

ATTENTIONAL GAIN CONTROL AND PATTERN MATCHING: THE 2/3 RULE

The STM reset and search process described above makes a paradoxical demand upon the processing dynamics of F_1: the *addition* of new excitatory top-down signals in the pattern V to the bottom-up signals in the pattern I causes a *decrease* in overall F_1 activity (Figures 1a and 1b). This property is due to the *attentional gain control* mechanism, which is distinct from *attentional priming* by the top-down template V. While F_2 is active, the attentional priming mechanism delivers *excitatory specific learned* template patterns to F_1. Top-down attentional gain control has an *inhibitory nonspecific unlearned* effect on the sensitivity with which F_1 responds to the template pattern, as well as to other patterns received by F_1. The attentional gain control process enables F_1 to tell the difference between bottom-up and top-down signals. In Figure 1a, during bottom-up processing, a suprathreshold node in F_1 is one which receives both a specific input from the input pattern I and a nonspecific attentional gain control input. In Figure 1b, during the matching of simultaneous bottom-up and top-down patterns, attentional gain control signals to F_1 are inhibited by the top-down channel. Nodes of F_1 must then receive sufficiently large inputs from both the bottom-up and the top-down signal patterns to generate suprathreshold activities. Nodes which receive a bottom-up input or a top-down input, but not both, cannot become suprathreshold: mismatched inputs cannot generate suprathreshold activities. Attentional gain control thus leads to a matching process whereby the addition of top-down excitatory inputs to F_1 can lead to an overall decrease in F_1's STM activity. Since, in each case, an F_1 node becomes active only if it receives large signals from two of the three input sources, this matching process is called the 2/3 Rule. Simple input environments exist in which code learning is unstable if the 2/3 Rule is violated[3,4]. Below are summarized the equations for the simplest ART network, which is called ART 1. Mathematical properties of ART 1 are also summarized.

NETWORK EQUATIONS: INTERACTIONS BETWEEN SHORT TERM MEMORY AND LONG TERM MEMORY PATTERNS

The STM equations for F_1 and F_2 and LTM equations for the bottom-up and top-down adaptive filters will now be described in dimensionless form, where the number of parameters is reduced to a minimum.

A. STM Equations

The STM activity x_k of any node v_k in F_1 or F_2 obeys a membrane equation of the form

$$\epsilon \frac{d}{dt} x_k = -x_k + (1 - Ax_k)J_k^+ - (B + Cx_k)J_k^-, \tag{1}$$

where J_k^+ is the total excitatory input to v_k, J_k^- is the total inhibitory input to v_k, and all the parameters are nonnegative.

Nodes in F_1 are denoted by v_i, where $i = 1, 2, \ldots, M$. Nodes in F_2 are denoted by v_j, where $j = M + 1, M + 2, \ldots, N$. Thus by (1),

$$\epsilon \frac{d}{dt} x_i = -x_i + (1 - A_1 x_i)J_i^+ - (B_1 + C_1 x_i)J_i^-$$ (2)

and

$$\epsilon \frac{d}{dt} x_j = -x_j + (1 - A_2 x_j)J_j^+ - (B_2 + C_2 x_j)J_j^- .$$ (3)

The excitatory input J_i^+ to the ith node v_i of F_1 in equation (2) is a sum of the bottom-up input I_i, the top-down template input V_i, and the nonspecific gain control input G. The top-down template input is the sum of all signals from F_2 nodes, via the adaptive filter:

$$V_i = D_1 \sum_j f(x_j) z_{ji},$$ (4)

where $f(x_j)$ is the signal generated by activity x_j of node v_j and z_{ji} is the LTM trace in the top-down pathway from v_j to v_i. Each gain control input is given by:

$$G = \begin{cases} G_1 & \text{if } I \text{ is active and } F_2 \text{ is inactive} \\ 0 & \text{otherwise.} \end{cases}$$ (5)

Setting

$$J_i^- = 1,$$ (6)

embodies the assumption that when no inputs are being processed ($J_i^+ = 0$), F_1 nodes are maintained at a tonic subthreshold level; that is, $x_i < 0$. When I becomes active while F_2 is inactive,

$$\epsilon \frac{dx_i}{dt} = -x_i + (1 - A_1 x_i)(I_i + G_1) - (B_1 + C_1 x_i)$$ (7)
$$= (I_i + G_1 - B_1) - x_i(1 + A_1(I_i + G_1) + C_1).$$

In the dimensionless equations, $0 \le I_i \le 1$. The 2/3 Rule requires that v_i become active when $I_i = 1$ but remain inactive when $I_i = 0$. The output threshold of each F_1 node v_i equals 0. Thus by (7), v_i becomes active iff $I_i + G_1 > B_1$. Therefore implementation of the 2/3 Rule when F_2 is inactive places constraints (8) on the strength of the gain control signal:

$$G_1 < B_1 < 1 + G_1.$$ (8)

At F_2, the excitatory input J_j^+ in equation (3) is the sum of a positive feedback signal $g(x_j)$ from v_j to itself and the bottom-up adaptive filter input T_j. The bottom-up input is the sum:

$$T_j = D_2 \sum_i h(x_i) z_{ij},$$ (9)

where $h(x_i)$ is the signal emitted by the F_1 node v_i and z_{ij} is the LTM trace in the pathway from v_i to v_j. Thus

$$J_j^+ = g(x_j) + T_j. \tag{10}$$

Input J_j^- adds up negative feedback signals $g(x_k)$ from all the other nodes in F_2:

$$J_j^- = \sum_{k \neq j} g(x_k). \tag{11}$$

Taken together, the positive feedback signal $g(x_j)$ in (10) and the negative feedback signal J_j^- in (11) define an on-center off-surround feedback interaction which contrast-enhances the STM activity pattern Y of F_2 in response to the input pattern T.

The parameters of F_2 can be chosen so that this contrast-enhancement process enables F_2 to choose for STM activation only the node v_j which receives the largest input T_j[7]. Then when parameter ϵ is small, F_2 behaves approximately like a binary switching, or choice, circuit:

$$f(x_j) = \begin{cases} 1 & \text{if } T_j = \max\{T_k\} \\ 0 & \text{otherwise.} \end{cases} \tag{12}$$

In the choice case, the top-down template in (4) obeys

$$V_i = \begin{cases} D_1 z_{ji} & \text{if the } F_2 \text{ node } v_j \text{ is active} \\ 0 & \text{if } F_2 \text{ is inactive.} \end{cases} \tag{13}$$

In the choice case, then, when I is active and the F_2 node v_j is active,

$$\epsilon \frac{dx_i}{dt} = -x_i + (1 - A_1 x_i)(I_i + D_1 z_{ji}) - (B_1 + C_1 x_i)$$
$$= (I_i + D_1 z_{ji} - B_1) - x_i(1 + A_1(I_i + D_1 z_{ji}) + C_1). \tag{14}$$

In the dimensionless equations, $0 \leq z_{ij} \leq 1$. The 2/3 Rule requires that v_i remain active when $I_i = 1$ and $z_{ji} = 1$, but become inactive when either $I_i = 0$ or $z_{ji} = 0$. By (14), x_i remains positive iff $I_i + D_1 z_{ji} > B_1$. Thus implementation of the 2/3 Rule when F_2 is active places constraint (15) on the strength of the patterned input signals:

$$\max\{1, D_1\} < B_1 < 1 + D_1. \tag{15}$$

The 2/3 Rule implies that if the top-down LTM trace z_{ji} becomes smaller than some critical valve $\bar{z}$, then when v_j is active, v_i will be inactive even if $I_i = 1$. That is, the feature represented by the F_1 node v_i will drop out of the critical feature pattern coded by v_j. By (14) and (15),

$$\bar{z} = \frac{B_1 - 1}{D_1}. \tag{16}$$

B. LTM Equations

The LTM trace of the bottom-up pathway from v_i to v_j obeys a learning equation of the form

$$\frac{d}{dt}z_{ij} = K_1 f(x_j)[-E_{ij}z_{ij} + h(x_i)],\tag{17}$$

where

$$h(x_i) = \begin{cases} 1 & \text{if } x_i > 0 \\ 0 & \text{if } x_i \leq 0. \end{cases}\tag{18}$$

In (17), term $f(x_j)$ is a postsynaptic sampling, or learning, signal because $f(x_j) = 0$ implies $\frac{d}{dt}z_{ij} = 0$. Term $f(x_j)$ is also the output signal of v_j to pathways from v_j to F_1, as in (4).

The LTM trace of the top-down pathway from v_j to v_i also obeys a learning equation of the form

$$\frac{d}{dt}z_{ji} = K_2 f(x_j)[-E_{ji}z_{ji} + h(x_i)].\tag{19}$$

In the present model, the simplest choice of K_2 and E_{ji} was made for the top-down LTM traces:

$$K_2 = E_{ji} = 1.\tag{20}$$

A more complex choice of E_{ij} was made for the bottom-up LTM traces in order to generate the *Weber Law Rule*, which is needed to achieve direct access to codes for arbitrary input environments after learning self-stabilizes. The Weber Law Rule requires that the positive bottom-up LTM traces learned during the encoding of an F_1 pattern X with a smaller number $| X |$ of active nodes be larger than the LTM traces learned during the encoding of an F_1 pattern with a larger number of active nodes, other things being equal. This inverse relationship between pattern complexity and bottom-up LTM trace strength can be realized by allowing the bottom-up LTM traces at each node v_j to compete among themselves for synaptic sites. The Weber Law Rule can also be generated by the STM dynamics of F_1 when competitive interactions are assumed to occur among the nodes of F_1.

Competition among the LTM traces which abut the node v_j is modelled by defining

$$E_{ij} = h(x_i) + L^{-1}\sum_{k\neq i} h(x_k)\tag{21}$$

and letting $K_1 = $ constant. It is convenient to write K_1 in the form $K_1 = KL$. A physical interpretation of this choice can be seen by rewriting (17) in the form

$$\frac{d}{dt}z_{ij} = Kf(x_j)[(1 - z_{ij})Lh(x_i) - z_{ij}\sum_{k\neq i} h(x_k)].\tag{22}$$

By (22), when a postsynaptic signal $f(x_j)$ is positive, a positive presynaptic signal from the F_1 node v_i can commit receptor sites to the LTM process z_{ij} at a rate $(1 - z_{ij})Lh(x_i)Kf(x_j)$. In other words, uncommitted sites—which number

$(1 - z_{ij})$ out of the total population size 1—are committed by the joint action of signals $Lh(x_i)$ and $Kf(x_j)$. Simultaneously signals $h(x_k)$, $k \neq i$, which reach v_j at different patches of the v_j membrane, compete for the sites which are already committed to z_{ij} via the mass action competitive terms $-z_{ij}h(x_k)Kf(x_j)$. In other words, sites which are committed to z_{ij} lose their commitment at a rate $-z_{ij}\sum_{k \neq i} h(x_k)Kf(x_j)$ which is proportional to the number of committed sites z_{ij}, the total competitive input $-\sum_{k \neq i} h(x_k)$, and the postsynaptic gating signal $Kf(x_j)$.

C. STM Reset System

A simple type of mismatch-mediated activation of A and STM reset of F_2 by A were implemented for binary inputs. Each active input pathway sends an excitatory signal of size P to the orienting subsystem A. Potentials x_i of F_1 which exceed zero generate an inhibitory signal of size Q to A. These constraints lead to the following Reset Rule.

Population A generates a nonspecific reset wave to F_2 whenever

$$\frac{|X|}{|I|} < \rho = \frac{P}{Q} \tag{23}$$

where I is the current input pattern, $|X|$ is the number of nodes across F_1 such that $x_i > 0$, and ρ is called the *vigilance parameter*. The nonspecific reset wave successively shuts off active F_2 nodes until the search ends or the input pattern I shuts off. Thus (12) must be modified as follows to maintain inhibition of all F_2 nodes which have been reset by A during the presentation of I:

$$f(x_j) = \begin{cases} 1 & \text{if } T_j = \max\{T_k : k \in \mathbf{J}\} \\ 0 & \text{otherwise} \end{cases} \tag{24}$$

where $\mathbf{J}$ is the set of indices of F_2 nodes which have not yet been reset on the present learning trial. At the beginning of each new learning trial, $\mathbf{J}$ is reset at $\{M + 1 \ldots N\}$. As a learning trial proceeds, $\mathbf{J}$ loses one index at a time until the mismatch-mediated search for F_2 nodes terminates.

THEOREMS WHICH CHARACTERIZE THE GLOBAL DYNAMICS OF THE ART 1 SYSTEM

A series of theorems[4] analyze the global dynamics of the ART system. The theorems are proved for the case that the input patterns are binary and that "fast learning" occurs, i.e., that the LTM traces approach their equilibrium values on each trial. With these hypotheses, the learning process is shown to self-stabilize. That is, after a finite number of trials, the learned critical feature pattern associated with each F_2 node remains constant. Thereafter, each input directly accesses that category whose critical feature pattern matches it best. This self-stabilization property does *not* require the assumption that plasticity is turned off, i.e., that K_1 in (17) and K_2 in (19) approach 0 after some finite interval. The length of time needed for the code to self-stabilize depends only upon the complexity of the set of input patterns, and is not set externally or *a priori*.

The theorems further specify details of system dynamics. For example, each LTM strength $z_{ij}(t)$ and $z_{ji}(t)$ is shown to oscillate at most once as learning proceeds. This occurs despite the fact that, in a complex input environment, many searches and category recodings may occur before the system self-stabilizes. Thus the learning process is remarkably stable. Also, given an arbitrary learning history, the order of search elicited by any input is characterized. The order of search is determined by bottom-up F_2 inputs T_j. Note, however, that the sum T_j depends upon both the pattern of STM activity across F_1 and the strengths of all the bottom-up LTM traces z_{ij}. Fluctuations which occur in these STM and LTM values could, in principle, destabilize the system as follows. First, the initial choice of an F_2 node depends only upon the F_1 (STM) activity pattern generated by I and the system's prior learning (LTM) history (Figure 1a). However, once F_2 becomes active, read-out of its template alters F_1 activity (Figure 1b). This read-out can dramatically alter the distribution of T_j values. However, the theorems guarantee that the original F_2 choice is confirmed by template read-out, so search proceeds as in Figure 1. Once search ends, however, learning alters both the pattern of F_1 STM activity, via changes in the top-down LTM traces, and the F_2 input function T_j, via the bottom-up LTM traces. The theorems also guarantee that the F_2 choice is confirmed by learning. In sum, F_2 reset can occur only via the orienting subsystem, which is activated by a mismatch between the input pattern and the critical feature pattern of an active F_2 node. While the order of search depends upon the entire coding history of the network, the decision to end the search depends upon the matching criterion as determined by the vigilance parameter ρ.

The size of ρ determines how coarse the learned recognition code will be. A small value of ρ leads to coarse recognition categories, whereas a large value of ρ leads to fine recognition categories. Environmental disconfirmation can increase ρ, thereby enabling the network to learn finer distinctions than it previously could. Using such a scheme, an alphabet of 26 letters can be classified in no more than 3 learning trials, at any level of vigilance.

REFERENCES

1. S. Grossberg, *Biol. Cyb.* **23**, 187-202 (1976).
2. G.A. Carpenter and S. Grossberg, *Proc. of the Third Army Conf. on Applied Math. and Comp.*, ARO Report 86-1, 37-56 (1985).
3. G.A. Carpenter S. Grossberg. In J. Davis, R. Newburgh, and E. Wegman (Eds.), **Brain structure, learning, and memory**. AAAS Symposium Series (1986).
4. G.A. Carpenter and S. Grossberg, *Comp. Vis., Graphics, and Img. Proc.* (1986).
5. S. Grossberg, *Psych. Rev.* **87**, 1-51 (1980).
6. S. Grossberg. In R. Karrer, J. Cohen, and P. Tueting (Eds.), **Brain and information: Event related potentials** (New York Academy of Sciences, N.Y., 1984).
7. S. Grossberg, *Stud. Appl. Math.* **52**, 217-257 (1973).

HIGH ORDER CORRELATION MODEL FOR ASSOCIATIVE MEMORY

H. H. Chen, Y. C. Lee[*], G. Z. Sun, and H. Y. Lee
University of Maryland, College Park, Md. 20742

Tom Maxwell
Sachs Freeman Associates/Naval Research Laboratory

C. Lee Giles
Air Force Office for Scientific Research

ABSTRACT

A neural network model of associative memory with higher order learning rule is presented. The new model could be fashioned to either the auto-associative or the multiple associative mode. Energy function, asynchronous or synchronous dynamics can be constructed. The retrieval of the stored patterns or pattern sets from an incomplete input is monotonic guaranteed by a convergence theorem. The higher-order correlation model shows dramatic improvement in its storage capacity in comparison to the conventional binary correlation model. It also opens up the possibility of storing spatial-temporal patterns and symmetry invariant patterns.

I. INTRODUCTION

Most of the conventional neural network models of associative memory are based on Hebb's associative conditioning rule:[1] when the firing of neuron cell A repeatedly causes the firing of the neuron B. The connection weight between cells A and B is increased. In these models, the Hebb's rule is incorporated into a binary correlation memory matrix[2,3]

$$T_{ij} = \sum_{\alpha=1}^{M} \xi_i^\alpha \xi_j^\alpha \tag{1}$$

where ξ_i^α assumes the value +1 or -1 representing the state of the i'th neuron in the αth pattern stored.

The most successful of these models is due to Hopfield[2], who introduced an energy function

$$E = - \sum_{i,j} S_i [T_{ij} S_j - 2\theta] \tag{2}$$

and an iterative dynamics

[*]Also at Center for Nonlinear Studies, Los Alamos National Laboratory.

$$S_i = W(\sum_j T_{ij} S_j - \theta) \tag{3}$$

where θ is a threshold value and W is a logical threshold function yielding a value $+1$ (or -1) depending on the positiveness (or the negativeness) of its argument.

Because of the symmetric form of T_{ij}, and the asynchronous updating of the neuron states (one at a time), Hopfield showed that the orbit of the dynamics is essentially a monotonic downhill slide on the energy surface in a high-dimensional phase space of the neuron states. The convergence theorem can be shown mathematically as

$$\Delta_i E = - \sum_j \Delta S_i (T_{ij} S_j - \theta) \leq 0 . \tag{4}$$

The final states of the dynamics are the minimum energy states of the system. When the number of stored patterns M is limited to[4,5]

$$M \lesssim \frac{N}{4 \log N} = M_c \tag{5}$$

where N is the size of the pattern (number of neurons). One can show statistically that the stored patterns are just these local minimum energy states. When M exceeds the capacity M_c, a lot of these minimum energy states turn into spurious states and the network's ability to store memory patterns is drastically reduced. M_c is a rather small number in comparison to the maximum number of patterns the network can represent (2^N). The problem is much worse in the storage of partially coherent patterns. For example, it is virtually impossible to store the twenty six English alphabet into a single neural network, of pixel size $N = 5 \times 7 = 35$. (see Fig. 1). Furthermore, Hopfield model works on the symmetrical correlation matrix (1). This restricts the model to auto-associative recalls only. In order to perform hetero-associative task, Hopfield introduced an asymmetric correlation matrix but did not provide a corresponding energy function and therefore could not retrieve the heterally associative memory pairs completely.

In this paper, we introduce a new model of associative memory utilizing higher-order correlations.[6] It not only increases dramatically the memory capacity but also allows the recall of multiply associated pattern sets. Energy function is constructed and iteration dynamics introduced. We found that in the case of even-order correlations synchronized dynamics can be implemented with monotonically convergent behavior. This technical aspect would increase the processing speed tremendously when running the model on a parallel machine.

II. THE HIGH-ORDER-CORRELATION MODEL

In a neural network model, the basic memory storage element is the connection weight between neurons (synapses). The more the number of connections, the larger the capacity of memories. For a Hebbian-type model, the memory matrix is binary, we have only C_2^N = N(N-])/2 number of connections. However, if we allow higher-order correlations, say mth order correlation, then the number of connections would increase to C_m^N = N(N-1)...(N-m+1)/m! which is enormously large when N itself is large. For example, in the case of one hundred neurons N = 100, Hopfield and others showed that the Hebbian model can store stably about five patterns. On the other hand, our simulation shows that the number increased to ∿500 for a triple correlation model.

Our model replaces the binary correlation matrix (1) by a multiply associated high order correlation tensor,

$$T_{i_1 i_2 \cdots j_1 j_2 \cdots k_1 k_2 \cdots} = \sum_\alpha \xi_{i_1}^\alpha \xi_{i_2}^\alpha \cdots \zeta_{j_1}^\alpha \zeta_{j_2}^\alpha \cdots \eta_{k_1}^\alpha \eta_{k_2}^\alpha \cdots \quad (6)$$

where ξ^α, ζ^α, η^α... belong to the α-th stored pattern set. They are multiply associated in the sense that one could recall any pattern from this set by an input of partially correct remainder patterns in the set. To facilitate this recall process, we follow Hopfield by introducing an energy (Lyapunov) function

$$E = \sum_{\substack{i_1, i_2 \cdots \\ j_1, j_2 \cdots \\ k_1, k_2 \cdots}} S_{i_1} S_{i_2} \cdots t_{j_1} t_{j_2} \cdots u_{k_1} u_{k_2} \cdots T_{i_1 i_2 \cdots j_1 j_2 \cdots k_1 k_2 \cdots} \quad (7)$$

In this expression we have set the threshold to zero for convenience. Note also that we have multi-slabs of neurons ($S_i, t_j u_k \cdots$) with connections among neurons in different slabs as well as among neurons within the same slab. Different slabs may have different size. $N_s \neq N_t \neq N_u \cdots$ in general.

The dynamical evolution of the neurons is governed by

$$S_{i_1} = W(\sum_{\substack{i_2 \cdots \\ j_1 j_2 \cdots \\ k_1 k_2 \cdots}} T_{i_1 i_2 \cdots j_1 j_2 \cdots k_1 k_2 \cdots} S_{i_2} \cdots t_{j_1} t_{j_2} \cdots u_{k_1} u_{k_2} \cdots \quad (8a)$$

$$t_{j_1} = W(\sum_{\substack{i_1 i_2 \cdots \\ j_2 \cdots \\ k_1 k_2 \cdots}} T_{i_1 i_2 \cdots j_1 j_2 \cdots k_1 k_2 \cdots} S_{i_1} S_{i_2} \cdots t_{j_2} \cdots u_{k_1} u_{k_2} \cdots \quad (8b)$$

$$u_{k_1} = W(\sum_{\substack{i_1 i_2 \cdots \\ j_1 j_2 \cdots \\ k_2 \cdots \\ \vdots}} T_{i_1 i_2 \cdots j_1 j_2 \cdots k_2 \cdots} S_{i_1} S_{i_2} \cdots t_{j_1} t_{j_2} \cdots u_{k_2} \cdots) \quad (8c)$$

The updating of the neurons can be done either asynchronously or synchronously. In the latter case, all the neurons are updated simultaneously whereas in the former case only one neuron is updated at a time. Both dynamics essentially converge. We can establish a convergence theorem as the following.

(A) Asynchronous dynamics converge if the correlation tensor $T_{i_1 i_2 \cdots j_1 j_2 \cdots k_1 k_2 \cdots}$ does not contain the diagonal terms. That is $T = 0$ whenever $i_\alpha = i_\beta$, or $j_\alpha = j_\beta$ or $k_\alpha = k_\beta \cdots$

(B) Synchronous dynamics converge if the correlations in each slab is even. We should retain the diagonal terms in the correlation tensor in this case.

We demonstrate the validity of this theorem in the case of auto-associative recall. The proof of general multiply associative recall is similar and will not be presented here.

For auto-associative recall, we have the correlation tensor given as

$$T_{ijk\ell\ldots} = \sum_\alpha \xi_i^\alpha \xi_j^\alpha \xi_k^\alpha \xi_\ell^\alpha \cdots$$

$$ \quad (9)$$

$$E = - \sum_\alpha (\vec{S} \cdot \vec{\xi}^\alpha)^n$$

where n is the order of correlations. Consider first case (A), the asynchronous scheme. We have a change in energy induced by updating a single neuron S_i

$$\Delta_i E \equiv E(\vec{S} + \Delta S_i) - E(\vec{S})$$

$$ \quad (10)$$

$$ = - (\Delta S_i) \sum_\alpha (\vec{S} \cdot \vec{\xi}^\alpha)^{n-1} \xi_i \leq 0$$

The last inequality follows from the updating rule (8). Since the energy is bounded, Eq. (10) guarantees the evolution of the system toward its closest minimum energy state.

As of the synchronous scheme, we set $n = 2q$ for even order correlations and find

$$\Delta E = E(\vec{S} + \Delta \vec{S}) - E(\vec{S}) = - 2q \sum_\alpha (\vec{\xi}^\alpha \cdot \vec{S})^{2q-1} (\vec{\xi}^\alpha \cdot \Delta \vec{S})$$

$$ - \sum_\alpha \sum_{\ell=2}^{2q} C_\ell^{2q} (\vec{\xi}^\alpha \cdot \vec{S})^{2q-\ell} (\vec{\xi}^\alpha \cdot \Delta \vec{S})^\ell$$

$$ < \sum_\alpha \sum_{\ell=2}^{2q} C_\ell^{2q} \left(\frac{\vec{\xi}^\alpha \cdot \Delta \vec{S}}{\vec{\xi}^\alpha \cdot \vec{S}} \right)^\ell (\vec{\xi}^\alpha \cdot \vec{S})^{2q} .$$

In order to show that $\Delta E < 0$, it is sufficient to prove that the real polynomial

$$f_q^{(x)} = \sum_{\ell=2}^{2q} C_\ell^{2q} x^\ell = (1+x)^{2q} - (1+2qx)$$

is non-negative.

Apply mathematical induction, we have obviously

$$f_1(x) = x^2 \geq 0.$$

For general q, we have

$$f_{q+1}(x) - f_q(x) = (1+x)^{2q+2} - (1+x)^{2q} - 2x$$

$$= \left[(1+x)^{2q} - 1\right]\left[(1+x^2) - 1\right] + x^2$$

$$\geq \left[(1+x)^{2q} - 1\right]\left[(1+x)^2 - 1\right]$$

$$\geq 0$$

q.e.d.

SIMULATION AND DISCUSSIONS

Numerical simulations generally confirm the prediction of higher capacity for pattern storages with the high-order correlation model. In the case of random patterns, we have done extensive simulations with thirty neurons. Our results show that if we require the probability of successfully retrieving any stored pattern be greater than ninety percent, then the maximum number of stored patterns increase dramatically with the order of correlations. At second order (i.e. binary correlation) we can store only two patterns. At fourth order, we can store forty patterns. At sixth order, we can store 150 patterns. At eighth order, we can store 250 patterns. The improvement is more dramatic when the number of neurons is large. For example, with 100 neurons, the second order model can store 5 patterns whereas the triple-correlation model can store 500 patterns, an increase of almost 100 times. In real applications, the stored patterns are not fully random. For example, there are large correlations among the 26 English alphabets. If we try to store these alphabets into a binary correlation network with 35 neurons, we found that the maximum number is only two and the minimum order we have to use to store all 26 of them is six. (see Fig. 2). The increase of storage capacity allows us to perform many tasks that are not possible with the binary model. We show two examples here. The first one is to store spatio-temporal memory. A string of spatial patterns would be recalled in definite order if an initial pattern is given as the cue. This can be accomplished in many ways.[7] The simplest such system would use a hetero-associative network with feedback. For example, if we would like to store the English alphabet in its natural order into the network, we simply store 25 associative alphabet pairs AB, BC, CD...YZ, into the memory. The initial input could be a defective A. The network would call out a perfect B and in the meantime restore A to its perfect form. The output B is then feedback to the network as the new input. It will then call out C and so on. In Fig. 3, we show that the minimum order of correlation needed to do this is 6×6. Another example of the necessity of high order correlation is the storage of symmetry invariant patterns. A typical case we studied is the storage of shift invariant patterns. To implement shift invariance one has to store all the patterns that are shifted with respect to one another. It thus requires a large capacity and therefore excludes the possibility of using the binary correlation model. In our simulation, we used a triple correlation model. Shift invariant storage implies $T_{i,j,k} = T_{i+\ell,\ j+\ell,\ k+\ell}$ for any ℓ. Average over the translation group, we have

$$<T_{ijk}> = \sum_{\ell} T_{i+\ell,\ j+\ell,\ k+\ell}$$

$$= \sum_{m} T_{m,\ m+\Delta j,\ m+\Delta k} \equiv T_{\Delta j, \Delta k}$$

The last equality means that we need only N^2 instead of the full N^3 elements to store the shift invariant patterns. Therefore symmetry has been used to reduce the memory space. On the other hand, it also implies the reduced ability to distinguish different shift invariant patterns for a memory tensor. Figure 5 shows that random initial patterns would converge to a shifted version of the stored pattern that is closest in Hemming distance to the input pattern.

Other applications of high-order correlations model in learning are discussed in another paper appeared in this proceeding.[8,9]

ACKNOWLEDGMENT

This work is supported by NSF, DOE, and AFOSR.

REFERENCES

1. D. O. Hebb, The Organization of Behavior (Wiley, N.Y. 1949).
2. J. J. Hopfield, Neural Networks and Physical Systems with Emergent Collective Computational Abilities, Proc. Nat. Acad. Sci. USA 79, 2554 (1983).
3. J. A. Anderson, Cognitive and Psychological Computation with Neural Models, IEEE Trans. Sys. Man. and Science, SMC-13, 799 (1983).
4. S. S. Venkatesh and D. Psalties, Information Storage and Retrieval in Two Associative Nets, Caltech preprint (1985).
5. D. J. Amit, H. Gutfreund, and H. Sompolinsky, Physical Rev. Lett. 55, 1530 (1985).
6. D. Psalties, paper in this proceeding.
7. T. Kohonen, Self-Organization and Associative Memory, Springer verlag (1984).

8. Y. C. Lee, G. Doolen, H. H. Chen, G. Z. Sun, T. Maxwell, H. Y. Lee, and L. Giles, Machine Learning Using a Higher Order Correlation Network, Proceeding of the "Evolution, Games, and Learning" Conference held at Los Alamos, May 20 - 24, 1985.

9. T. Maxwell, Lee Giles, Y. C. Lee, and H. H. Chen, Nonlinear Dynamics of Artificial Neural Systems, Paper presented in this meeting.

Fig. 1a) Storage of only three alphabets into
a binary correlation network is not successful.
The left letter in each pair is the input and
the right one the output. All the alphabets in
this work have pixel size 5 × 7 = 35.

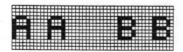

Fig. 1b) The binary correlation network can
store two letters perfectly.

94

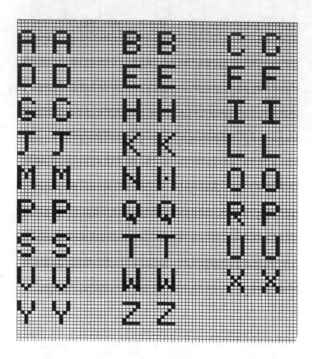

Fig. 2a) In a 5 × 5 order correlation model, it is still unable to store all 26 letters into the network. Four mistakes can be seen at C, G, N, and R.

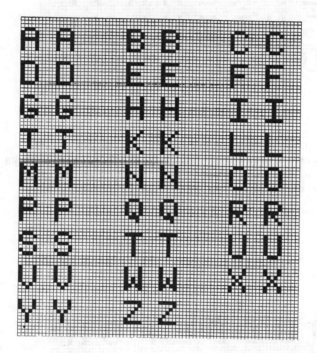

Fig. 2b) Perfect storage of 26 letters into
a 6 × 6 order network.

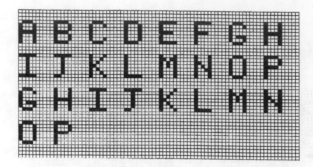

Fig. 3a) An unsuccessful attempt to recall the 26 letters sequentially, the sequence took a wrong turn after the letter P to G instead of Q. The order of correlation is 5 × 5.

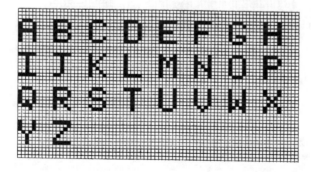

Fig. 3b) The network successfully recalled the alphabet sequence. The order of correlation is 6 × 6.

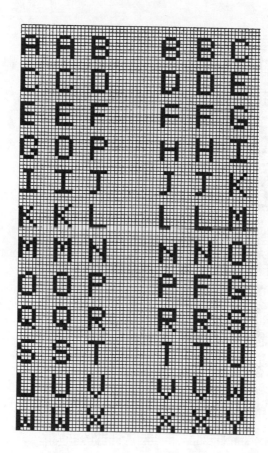

Fig. 4a) Hetero-associative recall of pattern
pairs. The input is shown on the first and fourth
column. They are at two Hemming distance away
from the exact form of the key pattern which should
appear at the second and fifth column. The order of
correlation in this figure is 6 × 6. There are two
unsuccessful recall for input G and P.

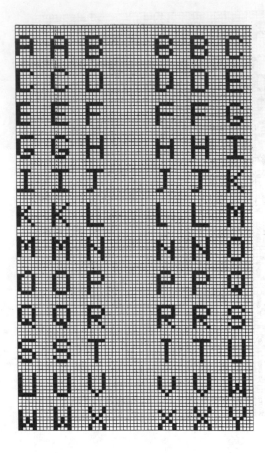

Fig. 4b) Fully successful recall of hetero-associative pattern pairs from imperfect input. The order of correlation is 8 × 8.

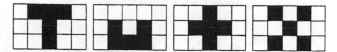

Stored patterns with shift invariance and
periodic boundary condition.

initial random patterns   final patterns

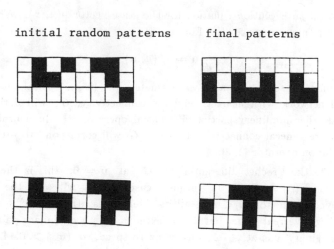

Fig. 5. Shift invariant associative memory.

Coupled Mode Theory for Neural Networks: The
Processing Capabilities of Nonlinear Mode-Mode
Interactions at Cubic and Higher Order

Marcus Cohen

Computing Research Laboratory
and the
Department of Mathematical Sciences
New Mexico State University
Las Cruces, NM, USA 88003
phone number: (505) 646-3901

1. Introduction and Mathematical Method

Continuum models of neural networks biased near threshold and of optical resonators with gain and nonlinear polarizations exhibit many phenomena which are common to all dispersive nonlinear media near a critical point. These are best analyzed using bifurcation theory; the technique is as follows:

1) Start with an evolution equation for the state vector $\mathbf{u}(\mathbf{x}, t)$. Write it as a flow on a function space H: For $\mathbf{u} \in H$,

$$\frac{d}{dt}\mathbf{u} = G\,(\mu, \mathbf{u}), \tag{1}$$

depending on some excitability parameter μ. In an optical medium or a neural network with nearest neighbor connectivity, G will derive from a second order nonlinear partial differential operator [1]. In neural networks with more general connectivity kernels, G will correspond to a thresholded convolution operator [2, 3].

2) Compute the Frechet differential of G at $\mathbf{u} = 0$; this is the linearized operator $L\,(\mu) \equiv G_{\mathbf{u}}(\mu, 0)$. In many cases, $L\,(\mu)$ will obey the Hopf spectral hypothesis: for $\mu < \mu_c$, all its eigenvalues $\lambda(\mu) = \alpha(\mu) + i\,\omega(\mu)$ will have negative real parts, but as μ exceeds μ_c a conjugate pair will cross the imaginary axis at $\pm i\,\omega_0$ with nonzero speed, $\alpha'\,(\mu_c) > 0$. Let $2n$ be the dimension of the eigenspace Y_C of $L\,(\mu_c)$ corresponding to eigenvalues $\pm i\,\omega_0$.

3) Recall [4] that under these conditions the system (1) possesses a locally invariant, attracting, $2n + 1$-dimensional *center manifold* $\mathbf{C} \subset H \times \mu$ tangent to Y_C at $\mathbf{u} = 0, \mu = 0$. Let $\mathbf{C}_\mu$ be a constant μ slice of $\mathbf{C}$. Split $H = \mathbf{C}_\mu \oplus \overline{\mathbf{C}}_\mu$ by writing $\mathbf{u} = \mathbf{v} + \mathbf{w}, \mathbf{v} \in \mathbf{C}_\mu, \mathbf{w} \in \overline{\mathbf{C}}_\mu$, and split 1) according to the projections P on $\mathbf{C}_\mu$ and $Q \equiv I - P$ on $\overline{\mathbf{C}}_\mu$:

$$\frac{d\,\mathbf{v}}{dt} = PG\,(\mu,\,\mathbf{v} + \mathbf{w}) \tag{2}$$

$$\frac{d\,\mathbf{w}}{dt} = QG\,(\mu,\,\mathbf{v} + \mathbf{w}) \tag{3}$$

4)   Note that $QG\,_\mathbf{w}(\mu,\,0)$ restricted to $\overline{C}_\mu$ has only negative eigenvalues. As $t \to \infty$, $\dfrac{d\,\mathbf{w}}{dt} \to 0$ and the implicit function theorem guarantees that (3) may be solved for $\mathbf{w} = \mathbf{w}(\mu,\,\mathbf{v})$. Plug this solution into (2) to obtain

$$\frac{d\,\mathbf{v}}{dt} = PG\,(\mu,\,\mathbf{v} + \mathbf{w}(\mu,\,\mathbf{v})) \equiv F\,(\mu,\,\mathbf{v}), \tag{4}$$

the *bifurcation equations* for system (1).

In practice, it is convenient to choose a basis of complex eigenfunctions $\{\phi_j\}$, $j = -n, \ldots, n$, with $\phi_{-j} = \overline{\phi}_j$ for Y_C and parametrize $\mathbf{C}_\mu$ about $\mathbf{u} = 0$, $\mu = \mu_c$ by coordinates z_j on its tangent space Y_C. We may thus write

$$\mathbf{v}(t) = \sum_{j=-n}^{n} z_j(t)\phi_j + \mathbf{y}\!\left[\sum_{j=-n}^{n} z_j(t)\phi_j\right]. \tag{5}$$

5)   Expand $F\,(\mu,\,\mathbf{v})$ in (4) as a Taylor series in the Frechet derivatives of F with respect to $\mathbf{v}$:

$$F\,(\mu,\,\mathbf{v}) = A\,(\mu) + B\,(\mu)(\mathbf{v},\,\mathbf{v}) + C\,(\mu)(\mathbf{v},\,\mathbf{v},\,\mathbf{v}) + \cdots \tag{6}$$

where $A\,(\mu) = G\,_\mathbf{u}(\mu,\,0)$, $B\,(\mu) = G\,_{\mathbf{uu}}(\mu,\,0)$ etc. are tensors over Y_C.

In "impressionable" materials, $B\,(\mathbf{v},\,\mathbf{v})$, $C\,(\mathbf{v},\,\mathbf{v},\,\mathbf{v})$, etc. are delayed non-linear functionals of $\mathbf{v}$. Such terms are called *memory nonlinearities*; the most common example is a contribution to $C\,(\mathbf{v},\,\mathbf{v},\,\mathbf{v})$ of the form

$$D\,\{\,|\,\mathbf{v}\,|^{\,2};\,T\,\}\mathbf{v} = ce^{-\beta T}\left[\int_{-\infty}^{T} e^{\beta T'}\,|\,\mathbf{v}(T')\,|^{\,2}dT'\,\right]\mathbf{v}. \tag{7}$$

In neural nets, such terms arise through persistent membrane polarizations or synaptic facilitation in the wake of continued depolarization; in ferroelectric materials they arise through long-lived intensity-dependent charge separation. (Visco-elastic, magnetic, and photographic materials also possess such memory nonlinearities.)

6)   Write the projection of (4) onto Y_C using the eigenfunction expansion (5) as

$$\frac{dz_j}{dt} = F_j\,(\mu,\,z_{-n},\,\ldots,\,z_n\,). \tag{8}$$

The goal is to solve (8) for $z_j(t)$. Using (6), we must account for all possible ways combinations of the amplitudes $z_l,\ z_m,\ \cdots$ of modes $\phi_l,\ \phi_m,\ \cdots$ may couple to produce growth of the amplitude z_j of mode ϕ_j. One such coupling occurs through (7). For example, with $\phi_j = e^{i\,k_j\cdot x}$, using (5) we may write

$$D\{\,|\,v\,|^2;\,T\,\}v = \left(\sum_{l,m=-n}^{n} K_{lm}\,e^{i(k_l - k_m)\cdot x}\right)\left(\sum_{p=-n}^{n} z_p\,e^{i\,k_p\cdot x}\right), \qquad (9)$$

where the term $\sim e^{i(k_l - k_m)\cdot x}$ is called an *interference grating* and

$$K_{lm} = ce^{-\beta t}\int_{-\infty}^{t} e^{\beta t'}\,z_l(t')\bar{z}_m(t')\,dt' \qquad (10)$$

is its *modulation depth*. When

$$k_l - k_m + k_p = k_j \qquad \text{(Bragg conditions)}, \qquad (11)$$

e.g. when $k_p = k_m$ and $k_l = k_j$, the cubic nonlinearity $C(v, v, v)$ couples excitation back into mode ϕ_j. This is the basis for all volume holographic phenomena.

7) To solve (8), use the method of *singular perturbations* [5]. Introduce a small amplitude parameter ϵ and a slow time scale $T = \epsilon^2 t$, so $z_j(t) \to z_j(\epsilon, t, T)$, and

$$\frac{d}{dt} \to \frac{\partial}{\partial t} + \epsilon^2\frac{\partial}{\partial t}. \qquad (12)$$

$$z_j(\epsilon, t, T) = \epsilon z_j^{(1)} + \epsilon^2 z_j^{(2)} + \epsilon^3 z_j^{(3)} \cdots \qquad (13)$$

and

$$\mu - \mu_c = \epsilon\mu_j^{(1)} + \epsilon^2\mu_j^{(2)} + \epsilon^3\mu_j^{(3)} \cdots. \qquad (14)$$

Insert expansions (12), (13), (14), and (6) into (8), equate terms in like powers of ϵ, and invoke the Fredholm alternative [5] for each eigenfunction ϕ_j at each order $O(\epsilon^n)$ to obtain, at $O(\epsilon)$:

$$z_j^{(T)} = A_j(T)e^{i\omega_0 t} \qquad (15)$$

and at $O(\epsilon^3)$:

$$\frac{dA_j}{dT} = (\lambda + 2\sum_{l=1}^{n} \kappa_{ll})A_j + 4\sum_{l=1}^{n}\kappa_{jl}A_l \qquad (16)$$

$$- 6g\sum_{l=1}^{n} |A_l|^2 A_j + 3g\,|A_j|^2 A_j - 3\kappa_{jj}A_j$$

or, in vector notation with $V = (A_1(T),\ \cdots,\ A_n(T))$,

$$\frac{dV}{dT} = (\lambda + 2\,Tr\,\kappa)V + 4\kappa V - 6g\,||\,V\,||^2 V + f, \qquad (17)$$

where

$$f_j = 3g \mid A_j \mid^2 A_j - 3\kappa_{jj} A_j.$$

(We have assumed that $z_{-j} = \overline{z}_j$, i.e., that the boundary conditions produce standing waves $z_j \phi_j + \overline{z}_j \phi_j$ as the normal modes of the cavity.) (16) are the *coupled mode equations* for the slowly evolving amplitudes $A_j(T)$ of each of the normal modes $e^{i\omega_0 t} \phi_j(\mathbf{x})$. If we could solve them, (5) would read (to $\mathbf{O}(\epsilon^3)$):

$$\mathbf{v}(t, T, \mathbf{x}) \cong e^{i\omega_0 t} \sum_{j=1}^{n} A_j(T) \phi_j(\mathbf{x}) + c.c. \tag{18}$$

Then $\mathbf{w} = \mathbf{w}(\mu, \mathbf{v})$ and $\mathbf{u} = \mathbf{v} + \mathbf{w}$ complete the approximate solution for $\mathbf{u}(t, T, \mathbf{x})$ to $\mathbf{O}(\epsilon^3)$; this is as far as most humans dare proceed in such perturbation expansions.

The coupled mode equations have a generic structure that depends only on the lowest order terms in the Taylor expansion of $F(\mu, \mathbf{v})$ in (6). This is the underlying mathematical explanation for the similarity of phenomena that unfold near a critical point μ_c in neural networks, optical media, and indeed in any system which supports small amplitude short wavelength bifurcating waves [8].

2. Phenomena Encompassed by Coupled Mode Theory

We mention here some of the phenomena predicted by coupled mode theory; please see [5, 6, 7] for a more complete description.

I) Noncoherent to Coherent Conversion

A sensory input to a neural network may be delivered as a nonhomogeneous depolarization $\eta(\mathbf{x})$ at $T = 0$; similarly, images may be input into an optical cavity as an initial intensity distribution $\eta(\mathbf{x})$. In either case, via the projection operator P on the center manifold, $\eta(\mathbf{x})$ is decomposed into an initial superposition of cavity modes $\phi_j = e^{i\mathbf{k}_j \cdot \mathbf{x}}$:

$$\eta(\mathbf{x}) \rightarrow \sum_{j=-n}^{n} A_j(0) e^{i\mathbf{k}_j \cdot \mathbf{x}}, \tag{19}$$

$$A_j(0) = \langle \phi_j, \eta \rangle = \int \phi_j^{\dagger}(\mathbf{x}) \eta(\mathbf{x}) d\mathbf{x}. \tag{20}$$

A short but intense initial exposure imprints the interference gratings $\sum_{j=-n}^{n} \sum_{l=-n}^{n} \kappa_{jl} e^{i(\mathbf{k}_j - \mathbf{k}_l) \cdot \mathbf{x}}$ in the medium, with $\kappa_{jl} = CA_j(0)\overline{A}_l(0)$ (see [4]). The matrix $\kappa = \mid \kappa_{jl} \mid$ is known as the *modal autocorrelation matrix* of the initial modal vector $\mathbf{V}(0) = (A_{-n}(0), \cdots, A_n(0))$; clearly, $\kappa = C\mathbf{V}(0)\mathbf{V}^{H}(0)$, where H denotes conjugate transpose. C is in general complex; its phase determines the gain/index ratio of the grating, while its modulus gives the modulation depth, which is a function of exposure time, and of time after exposure (for convenience, we fix $\| V(0) \|^2 = 1$). We assume that the medium is capable of storing multiplexed gratings in its volume; thus, in general, $\kappa = \sum_{I=1}^{M} C_I V_I V_I^H$ where the M vectors V_I have been stored at different times in the medium. Vectors stored in the recent past tend to have greater modulation depths C_I than "old memories", due to a decay term in C giving the "wash out" rate of gratings due to diffusion.

II) Recovery of a Stored Vector From Partial Input

To determine the temporal response $V(T) = (A_{-n}(T), \cdots, A_n(T))$ of the medium with all these gratings stored to newly input images, we employ its coupled mode (CM) equations.

We ask of the CM equations the following two questions:

1) Will the saturated (output) state $V(\infty)$ resemble *any* of the patterns V_I stored in κ, or will it be some hopeless nonlinear mixture of stored memories?

2) How will $V(\infty)$ depend on the input $V(0)$?

The answers to these questions are contained in the solutions of the CM equations; these depend on the values of the parameters C, g, and λ.

For one stored vector V, note first that V is the eigenvector of $\kappa = C V V^H$ with maximal eigenvalue. It can be shown from the CM equations that for C/g real, the system reconstructs V as the stable saturated output state, whatever the input (1 globally stable basin of attraction). The reconstruction is *faithful* (i.e., the output state is exactly colinear with the stored eigenvector V), for one value of λ only -- $\lambda = 0$.

For M stored vectors, $\kappa = \sum_{I=1}^{M} C V_I V_I^H$, with C/g real there are M stable equilibria, reconstructing each of the M stored vectors, i.e., there are M basins of attraction, one surrounding each eigenvector V_I of κ. The reconstruction is faithful for $\lambda = 2(1 - M)C$, provided that no two of the V_I share any modes ϕ_j in common (i.e., the V_I are *absolutely orthogonal*, $\sum_{j=1}^{n} |A_j^I||A_j^J| = \delta^{IJ}$). This behavior may be understood by noting that each of the V_I is an eigenvector of κ with eigenvalue C and therefore may grow linearly. As it grows into the nonlinear region, each eigenvector damps itself less than it is damped by the others. This competition among eigenvectors for saturation produces the multiple basins of stability, which happen to lie along the eigenvectors for $\lambda = 2(1-M)C$ (see Figure 1).

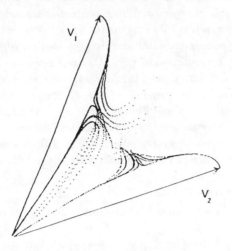

Figure 1:   **Dynamical Behavior of the system.** The system has been trained by exposure to the modal vectors $V_1^H = (1,4,1,0,0,0)$ and $V_2^H = (0,0,0,4,1,1)$, imprinting the modal autocorrelation matrix $\kappa = \dfrac{1}{36}(V_1 V_1^H + V_2 V_2^H)$. The state vector $V(T) = \sum A_j(T)\phi_j$ evolves according to the *coupled-mode equations* (2). Input patterns η set the initial components of V by $<\eta, \phi_j> = A_j(0)$. The simulation shows the evolution of the 6 dimensional state vector V projected into three dimensional space; each dot represents a time step. For the parameter value $\lambda = -2Tr\,\kappa$, basins of attraction of the full nonlinear system (saturated output states) lie along the eigenvectors V_1 and V_2 of κ. Initial states η evolve to the output state V_I for which $<\eta, V_I>$ was maximal, corresponding to recognition of an input pattern η as belonging either to class $\{V_1\}$ or $\{V_2\}$.

III) Associative Memory

An input state (initial condition) $V(0)$ will stimulate growth of each eigenvector V_I according to its projection $V_I^H V(0)$ on V_I; the eigenvector that wins the competition for saturation will be the one that "most resembles" $V(0)$, i.e., for which $V_I^H V(0)$ is maximum. In particular, if $V(0)$ happens to reproduce, or nearly reproduce, a portion of V_I, it will kick the system into the basin of stability corresponding to reconstruction of the *entire* $\mathbf{V}_I$ as the saturated (output) state; this is *error-correcting content-addressable memory* (see Figure 1). If each eigenvector $\mathbf{V}_I$ is made up of two or more parts, $\mathbf{V}_I = \mathbf{U}_I + \mathbf{W}_I$ (corresponding to simultaneous presentation of two images to be associated in memory), presentation of each $\mathbf{U}_I$ reconstructs the corresponding member $\mathbf{W}_I$ of the associative pair; this is *associative memory*. In fact, the system implements an associative memory of the Kohonen/Anderson/Hopfield type [9, 10, 11]; however, it works on "Fourier" vectors in the Hilbert space H spanned by the critical cavity modes $\{\phi_j\}$ rather than on vectors whose components are "pixels" in physical space. We mention below some of the new features that arise a consequence of multiple scattering in the volume.

IV) Associative Chains and Cycles

Imagine that an *associative chain* of gratings $\kappa = (\mathbf{V}_1 + \mathbf{V}_2)$ $(\mathbf{V}_1 + \mathbf{V}_2)^H + \qquad (\mathbf{V}_2 + \mathbf{V}_3) \qquad (\mathbf{V}_2 + \mathbf{V}_3)^H + \cdots + \qquad (\mathbf{V}_{M-1} + \mathbf{V}_M)$ $(\mathbf{V}_{M-1} + \mathbf{V}_M)^H$ has been imprinted by overlapping pairwise exposure to $\mathbf{V}_{J-1} + \mathbf{V}_J$, $J = 1 \cdots M$. An example would be pairwise association of sequential frames in a movie. The chain could, of course, be closed into an *associative cycle* by imprinting one more association $(\mathbf{V}_M + \mathbf{V}_1)(\mathbf{V}_M + \mathbf{V}_1)^H$. The possibility arises that an input that partially reconstructs $\mathbf{V}_1$ may scatter sequentially off the stack of gratings composing κ to reconstruct $\mathbf{V}_1 \cdots \mathbf{V}_M$ in turn. If so, are patterns $\mathbf{V}_1 \cdots \mathbf{V}_M$ reconstructed *sequentially*, thus playing back the movie frame by frame, or does the energy spread out among all the frames at once?

To answer this question for a particularly simple test case, we will let $\mathbf{V}_1 = (1, 0, 0, \cdots, 0)$, $\mathbf{V}_2 = (0, 1, 0, \cdots, 0)$, $\mathbf{V}_j = (0, 0, \cdots, 1_j, 0) = e_j$, and write the coupled mode equations for $\dot{A}_1$ to $\dot{A}_M$ as

$$\frac{dA_j}{dT} = (\lambda + 2 \sum_{l=1}^{M} \kappa_{ll} + \kappa_{jj})A_j + 4\kappa_{jj-1}A_{j-1} \tag{21}$$

$$+ 4\kappa_{jj+1}A_{j+1} - 6gEA_j + 3g |A_j|^2 A_j,$$

where $E = \sum_{l=1}^{M} |A_l|^2$ is the total "energy". It can be shown that for $\text{Re}(\lambda + 2Tr\,\kappa) \geq 0$, $\text{Re}(g) \geq 0$, as $T \to \infty$, E is approximately conserved during the motion; the system moves (asymptotically) on an attracting "constant energy surface". We now imagine that $M \to \infty$, so the $j = 1 \cdots M$ divisions become infinitesimally fine, and may be replaced by a continuous coordinate Z in "memory sequence space". The system (4) above is then replaced by a partial differential equation for $A(Z, T)$:

$$\frac{\partial A}{\partial T} = 4\frac{\partial^2}{\partial Z^2}(\kappa A) + \gamma A + 3g \mid A \mid^2 A \, , \tag{22}$$

where $\kappa = \kappa(Z)$ and $\gamma = \lambda + 2\int \kappa(Z)dZ + 9\kappa - 6gE$, and is (approximately) constant (since $E = \int \mid A(Z,T)\mid^2 dZ$ is approximately conserved as $T \rightarrow \infty$).

5) is a complex nonlinear diffusion equation; it has two basic kinds of solution. (We assume periodic boundary conditions, corresponding to a cyclic coordinate for Z space.)

a) *Diffusive:* κ, g, λ real, $\kappa > 0$, $g > 0$. For $\kappa(Z)$ constant, the energy tends to spread itself out evenly in Z; $A(Z,T) \rightarrow$ constant. This behavior is retained in the discrete case.

b) *Dispersive:* κ, g, λ imaginary, $\kappa = i\tilde{\kappa}$, $g = i\tilde{g}$, $\gamma = i\tilde{\gamma}$. 4) becomes

$$i\frac{\partial A}{\partial T} + 4\frac{\partial^2}{\partial Z^2}\tilde{\kappa} A + \tilde{\gamma} A + 3\tilde{g} \mid A \mid^2 A = 0, \tag{23}$$

a cubic Schroedinger equation for $A(Z,T)$. For $\tilde{\kappa}(Z)$ slowly varying, (6) possesses travelling pulse solutions (solitons) that sweep down the chain or around the cycle in Z space, reconstructing a narrow band of images at a time. Referred to our discrete index j, this corresponds to sequential reconstruction of images $A_1, \cdots, A_M$, with time T, replaying the movie of associated frames that were presented pairwise to imprint the gratings κ in the first place.

Computer simulations of (4) in the discrete case (M finite) show, in addition to the periodic solutions that correspond to the cycling pulses in Z space, both quasi-periodic and chaotic solutions as the value of λ is varied. It can be easily shown that for λ, κ_{jl}, and g all imaginary, that E is strictly conserved; the system may wander around on the energy surface without ever "falling into" an equilibrium state. Such behavior has indeed been observed experimentally in phase conjugating resonators (Dana Anderson, personal communication).

3. The Processing Capabilities of Higher Order Nonlinearities

If one were courageous enough to proceed to higher order than $\mathbf{O}(\epsilon^3)$, say to $\mathbf{O}(\epsilon^{n+2})$ in the perturbation scheme, one would pick up contributions on slower time scales $T_n = \epsilon^{n+1}t$ to the modal amplitudes. These have the form

$$\frac{dA_j}{dT_n} \sim T_j^{i_1 \cdots i_n} A_{i_1} \cdots A_{i_n}, \qquad \text{(summation convention)} \tag{24}$$

or

$$\frac{d}{dT_n}\mathbf{V} \sim T\,_1^n\,(\mathbf{V}_1, \ldots, \mathbf{V}_n\,), \tag{25}$$

where $T\,_1^n$ is an n-covariant 1-contravariant tensor over H. These represent couplings of n input vectors (patterns) to give one output vector. Such "higher order correlations" have been used to implement group-invariant pattern recognition [12]. For example, when the two arguments in $T\,_1^2(\mathbf{V}, \mathbf{V})$ are the same, if $T_j^{lm} = T_j^{l-m}$ the system responds only to differences in pattern elements and hence is shift-invariant [13].

We note here that such n-input - one output transfer functions may be used to represent $n \rightarrow 1$ branch nodes (or $(n-1) \rightarrow 2$ nodes, etc.) in a directed graph or tree. Such structures provide the skeletal substrate for search and inference (these ideas will be explored more fully in a forthcoming paper).

4. Conclusion

Generalizing the above considerations, we may state the basic rules which govern the unfolding of bifurcating waves in the supercritical neural network or optical resonator which has been "programmed" with a matrix of previous associations:

1) An initial input will stimulate stored associative chains and cycles in proportion to its projection on those chains or cycles. Each chain (or cycle) forms a gain *cascade* (a positive feedback avalanche or loop) through which the amplitude of waves may grow.

2) The cascade with the greatest net path product for the gain will tend to grow fastest in amplitude when stimulated.

3) Disjoint cascades compete for the energy of saturation and tend to lock each other out.

4) Chains and cycles may imbed into trees using $n \rightarrow m$ branch nodes.

These rules endow the system with some capabilities for implementing "higher" cognitive tasks such as goal seeking, sentence disambiguation, search and inference, and optimization problems by maximization of a path product.

5. References

[1] Cohen, M.S. (1984). Interacting Nonlinear Waves in a Neural Continuum Model: Associative Memory and Pattern Recognition, from *Wave Phenomena: Modern Theory and Applications*, C. Rogers and T.B. Moodie (eds.), North-Holland.

[2] Ermentrout, G.B, and Cowan, J.D. (1979). Temporal Oscillations in Neuronal Nets, *J. Math Biology*, **7**, pp. 265-286.

[3] Ermentrout, G.B, and Cowan, J.D. (1979). A Mathematical Theory of Visual Hallucination Patterns, *Biol. Cybernetics*, **34**, pp. 137-150.

[4] Hassard, B.D., Kazarinoff, N.D., and Wan, Y-H. (1981). *Theory and Applications of Hopf Bifurcation*, London Mathematical Society Lecture Note Series 41, Cambridge University Press.

[5] Cohen, M.S. (1986). Design of a New Medium for Volume Holographic Information Processing, *Applied Optics*, special issue on Optical Computing, **2** (to appear).

[6] Cohen, M.S. (1986). Self Organization, Association, and Categorization in a Phase-Conjugating Resonator, *proc. SPIE Conference on Optical Computing*, January (to appear).

[7] Cohen, M.S. (1986). Volume Holographic Information Processing in a Phase Conjugating Resonator, Computing Research Laboratory preprint, New Mexico State University, Las Cruces, NM.

[8] Newell, A. (1980). Bifurcation and Nonlinear Focusing, in *Pattern Formation and Pattern Recognition*, Springer series in Synergetics, **5**, pp. 244-265.

[9] Kohonen, T. (1984). *Self Organization and Associative Memory*, New York: Springer-Verlag.

[10] Anderson, J.A. (1983). Cognitive and Psychological Computation with Neural Models, *IEEE Transactions on Systems, Man, and Cybernetics*, **SMC-13**, 5, pp. 799-815.

[11] Hopfield, J. (1982). Neural Networks and Physical Systems with Emergent Collective Computational Abilities, *PNAS*, **79**, pp. 2554-2558.

[12] Lee, W.H., et al. (1986). (In this volume).

[13] Psaltis, D., Hong, J., and Venkatesh, S. (1986). Shift Invariance in Optical Associative Memories, *proc. SPIE Conference on Optical Computing*, January (to appear).

A MODEL FOR CORTICAL FUNCTION

M. E. Colvin, F. H. Eeckman and J. W. Tromp
University of California, Berkeley, CA 94720

ABSTRACT

We investigate a dynamical model of the cerebral cortex. This model is based on a hexagonally close packed array of discrete units that represent densely interconnected neurons (cortical columns). We assume only local interactions between these units. Under certain regimes this system will display long lived periodic oscillations. Moreover, when the system is trained using a Hebb-like modification it can act as a content addressable memory.

INTRODUCTION

An ideal model for the study of the brain would incorporate known anatomical and physiological facts while being simple enough to actually implement. Moreover, such a model would be general enough to allow investigation of various hypothesized neuronal interaction rules. The neurons of the cortex are known to be highly interconnected. Thus any model of cortical function that attempted explanation at the single neuron level would become exceedingly complex. Fortunately, there is physiological evidence that the cortex is organized into larger units than individual neurons.

Mountcastle[1] proposed a vertical organization into cortical columns. These columns are functional units, that receive a common input and have a common output. Inherent in the definition of the column is the concept of lateral pericolumnar inhibition. This lateral inhibition allows for selective processing (ie. feature extraction) of certain input signal parameters.

The anatomical work of Szentagothai[2] showed that the cortex is composed of modules of densely interconnected neurons. Again the importance of lateral inhibition was stressed. But Szentagothai also drew attention to the recurrent excitatory axon collaterals of pyramidal cells that excite cylindrical slabs of tissue at a distance away from the original focus. Based on the anatomical evidence, Palm[3] estimated that a strong excitation to a single spot would give rise to surround inhibition and longer range excitation (inverted Laplacian rule).

These ideas are incorporated into a model that we describe in the next section.

THE MODEL AND RESULTS

Following the work of Shaw[4] we assumed that each unit in our model represents a densly interconnected group of neurons. These units can have three discrete states corresponding to above average firing, average firing, and below average firing rates. This discretization of firing activity should qualitatively describe phenomena that depend on correlations of fluctuations of neural activity on nearby groups.

Operation of this model involves the periodic updating of an interconnected array of these units according to some interaction rule. Shaw[4] studied the dynamical behavior of a 6-membered ring of these units. In order to better represent the spatial structure of the cortex we consider a two dimensional hexagonal close-packed array of these elements. The rule used to update the units' states in our model is that derived by Shaw:

$$P_i(S) = \frac{g(S) \exp[\ B\ M_i\ S\ ]}{\sum_s g(s) \exp[\ B\ M_i\ s\ ]} \qquad (1)$$

where S is the state of the units (+1 : above average firing, 0 : average firing, −1 : below average firing). g(s) is the degeneracy of each of the s states and g(0) is much greater than g(+) or g(−) reflecting the fact that the non zero states represent rare large fluctuations. B is the level of noise in the system (B approaching infinity means no noise). The "energy field" M_i is given by equation (2).

$$M_i = \sum_j [\ V_{ij}\ S(T-1) + W_{ij}\ S(T-2)\ ] - \Theta_i \qquad (2)$$

where Θ_i is the threshold for activation of cell i, and V and W are the nearest neighbour and next nearest neighbour interaction strengths respectively. The update feature with two time steps models the finite speed of neural propogation through the cortex.

In the implementation of this interaction rule, equation (1) can be used deterministically (where the unit is assigned the state with highest probability) or stochastically (where the assignment is based on a random number weighted by the state probability).

In order to allow the model to "learn" we have implemented a Hebb type mechanism for modification of the interaction strengths, V and W. This algorithm has the form:

$$\Delta V_{ij} = \varepsilon\ S_i(T)\ S_j(T-1) \qquad (3)$$

$$\Delta W_{ij} = \varepsilon\ S_i(T)\ S_j(T-2) \qquad (4)$$

where ε is the modification strength (typically 5-10 % of the initial weight).

112

We initially investigated an interaction rule inspired by the work of Mountcastle. For this interaction rule the nearest neighbour weights were +1 and the next-nearest neighbour weights were -0.5. The net effect of such a rule is that a cluster of 7 units represents a cortical column which, when activated, is surrounded by an inhibitory region. We studied the behavior of this interaction rule on hexagonal segments of the model with 7, 9, and 15 units on a side (corresponding to a total of 127, 217, and 631 respectively). The gross behavior of this system was the same for all starting conditions: the system would oscillate between largely below average firing (s =-1) and above average firing (s=+1) until after 40 - 80 time steps the system was flipping between all -1 and all +1 states with a period of 2.

Next we added a Hebb type modification rule for the interaction weights (ϵ = 0.05). This had a dramatic effect on the behavior of the system. From any starting pattern the system would slowly oscillate until after 40-80 time steps it would converge to a fixed activity pattern.

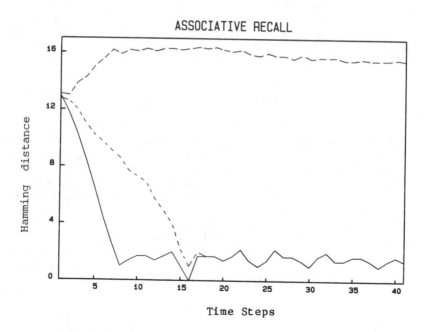

Figure (1): Demonstration of associative recall. For a description of the curves, see text.

An interesting result is that when operated in this mode the system acts as a content-addressable memory. This works as follows. With a given initial pattern (called the training pattern) we allow the network to evolve with the Hebb algorithm on. A stable state is reached within 30–50 time steps (called the stored pattern). If we now run the network with the interaction strengths determined by the training run, and start with the same or a close pattern the network converges to the stored pattern in 10–20 steps. There are small residual fluctuations due to the stochastic nature of the updating.

We measure closeness of a given pattern to the stored pattern using the Hamming distance (bit difference) between the two patterns. Perfect recall corresponds to the Hamming distance approaching zero or a small neighbourhood around zero and remaining there. Attraction to another memory state corresponds to the Hamming distance approaching some finite non zero value. Associative recall is demonstrated in figure (1). In the solid curve, the recall pattern is identical to the training pattern, and convergence is rapid. In the short dashed curve, a similar but not identical recall pattern is used and slower convergence to the stored memory is seen. Finally, the long dashed curve shows that when the recall pattern is sufficiently different from the training pattern, the stored pattern is not evoked.

Despite the interesting behavior of this system the convergence to a fixed pattern of activity is not realistic. Thus, as an attempt to stabilize without totally damping the systems dynamics behavior we introducted a fatigue rule. A fatigue rule can be justified from the available physiological evidence. It is a well known fact that synapses adapt to continuous high levels of input, and attenuate their response.

The effect of this fatigue rule was that from any initial pattern the system displayed long-lived, well behaved oscillations with a period of 10 time steps. Since the time scale of the oscillations is much longer than the time scale of the rule, these oscillations are a true emergent property of the network.

We next investigated an interaction rule motivated by the work of Palm. For this interaction rule the nearest neighbour weights were set to −1 and the next nearest neighbour weights to +.5. The behavior of this system was very different than that using the first rule. For all initial patterns the system immediately went into rapid oscillation between all −1 and all +1 states.

In order to damp this clearly uninteresting behavior we again introduced a fatigue rule of the type described above. The system then produced long lived regular oscillations. These oscillations could be decomposed into a low frequency wave with period 25 and a high frequency noise of period 2.

DISCUSSION

We have presented results showing that our model cortex can store stable spatial patterns of activation. However, activity in the cortex is always fluctuating, so memories should correspond to spatio-temporal patterns of activity. Use of the extra dimension of time in the patterns will yield a much greater memory storage capacity. A new definition of closeness of memories is also required. By starting with a very small hexagonal model, we hope to be able to identify repeating patterns of finite length. The memory distance can be generalized to be the sum of the Hamming distances for all time steps over the cycle. Then for time steps larger than the cycle time we can observe convergence of memories as before.

We also want to investigate the effect of dynamic input on the functioning of the model. The cortex of course is always receiving input from the outside world, and the most interesting future results will involve understanding how external inputs can affect the dynamics of recall, and cause transitions between stored memories.

The model as presented only represents activity on the surface of a small area of the cortex. A full scale representation of the cortex would require a large set of cortical segments doing local computation and interconnected by long range collaterals.

REFERENCES

[1] V. B. Mountcastle, The Mindful Brain (MIT Press, Cambridge, MA, 1978), p. 1.
[2] J. Szentagothai, Brain Res. 95: 475 (1975).
[3] G. Palm, Neural Assemblies (Springer, Berlin, 1982).
[4] G. L. Shaw, D. J. Silverman and J. C. Pearson, Proc. Natl. Acad. Sci. USA 82 : 2364 (1985) Biophysics.

FLOW-OF-ACTIVATION PROCESSING:
PARALLEL ASSOCIATIVE NETWORKS (PAN)

Claude A. Cruz-Young, William A. Hanson, Jason Y. Tam
IBM Palo Alto Scientific Center
1530 Page Mill Road, MS/35B, Palo Alto, California 94043

ABSTRACT

Parallel associative networks, or "PANs", are an information processing mechanism based on formal neuron networks. The IBM Palo Alto Scientific Center is exploring the nature and possible computing applications of PANs. This work has led to the creation of a versatile experimentation environment called the "PAN workstation". The workstation includes a parallel PAN emulation processor (the Network Emulation Processor, or "NEP") which greatly speeds up the processing of large PAN nets. Our initial application studies have dealt with low complexity functions such as binary and graded logic gates, as well as applications in robotics vision and expert systems. Our current research is in the area of PAN-based knowledge representation and inference mechanisms.

PARALLEL ASSOCIATIVE NETWORKS (PAN)

Since 1982, the IBM Palo Alto Scientific Center has been studying a novel information processing mechanism which we call parallel associative networks, or "PANs". This mechanism is a particular type of formal neuron network, meaning that the PAN network elements embody our view of the significant information processing properties of biological nerve nets. This was done at a level of abstraction above detailed single-cell models (e.g. detailed cell-membrane dynamics), but below that of multi-cell behavioral descriptions. Our intent has been to examine the computational capabilities of large networks of PAN processing elements ("nodes" and "links"). This work is documented in detail in the references to this paper, which are available from the authors.

A parallel associative network consists of a set of simple processors called nodes, connected by unidirectional communication channels called links. All nodes operate as non-linear summing junctions, and all links transmit state information between nodes. However, the precise behavior of each node and each link is determined by the values of an associated set of node and link "parameters". The design of a PAN network includes specification of these parameters, as well as the definition of the network topology (node/link relationships).

Each node continuously computes an associated formal variable called its "activation level" (a numeric variable in the range 0.0 to 1.0, inclusive). It does so by summing activation flowing to it over links from its set of source nodes. A node's resultant output activation level constitutes a "short-term memory" which decays passively over time.

Each link has an associated "link weight" which scales activation flowing across the link. Links with positive weights are said to be "excitatory", since they drive the downstream node toward its firing threshold. Conversely, links with negative weights are called "inhibitory".

Links are not simply passive transmission channels for activation. Each link is adaptive, in that its link weight can vary over time as a function of the activation levels in the source and sink nodes of that link. Thus, modification of each link weight is locally controlled. In our current PAN model, links incorporate a "Hebb synapse" model. Through this mechanism, the modulus of a link increases if its source node becomes highly active, and then its sink node also becomes active. The intent of this mechanism is to capture possible correlated activation patterns of the network. Such correlations may reflect causal or non-causal rela-

116

tionships to which the PAN may respond. The time varying component of link weights constitutes a "long-term memory" which decays passively over time. The decay rate for long-term memory is typically much slower than that for node activation levels.

PAN nets are basically pattern processors, where "pattern" means a vector of activation levels across the constituent nodes of a network at a given instant of time. The basic PAN functional units, or "circuits", appear in Figure 1.

The function of an "encoder" is to detect the presence of a particular activity pattern. The activity pattern to which a particular encoder is tuned is determined by the current values of its link weights. These weights (possibly time varying) can be interpreted as a "recognition template". A vector of input nodes may fan-in to a set of encoders, all concurrently assessing the degree of match between the current input vector, and the "recognition template" associated with each encoder. Local competitive mechanisms may be used to ensure that one or a few match nodes become highly active, while all other nodes are nearly totally inactive.

Decoders are used to produce a particular output pattern across a vector of output nodes. Here, activity in one input node flows across a fanned-out set of links leading to output nodes. The weights on these links comprise a pattern "production template", such that when the input node becomes active, the output activity vector becomes a scaled version of the link weights of the decoder.

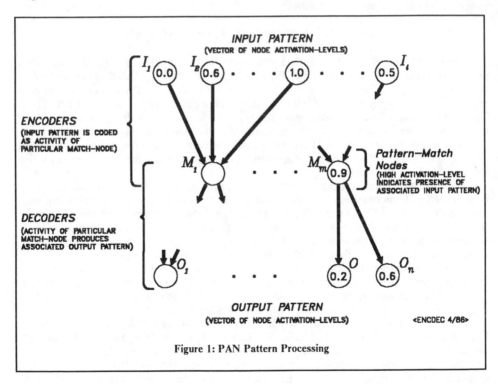

Figure 1: PAN Pattern Processing

As shown in Figure 1, encoders may be cascaded into decoders. This allows construction of state machines in which the state vectors may be inexact, or "fuzzy". Such a state machine is deterministic, but has many possible state trajectories.

Encoders and decoders operate upon node activation levels, or PAN short-term memory. In addition, there is an operation that mediates transfer from short- to long-term memory: pattern storage. Through this operation, network activation levels cause changes in the net's link weights. This is accomplished by the Hebb synapse mechanism residing in each link. Pattern storage allows a PAN's activation patterns to affect the net's future behavior.

As described above, parallel associative networks are made up of very simple elements. Nonetheless, the implications of this information processing approach are far-reaching. Consider the nature of the prevalent von Neumann digital computer architecture. Here, the system's state is encoded in the contents of system memory, which is a passive repository whose contents are manipulated by the system's single processor. State transitions then correspond to changes in the system's memory. The system's transition rules are encoded in a precisely defined "program" (i.e. in the contents of program memory, plus a sequential instruction fetch-and-execute cycle). In comparison, a PAN's state is encoded in two places: in node activation levels (short-term memory), and in link weights (long-term memory). Since flow of activation across links can alter link weights, and link weight changes affect flow of activation, PAN memory is active. Each PAN node and link has memory, but each is also a simple processor. Both memory and processing are thus distributed. A PAN contains no explicit transition rules. Instead, the network topology and node/link parameters determine the network's behavior. Whereas a standard computer's activities are orchestrated by its serial fetch-and-execute cycle, a PAN's behavior is elicited by parallel flow of activation.

Without detailed exposition, our current (discrete time) PAN equations appear below. It should be noted that we have used several variants of these equations, and will probably continue to do so. We view parallel associative networks as a family of networks based on flow-of-activation processing, and possessing a local "association" mechanism which alters the degree of coupling of activity in various network nodes.

$$A(i, t+1) = \min\left[1, \; G(i) \; \times \; \Big(\delta(i) \times A(i,t) + \max[0, \; s(i,t) - \tau(i)]\Big)\right] \qquad (1)$$

$$s(i,t) = \sum_{j \in N(i)} \left[w(i,j) + \Delta w(i,j,t)\right] \times A(j,t) \qquad (2)$$

$$\Delta w(i,j,t+1) = \Delta(i,j) \times \Delta w(i,j,t) + \lambda(i,j) \times A(i,t+1) \times A(j,t) \qquad (3)$$

The parameters (static quantities) and variables (time varying quantities) in these equations are:

$A(i,t)$	Current activation level for node i; a number in the range 0 to 1, inclusive
$G(i)$	Gain for node i; a non-negative number
$\delta(i)$	Activation decay constant for node i
$s(i,t)$	Current driving activation for node i
$\tau(i)$	Activation threshold for node i
$N(i)$	Set of source nodes feeding node i
$w(i,j)$	Static link weight from source node j to sink node i
$\Delta w(i,j,t)$	Current dynamic link weight from source node j to sink node i
$\Delta(i,j)$	Decay constant for dynamic component of i,jth link weight
$\lambda(i,j)$	"Plasticity", or learning rate of i,jth link (Hebb synapse model)

The terms in these equations have the following interpretation:

$\delta(i) \times A(i,t)$	Activation decay component for node i; encodes passive "forgetting" of node's "short-term memory", or STM
$[w(i,j) + \Delta w(i,j,t)]$	Current effective total weight of i,jth link
$\Delta(i,j) \times \Delta w(i,j,t)$	Link weight decay component for i,jth link; encodes passive "forgetting" of link's "long-term memory", or LTM
$\lambda(i,j) \times A(i, t+1) \times A(j,t)$	Current dynamic link weight increment for i,jth link; reflects active LTM "learning" implemented by Hebb synapse model

PAN IMPLEMENTATION APPROACHES

The basic operations performed by PAN nodes and links are analog in nature. Unfortunately, these do not lend themselves to direct implementation using extant circuit technology. For example, there exists no analog circuit which can store continuous valued link weights with stability over months or years (though there are experimental means for doing this, such as optical storage media). Given these circumstances, it is reasonable to consider digital emulation of the PAN elements. The discrete time PAN equations require fairly modest numerical resolution (16-32 bit integers or floating point numbers), and lend themselves to efficient implementation on signal processing computers (i.e. on computers specialized for numerical operations, such as inner products). Moreover, the local processing characteristic of PANs allows a large network to be partitioned, with each sub-net being implemented on a distinct hardware processor. In standard multiprocessor computers, partitioning of application programs to run on a set of processors often presents severe resource contention problems (such as shared access to memory). In contrast, the local communication characteristic of PANs greatly simplifies partitioning of a network. The parallelism of a PAN can be readily exploited in parallel hardware, resulting in large performance gains. In addition, one can achieve a desired PAN update rate (i.e. number of total network state updates per second) by varying the number of processors in the emulation system.

We have constructed an intermediate-speed PAN emulation computer called the Network Emulation Processor, or "NEP". This is a cascadable parallel coprocessor for the IBM PC/XT and PC/AT (with IBM RT PC support planned). Each NEP can simulate about 4,000 nodes and 16,000 adaptive links, with at least 30-50 total network updates per second. Nodes can be exchanged for links and vice versa, and the processor speed is considerably increased if few of the links are made adaptive. PAN nets of arbitrary topology may be emulated, and up to 256 NEPs can be cascaded for implementation of networks containing up to 1M nodes and 4M links. Multiple NEPs are cascaded through a dedicated bus, and each NEP contains local interfaces for attachment of high data rate input/output (I/O) devices. These I/O devices (such as video scanners or actuators) interface the network with the surrounding environment. The NEP system is described at length in Reference 3. Here, we simply note that it allows one to explore PAN applications that could not otherwise reasonably be undertaken.

PAN WORKSTATION

Having established a model for the basic PAN elements, we seek to apply these "building blocks" in a variety of application domains. This requires a supportive experimental environment, in which large associative networks can be constructed, driven with meaningful stimulus signals, updated using the PAN equations, and have their evolving state displayed in an easily understood format. From our early mainframe based simulations (developed in APL), we have progressed to a special high performance PAN Workstation [5] developed by the authors at the IBM Palo Alto Scientific Center. This system is depicted in Figure 2.

The PAN Workstation consists of an IBM PC/XT or PC/AT equipped with a PGA (Professional Graphics Adapter) and optionally equipped with a NEP coprocessor (described above). The workstation provides an interactive environment for the development, test, debug, and analysis of networks. A major goal during the implementation was to maintain flexibility while simultaneously providing low level display of the network state. The workstation is capable of simulating the NEP coprocessor (at the cost of execution time), so that investigation of networks can be done even without the special purpose hardware.

The workstation is is a dual display system that uses a monochrome display for operator interaction and a high resolution graphics display for the presentation of the network state. The network state is portrayed on the graphics display by drawing a color-coded circle for each node in the network. Experience has shown that this method of presentation allows the

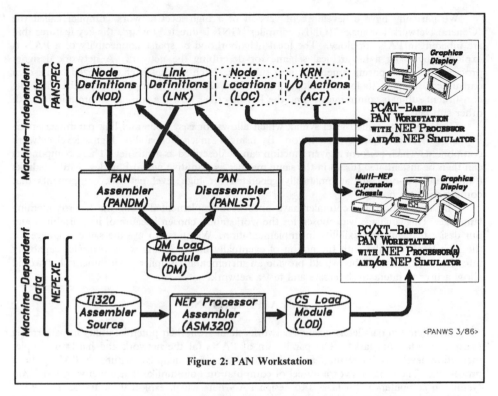

Figure 2: PAN Workstation

user to comprehend the enormous amount of information contained in the network state. The user can select various subsets of the entire network to be displayed. By using the cursor keys, the user can "traverse" the network to view detailed information on the node and link states. Alternatively, the workstation supports referring to nodes by name. The user can also alter the network state by forcing nodes to a desired value. There are three modes of network update execution: single cycle, multiple cycle displaying after each cycle ("animate"), and multiple cycle displaying after the last cycle ("n-cycle"). These options allow the user to trade processing time for detailed presentation of network state.

During a PAN session, the workstation maintains a log of all user commands. The user can also request the workstation to "trace" (exhaustively record) network state data in a disk file. These two data files (log and trace) can then be used by off-line performance analysis programs such a timing-trace plotters.

We are exerting a major effort to create a high level mechanism for describing networks and then executing them. Towards this end, the PAN Workstation has an easy-to-use network specification interface, called PANSPEC. PANSPEC consists of two ASCII files that contain simple tables of the node parameters, link parameters, and the network topology. The node definition file contains the parameters for each node in the network (threshold, gain, short-term memory decay rate, and initial activation level). The link definition file contains the topology and link parameters (source node, destination node, static weight, initial dynamic weight, long-term memory decay rate, and plasticity). These files provide a fixed target specification for the output of higher level compilers from which the workstation can operate. Subsequent changes in implementation of the network or changes in the workstation itself are thus isolated from higher level compilers.

We currently have a development version of a high level network compiler called the "General Network Language" (GNL) compiler. GNL is directed towards the key features that are inherent in PAN topology. The local neighborhood or sparse connectivity of a PAN is exploited by using a hierarchical scheme for describing the network. A network designer proceeds by decomposing a network into sub-nets. Each sub-net is characterized by its own constituent sub-nets, its local topology, and by its I/O. The local topology describes how the constituent sub-nets communicate, and the I/O describes how the sub-net communicates with other sub-nets.

GNL supports a high level syntax which allows for easy node and link parameter specification as well as topology description. By using a "macro" capability, higher level network components which perform a given function can be described as a sub-net. These components can then be instantiated (used) in the same manner as primitive network nodes. Multiple links (fan-in and fan-out) are automatically generated for high level network components that contain multiple nodes.

The workstation and compiler have both been implemented using the C programming language. This language was chosen for the workstation chosen because of its portability and our desire to maintain flexibility in implementation. We chose to use the same language for the high level compiler with the notion of eventually incorporating some compiler capability into the workstation. This would provide an incremental compiler environment that would allow a user to interactively create and test a network.

PAN APPLICATIONS

Our interest in parallel associative networks stems from their potential utility in performing many computational tasks. The parallelism of PANs (at the network and hardware implementation levels) is interesting, as is the distributed and adaptive nature of PAN pattern processing. These traits favor a model of computation quite unlike that underlying serial von Neumann computers. Our task is to design PAN nets which exploit these traits.

To date, we have used PANs to construct "circuits" which act as Boolean and graded logic gates, as well as state machines. We have used PANs to develop local operators (such as $\nabla^2 G$) required by David Marr's model of low level visual processing. This work is been carried forward by the IBM Los Angeles Scientific Center. We have also used PANs to implement a parallel control structure for an expert system (ANEX) similar to Stanford University's EMYCIN system. This work is described in detail in the references. Our current work is in the area of knowledge representation and inference mechanisms. We are testing our belief that PANs will prove to be a useful mechanism on which to build these functions.

REFERENCES

1. C. A. Cruz-Young and H. J. Myers, IBM Palo Alto Scientific Center Report #G320-3474, *Associative Networks* (1982).
2. C. A. Cruz-Young and H. J. Myers, IBM Palo Alto Scientific Center Report #G320-3446, *Associative Networks II* (1983).
3. C. A. Cruz-Young and J. Y. Tam, IBM Palo Alto Scientific Center Report #G320-3475, *NEP: An Emulation-Assist Processor for Parallel Associative Networks* (1983).
4. J. M. Oyster, W. Broadwell and F. Vicuna, IBM Los Angeles Scientific Center Report #G320-2777, *Associative Network Applications to Robot Vision* (1986).
5. W. A. Hanson, IBM Palo Alto Scientific Center Report (Number to Be Assigned), *The PAN Workstation* (1986).

Neural Network Refinements and Extensions

John S. Denker
AT&T Bell Laboratories, Holmdel, New Jersey 07733

ABSTRACT

This paper contains some ideas for tidying up and extending the standard neural network model. It includes some fundamental generalities as well as some arcane technicalities. One central theme is how to make a network that is as close to ideal as possible, given various restrictions on the available fabrication resources. Another theme is how to understand not only the statics but also the dynamics of the network.

Key phrases: Positive vs. Negative Synapses; Constant Conductance Synapse; Virtual Ground; Inertia; Complex Synapses; Zero Diagonal.

Note: This paper is intended for experts in the field. Space does not permit including a lot of introductory or philosophical material. I must assume that the reader has read my previous paper[1], which introduces the ideas and notation.

1 Separate T+ and T−

One refinement that was tacitly adopted in the previous paper was the separation of the T matrix into two pieces, T^+ and T^-. This was motivated by the fact that the two are logically separated when one actually builds a circuit. The interaction with other neurons depends on $T \equiv T^+ - T^-$, while γ_i (the total Thevenin conductance of the dendrite) and hence the settling time of the neuron depend on the sum $T^+ + T^-$. The reason that no one bothered to make the distinction before is that when T^+ and T^- are disjoint (i.e. they are never both non-zero at the same (i, j) crosspoint in the matrix) then it is possible to write the impedance in terms of abs(T). This glib and compact notation breaks down if we put synapses of both signs at the same crosspoint, which is in fact an interesting thing to do — see below.

2 Virtual Ground Neurons

Another refinement is what I call *virtual ground neurons*, as opposed to the standard model, which I call *floating dendrite neurons*. The circuit for such a model neuron is shown in figure 1. The circuit equation is

$$U_i = \sum_j (T_{ij}^+ V_j - T_{ij}^- V_j)$$
$$U_i - C_i \dot{V}_i - A^{-1}(V_i) = 0 \tag{1}$$

where U_i is now the dendrite *current*, whereas in the floating dendrite model the symbol U_i was used to represent the dendrite voltage. The transfer function $A(U)$ now has dimensions of transimpedance, whereas in the floating dendrite model $A()$ was a simple voltage gain.

In this figure, it is easy to see one advantage of this refined model: the current contributed by the matrix $(T_{ij} V_j)$ enters the equation on the same footing as the current contributed by

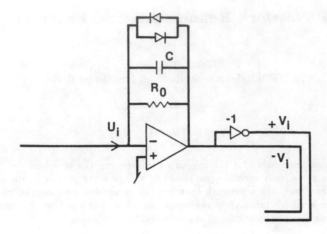

Figure 1: Virtual Ground Neuron

the inverse gain $A^{-1}(V)$. Both represent currents feeding into virtual ground at the summing node. It also becomes quite clear how to synthesize simple but interesting inverse transfer functions. For example, if we omit the feedback resistor in figure 1, the pair of crossed diodes implements $U = \sinh(V)$. Similarly, if we omit the diodes instead, we get $U = V/R_0$, i.e. $A(U) = R_0 U$.

As in the previous paper, I will use a Hopfield potential energy of the general form

$$PE = -\frac{1}{2}V_i T_{ij} V_j - \theta_k V_k - \sum_k \gamma_k \int_0^{V_k} A_k^{-1}(V)dV \tag{2}$$

and differentiate it according to the principle of virtual work

$$F_i \equiv -\frac{\partial PE}{\partial V_i} = T_{ij}V_j + \theta_i - \gamma_i A_i^{-1}(V_i) \tag{3}$$

and convert force to velocity using some viscous mobility

$$\frac{dV_i}{dt} = \mu_i F_i \tag{4}$$

Now a difference appears. The equations of motion on the energy surface are equivalent to the circuit equation if we make some new identifications:

$$
\begin{aligned}
T_{ij} &= T_{ij}^+ - T_{ij}^- \\
\theta_i &= \sum_k S_{ik} X_k \\
\gamma_i &= 1 \\
\mu_i &= 1/C_i
\end{aligned}
\tag{5}
$$

In this case, the equations for γ and μ are much simpler than in the case of the floating-dendrite model.

2.1 Simplified Mu

The mobility μ no longer depends on where in the hypercube the particle sits. We now have "evenly distributed honey". For some applications, this might not make a difference. The locations and energies of the fixed points will be the same no matter what μ is.

On the other hand, μ does affect the dynamics. In typical applications, one chooses the T_{ij} values to implement some desired energy surface $PE(V)$. In a floating dendrite network, you know that the vector will always move to lower energy values, but you don't know very much about its specific direction of motion. It prefers to move along the directions of highest mobility. The virtual ground network has the advantage that the vector moves straight down the gradient of the chosen energy surface. Therefore, if you are concerned with the dynamics of the network (as opposed to being concerned only with final states) then you might prefer to use virtual ground neurons.

Aside: In the floating dendrite model, it is in principle possible to choose a new set of T_{ij} values so that the dynamical vector would descend straight down the gradient of $PE(V)$, but the T_{ij} values would not be related to your original PE in a simple way.

2.2 Simplified Gamma: Gain

The simplified Thevenin impedance γ has important consequences too. In the virtual ground model model, a particular axon voltage across a particular synapse conductance makes a definite, predictable contribution to U_i. In the floating dendrite model, though, the effect of one synapse is voltage-divided with γ_i, so U_i depends on what synapses and other conductances are connected to that dendrite. Making or breaking a synaptic connection changes the effective strength of all the other synapses on that dendrite. Physically this comes about because of the current that flows backwards through the synapses, from dendrite to axon, proportional to the dendrite voltage. Once again, you have a situation where you can't just choose an energy function and use it to determine circuit component values in a simple way.

There are three main ways of dealing with the side effects that come from changes in the synapse resistors:

Large g^*: One way is to add to each dendrite a very large conductance g^* to ground. Then γ will be dominated by g^*, and will be effectively constant. This option is mathematically equivalent to the virtual ground design (since it holds the dendrite voltage near zero) but it is not optimal from a circuit engineering viewpoint, since it requires you to throw away practically all of your dendrite signal into a resistor to ground. It is pretty obvious that holding the dendrite at ground using a transimpedance amplifier is better than holding it at ground with a resistor.

Regulated g^*: Another non-simple way is to put a conductance g^* to ground that is just big enough to bring the γ up to some standard value. This is better than using a large g^*, since you don't throw away as much of your signal. It has the slight bug that every time you change a T_{ij} element, you have to change g^* to compensate. In the literature g^* is always shown as a special resistor to ground somewhere, which means that the learning rule cannot be local. This is not a bug if you are producing a large number of identical network chips, but might be a problem if you tried to implement changeable synapses using a cellular automaton.

Locally Regulated g^*: Here is a scheme for keeping γ at a pre-determined value by means of local changes at each crosspoint: Make each synapse have a regulated conductance, which can be connected either to ground or to an axon. If you want to be really fancy, you can build a voltage divider at the crosspoint by choosing the following T_{ij} conductances:

$$
\begin{aligned}
T_{ij}^+ &= \frac{\alpha+1}{2}g \\
T_{ij}^- &= \frac{\alpha-1}{2}g \\
g &= T_{ij}^+ + T_{ij}^- = \text{const} \\
V &= V_i\frac{T_{ij}^+ - T_{ij}^-}{T_{ij}^+ + T_{ij}^-} = \alpha V_i
\end{aligned}
\tag{6}
$$

These are just the equations for a voltage divider, with constant Thevenin conductance g and open circuit voltage V. V can be anywhere you want between V_i and $-V_i$. It contributes to T any amount you want in the interval $[-g, g]$, but contributes to γ a constant amount g. Notice that it does not require a ground wire.

Note that a lumped g^* might have to be fabricated by different techniques, since it might have to compensate for a very great number of missing synapses, while the locally regulated g^* never needs to compensate for more than one at a time. The compensation is distributed throughout the matrix.

2.3   Simplified Gamma: Timing

Another reason why it is highly desirable to regulate the conductance in the floating dendrite model is that the RC settling time of the neuron depends on γ. In the virtual ground model it depends only on the resistance of the feedback elements.

2.4   Miscellaneous Remarks

1) One drawback of the virtual ground neural network is that the feedback resistor must have a different conductance value than the T_{ij} elements, by a factor of order N.

2) An argument in favor of the virtual ground model is that the physically important variable is the same as the conceptually important variable. One thing that is obvious from the physical point of view, and which is driven home as soon as you try to simulate this system, is that the important variable is the charge on the capacitor. That is the initial condition that has to be specified. In the floating dendrite model, the dendrite voltage U is tied to the capacitor. Alas, the variable which goes to the outside world is the axon voltage V, and the energy is written in terms of V. It doesn't help to try to write E in terms of U. In the virtual ground model, these difficulties do not arise, because the capacitor voltage and the axon voltage are the same thing.

3) In the absence of hard data, one may speculate that this formulation is at least as sensible as previous ones for modelling the behavior of real, biological neurons. Many of the counter-intuitive properties of the earlier models came from the "back current", the discharge of the capacitor back through the synapse conductivities. In biological system, one imagines that the presynaptic neuron acts like a constant current source; i.e., the flow of neurotransmitters is a one way process.

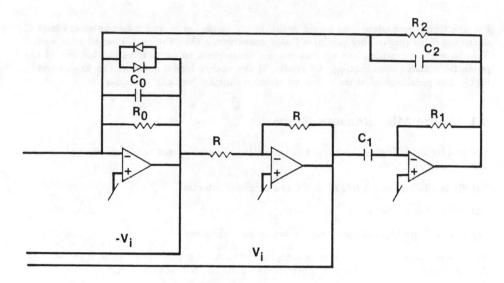

Figure 2: Neuron with Second Derivative

3 Second Derivatives & Inertia

The fact that the term in $\dot{V}$ enters equation 1 in such a simple way makes it easy to design a circuit that has a term in $\ddot{V}$ also.

The equation of motion for a network of such neurons is

$$\sum_j T_{ij}V_j - A^{-1}(V_i) - \eta \dot{V}_i - m\ddot{V}_i = 0 \tag{7}$$

where the damping is $\eta = R_1 C_1 / R_2$ and the mass is $m = R_1 C_1 C_2$. Note that one can choose the values of mass and damping independently. Of course A, η, and m can vary from neuron to neuron, but I have suppressed the subscript i.

The potential energy (equation 2) and the principle of virtual work (equation 3) remain the same. The response equation is modified to be

$$m\frac{d^2 V_i}{dt^2} + \eta \frac{dV_i}{dt} = F_i \tag{8}$$

Where η plays the role of $1/\mu$. In this way the energy and force laws reproduce the correct dynamics as given by equation 7. It is not surprising to find, though, that the potential energy is not a Lyapunov function for this system. The marble can pick up some kinetic energy as it rolls downhill, overshoot the minimum, and roll back up the opposite side of the pit. The unsurprising but still beautiful solution is to add a kinetic term:

$$E = PE + \frac{1}{2}\sum_i m_i \dot{V}_i^2 \tag{9}$$

It is easy to show that this total E is non-increasing and that the particle comes to rest at a point where E, and *ipso facto* PE, are at a local minimum.

For practical applications, one would probably try to choose m and η to implement critical damping. This presumably produces faster convergence than the over-damped case would. The descending marble picks up just enough momentum to carry it to the bottom of the potential without overshooting. Of course, if the energy landscape near the fixed point is highly non-paraboloidal, it isn't clear what one means by "critical" damping.

3.1 More Miscellaneous Remarks

1) Consider the following three types of interesting circuit features or complications in neural network models:

   a) T_{ij} is off-diagonal: it lets neuron i and neuron j interact

   b) $A^{-1}(V_i)$ is nonlinear

   c) $\dot{V}$ and $\ddot{V}$ are higher-order: they contain time derivatives

The present models do not mix these complications, but it is possible to imagine models that do. For instance, it is amusing to contemplate an off-diagonal mass tensor.

2) Note that it is not really necessary for a single biological nerve cell to embody all of the functions required in figure 2. Just as that figure uses an additional op-amp to carry out the second time derivative, you could imagine a neuron with an interneuron such that one cell performs the first derivative and the other performs the second derivative.

4   Complex Numbers

Another interesting generalization is a scheme to represent complex numbers. This can be implemented by using one neuron to represent each of the two phasor projections: one neuron carries the real part of the signal and the other carries the imaginary part. There are then 4 "real" synapses when a complex axon meets a complex dendrite. There are actually 8 crosspoints if each axon has a separate wire for T^+ and T^-. The strengths of the "real" synapses within one "complex" synapse must be chosen in accordance with the laws of complex multiplication, which is easy.

It is clear that any such pairing scheme can be thought of either as a network with N complex neuron-pairs, or as a network of $2N$ simple neurons. Even if the $N \times N$ network is Hermitian, the $2N \times 2N$ network will be systematically non-symmetric in an interesting way.

Such methods of embedding sophisticated processors within simpler ones may help explain the development and evolution of the brain, and may help us design algorithms to exploit artificial neural networks. They also imply that if experiments ever show that biological systems have non-symmetric synaptic connections, it does not necessarily follow that the neurons cannot be understood in terms of ideas like Hopfield's.

Introducing complex numbers is motivated partly by general mathematical curiosity, and partly by the hope that we can represent the Schrödinger equation. Quantum mechanical dynamics is an obvious way to try to generalize the classical dynamics in the Hopfield model. Also, as I have said before, the analogy between neural networks and spin glasses is useful, but not very deep. The analogy involves mostly the statics: ground state energies and fixed points. I would like to develop a network that is analogous to the quantum mechanical *dynamics* of a spin glass or some such thing.

This would give us a new way to escape from local minima in the energy surface — simulated tunneling, rather than simulated annealing. Complex numbers would also provide a natural way to produce eigenstates that are limit cycles, with a periodic time dependence. We know that there are all sorts of oscillator circuits and periodic signals in the brain.

The Schrödinger analogy has not progressed much beyond the speculation stage, and there are many problems. For one thing, the wavefunction of a particle is a vector *field*, whereas the vector V of neural voltages is only a vector *space*. Secondly, the Schrödinger equation equates a time derivative to an *energy*, whereas the neural circuit equates a time derivative to a *force*. This means we have to be clever; we cannot easily exploit simultaneously the analogy to physics and the analogy to the real-valued neural network.

5   Unipolar Synapses

It is awkward to build a network chip that contains both synapses of both signs. You apparently need twice as many synapse resistors, twice as many axon wires, and twice as many amplifiers. There are various ways to solve this problem. One way is to add a constant to T_{ij} and compensate for it later. Assuming T_{ij} is in the interval $[-1, +1]$,

$$
\begin{aligned}
U_i &= T_{ij}V_j \\
&= (T_{ij} + Z_{ij})V_j - Z_{ij}V_j \\
&= T'_{ij}V_j - \sum_j V_j
\end{aligned}
\tag{10}
$$

where $Z_{ij} = 1$ for all i, j. The compensation term is independent of i so it can be computed once and for all by an interneuron and then fed to all the regular neurons. Note that the synapses coming from the interneuron axon will typically be a factor of order N stronger than ordinary synapses, so you might want to think carefully about how you do it. You might perhaps provide a separate offset input to the amplifier. One should take pains to ensure that the interneuron does not introduce a big time delay.

Technical note: Suppose that T_{ij}^+ and T_{ij}^- are each chosen from the two values $\{0, +1\}$. Then T_{ij} will have *three* possible values, $\{-1, 0, +1\}$. Shifting the zero maps this onto $\{0, +1, +2\}$, which is good, since they all have the same sign. You don't have to build synapses where the zeroes are, but you do have to build synapses where the $+1$'s are. This means that that shifting the zero is more complicated than just throwing away the T_{ij}^- synapses. It isn't a serious complication; building two types of synapses ($+1$ and $+2$) is much easier than building two types of axons (inverting and non-inverting), and takes up less real estate on the chip. Also, one can revise the learning algorithm so it doesn't generate any $T_{ij} = 0$.

Remember that what we are worrying about here is the sign of T_{ij}; having bipolar signals on the V_j is not a problem.

Consider for a moment what would happen if we arbitrarily threw away the negative synapses T_{ij}^-. It would be possible to have stable states, in the following way. Divide the neurons into two groups. The positive neurons (i.e. that are supposed to be at $+1$ volt) have positive feedback connections to all of the other positive neurons, and the negative neurons have positive feedback connections to all of the other negative neurons. Both groups are stable. On the other hand, they are independent. Because there are no negative synapses, there is no way for a positive neuron to stabilize a negative neuron or vice versa. Suppose I started the network in a state where some of the neurons were positive, and the rest were at their "maybe" logic level. Then the the network would reconstruct all of the positive bits in the word, but it could not possibly reconstruct any of the negative bits. In fact, the system would quickly wind up in the "all +1" state, which is super-stable.

The solution is obvious: throw away the positive synapses instead. In the stable state, the neurons can still be divided into two groups, according to the sign of their voltage, but now the positive neurons reinforce the negative ones, and vice versa. Neither can get along without the other.

Obviously, one can easily put in a compensation interneuron in a negative-only synapse system, by choosing $Z_{ij} = -1$ in the above analysis. A compensated negative-only system is more tolerant than a compensated positive-only system against loss of compensation, and presumably it will be more tolerant of lesser errors too. I recommend this compensated negative-only system to all hardware builders.

A back-of-the-envelope calculation by Hopfield indicates that even in the absence of compensation, throwing away the positive synapses reduces the asymptotic capacity of a CAM by only a factor of $\pi/2$.

6   Zeroing the Diagonal of T

There is a stock phrase that appears in almost all papers in this field: After deriving some rule for setting up T_{ij}, people tend to say something like "and of course we must set $T_{ii} = 0$".

There is a case where it is important to zero the diagonal. In the closely related model where rather than having a differential equation, one has a discrete time series, and where furthermore one updates the neurons one at a time (and applies the changes before evaluating the RHS of the update equation for the next neuron) then the size of the time step that one may take depends (reciprocally) on T_{ii}.

When one deviates from this case, it pays to think a little bit. If T_{ii} is non-zero and one takes too big a step, it is possible for the energy to increase at each step — the model becomes unstable. This is discussed in more detail in the paper by Jeffrey and Rosner[2].

Secondly, if one updates more than one neuron at a time, the stable stepsize condition is more complicated, involving eigenvaules and Rayleigh quotients.

On the other hand, the differential equation model corresponds to infinitesimal time steps, and you can have any T_{ii} you like. Changing T_{ii} is completely equivalent to changing the gain-setting feedback resistor R_0.

References

1. J. S. Denker, to appear in Physica D.

2. W. Jeffrey and R. Rosner, this volume.

Stochastic Spin Models for Pattern Recognition

R. Divko and K. Schulten
Physik-Department
Technische Universität München, D-8046 Garching

Abstract

We exploit for the recognition of patterns the properties of physical spin systems to assume long range order and, thereby, to establish a global interpretation of patterns. For this purposes we choose spins which can take a discrete set of values to code for local features of the patterns to be processed (feature spins). The energy of the system entails a field contribution and interactions between the feature spins. The field incorporates the information on the input pattern. The spin-spin interaction represents 'a priori' knowledge on relationships between features, e.g. continuity properties. The energy function is choosen such that the ground state of the feature spin system corresponds to the best global interpretation of the pattern. The ground state is reached in the course of local stochastic dynamics, this process beeing simulated by the method of Monte Carlo annealing[1]. Our study is related to work presented in Refs. 2, 3.

Introduction

Spin systems are characterized by a set of values for the spin variable $S_{i,j}$, a lattice on which they are defined, and by an interaction energy. In the two-dimensional Ising model the spins take the values ± 1 and the interaction energy is defined by the Hamiltonian

$$E = -J \sum_{<(i,j),(k,l)>} S_{i,j} S_{k,l} - \sum_{(i,j)} H_{i,j} S_{i,j} \qquad (1)$$

where the brackets indicate summation over next neighbors. In the ferromagnetic case ($J > 0$) the first term, the exchange interaction, gives a negative contribution if neighboring spins point into the same direction. This term creates a tendency for an alignment of all the spins. The second term describes the interaction of the spins with a local magnetic field $H_{i,j}$ tending to align the spins locally with the field. The regularizing effect of the exchange interaction will be utilized in the following to solve pattern recognition tasks under the constraint that pattern features are expected to vary continuously.

Feature Spins

For the purpose of picture processing the spins are choosen to code local features of a pattern. Examples for attributes coded by such feature spins are intensities, disparities between corresponding points in a stereogram or edges of different directions. Several different types of feature spins, interacting with each other and with external fields, may be needed to solve a specific pattern recognition problem.

At finite temperatures the feature spin system shows fluctuations like its physical counterpart. Certain values of a feature spin at a certain lattice point are more probable than others. One may consider the value of a feature spin as the hypothesis that the picture has a certain local attribute at this point.

At high temperatures all hypotheses are equally probable. After carefully cooling down to low temperatures (simulated annealing[1]) the fluctuations eventually disappear. At zero temperature the feature spins take definite values indicating the final global hypothesis about a pattern.

The final hypothesis, i.e. the ground state, achieved by the system after cooling down to low temperatures depends on the interaction among the feature spins as well as on the interaction with the external field. The interaction among the feature spins contains an 'a priori' global knowledge on relationships to be expected to hold between the features of a pattern. Correct interpretations of patterns must meet certain constraints, e.g. the constraints of continuity, which have to be realized by the feature spin configurations in the final hypothesis. Such configurations can be achieved by a properly choosen interaction between the feature spins. For example a Potts model type interaction between intensity spins yields a smooth change of brightness. The external field serves to communicate the pattern to be processed to the system of feature spins. Examples for pattern attributes coded by the external fields are local brightness or edges of various directions.

STEREO VISION

Whereas a certain degree of depth vision can be obtained from perspective distortion or from hidden parts of a scene, full stereo vision is a result of binocular perception. The projections of an object in both eyes differ slightly from each other. This difference (disparity) allows the reconstruction of the three-dimensional information. Figure 1 shows an image pair of dot patterns appearing completely random when viewed monocularly. But when viewed one through each eye the two pictures fuse showing a three-dimensional structure (square hovering over the ground). The absence of higher level structures in the patterns shows that disparity alone can be used to obtain three-dimensional information from a stereogram.

Fig. 1. Julesz pattern[5] with 50% black dots. This random-dot stereogram of 50×50 pixels is generated by copying the rigth image from the left one, shifting a square-shaped region of 30×30 pixels sligthly to the left and filling the gap caused by the shift with a new random pattern.

To obtain stereo information the disparity of corresponding points in the two retina projections of an image must be determined. The problem is to assign correspondences between points of the two pictures. This is a difficult task because of the socalled 'false target problem'[4] occuring in its extreme in Julesz patterns[5]. Every black pixel could correspond to every other black pixel. To restrict all possible combinations of points from both pictures to physically plausible correspondences the following matching conditions must hold:
- Compatibility: Black dots can only match black dots and vice versa.
- Continuity: The physical feature disparity varies smoothly almost everywhere over the image.
- Uniqueness: Except in rare cases each point from one image can match only one point from the other image.

The following spin model is designed to find the correspondences between the pixels of both pictures of a random-dot stereogram and measure the disparities. This information will be contained in the ground state of the spin system.

A Spin Model for Stereo Vision

The feature coded by a spin is the disparity of corresponding pixels. A disparity spin with a value $S_{i,j} \in \{0, \pm1, \ldots, \pm N\}$ at lattice site (i,j) stands for the hypothesis of a correspondence between the pixel (i,j) in the right picture of the stereogram and the pixel $(i, j + S_{i,j})$ in the left picture shifted $S_{i,j}$ units to the rigth. Both pixels are assumed to correspond to the same original point of an object and to have the disparity $S_{i,j}$.

The Hamiltonian of the disparity spin system is split into two contributions

$$E_{total} = E_{exchange} + E_{field}. \tag{2}$$

These terms reflect the continuity and the compatibility conditions, respectively. The distance of an observer to a point on the surface of an object is a smoothly varying property. To achieve a corresponding property for the values of the disparity spin a Potts model interaction is choosen

$$E_{exchange} = -J \sum_{<(i,j),(k,l)>} F(S_{i,j}, S_{k,l}),$$

$$\tag{3}$$

$$F(S_{i,j}, S_{k,l}) = \begin{cases} 1, & \text{if } S_{i,j} = S_{k,l} \\ q, & \text{if } S_{i,j} = S_{k,l} \pm 1; \quad q < 1. \\ 0, & \text{else} \end{cases}$$

If the value of a disparity spin is $S_{i,j}$ and if this hypothesis is correct, the pixel (i,j) in the right picture and the pixel $(i, j + S_{i,j})$ in the left picture have identical surroundings. If the disparity hypothesis $S_{i,j}$ is wrong the surroundings may be completely different. Therefore, the comparison of the neighborhoods of the two points assumed to correspond to eachother indicates a possible correct match.

Whereas many features can (and for real pictures must) be used for comparison, we restrict ourselves in the present application to the most simple choice and compare the pixel intensities in a square shaped region only. Comparison is established by the following energy contribution

$$E_{field} = \sum_{(i,j)} G(S_{i,j}), \qquad G(S_{i,j}) = G_0 \sum_{k=i-w}^{k=i+w} \sum_{l=j-h}^{l=j+h} |H_{k,l+S_{i,j}}^{left} - H_{k,l}^{right}|. \tag{4}$$

Here $H_{i,j}^{left}$ and $H_{i,j}^{right}$ denote the intensity of the pixel (i,j) in the left and in the right picture of the stereogram and $G_0 = [(2h+1)(2w+1)]^{-1}$ is a normalization constant. Correct disparity spin configurations are characterized by low energy contributions.

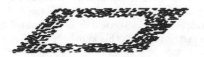

For the input pattern shown in Figure 1 the disparity field obtained is presented in Figure 2. Starting from the temperature $T = 1.5$ the annealing process was stopped at a low temperature $T = 0.01$. For the interaction parameters in (3) we have assumed the values $J = 2$ and $q = 0$ and for the maximal disparity the value $N = 5$.

Fig. 2. Ground state of the disparity spin system corresponding to the Julesz pattern in Figure 1.

A more complicated stereogram containing the three-dimensional information of an eight level pyramide is shown in Figure 3. The solutions of the disparity spin system ($N = 10$) with interaction parameters $J = 2$ and $q = 0.2$ are shown for three temperatures in Figure 4.

Fig. 3. Julesz pattern of an eight level pyramide.

At the high temperature $T = 0.8$ the fluctuations of the disparity spin values are large. This is demonstrated by a snapshot of the dynamics shown in Figure 4.a. Figure 4.b illustrates that at the intermediate temperature $T = 0.3$ the system still fluctuates; however, the disparity field already indicates the presence of different disparity planes. Figure 4.c shows that at the low temperature $T = 0.1$ the fluctuations almost disappeared and that the disparity spin system achieves the correct interpretation of the Julesz pattern, an eight level pyramide.

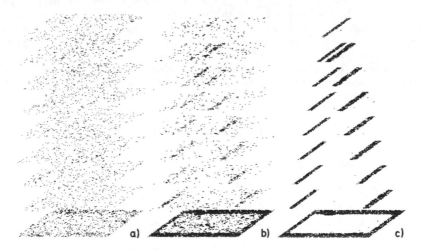

Fig. 4. Behaviour of the disparity spin system for the Julesz pattern in Figure 3.

A SPIN MODEL FOR PICTURE RESTORATION

Restoration of noisy pictures can be simplified if expected relations between picture attributes are known. As an example we consider a chessboard-like pattern as input for a picture restoring system. There are several 'a priori' qualities present in such pattern: the intensity in a square is constant, at a square's border are straight edges in vertical or in horizontal direction, edges are continuous. A system of feature spins instructed with this knowledge can restore noisy

chessboard patterns. Such system entails three kinds of feature spins
- Intensity spins which take the values ± 1 for black and white colours, respectively.
- Horizontal edge spins which take the values $+1$ for intensity changes from white to black, the value -1 from black to white and the value 0 in the case of an absence of any edges.
- Vertical edge spins which follow corresponding rules.

The intensity spins are defined on a square lattice. The edge spins are located between neighboring intensity spins.

The Hamiltonian of the picture restoring spin system can be written

$$E_{total} = E_{i-ifield} + E_{h-hfield} + E_{v-vfield} + E_{i-i} + E_{h-h} + E_{v-v} + E_{i-h} + E_{i-v} \quad (5)$$

where the indices i, h, v refer to intensity, horizontal and vertical edge spins, respectively, with the corresponding fields $ifield, hfield, vfield$. To implement the continuity condition for the intensity spins $I_{i,j} \in \{-1, +1\}$ an Ising-like interaction is assumed. To implement the continuity property of edges an attractive interaction in the proper direction is employed for the horizontal edge spins $H_{i,j} \in \{-1, 0, +1\}$ as well as for the vertical edge spins $V_{i,j} \in \{-1, 0, +1\}$.

$$E_{i-i} = - J_i \sum_{<(i,j),(k,l)>} I_{i,j} \, I_{k,l} \quad (6)$$

$$E_{h-h} = - J_h \sum_{(i,j)} \delta(H_{i,j+1}, H_{i,j}); \qquad E_{v-v} = - J_v \sum_{(i,j)} \delta(V_{i+1,j}, V_{i,j}) \quad (7)$$

To obtain compatibility between the hypotheses of edge spins and intensity spins an interaction energy favoring consistent configurations is added

$$E_{i-h} = - J_{i-h} \sum_{(i,j)} T(I_{i,j}, I_{i+1,j}, H_{i,j}), \qquad E_{i-v} = - J_{i-v} \sum_{(i,j)} T(I_{i,j}, I_{i,j+1}, V_{i,j}). \quad (8)$$

Here T contains the compatibility condition listed in Table 1.

i	j	k	T
1	1	0	1
-1	-1	0	1
1	-1	-1	1
-1	1	1	1
1	1	1	0
-1	-1	1	0
1	1	-1	0
-1	-1	-1	0
1	-1	0	0
-1	1	0	0
1	-1	1	0
-1	1	1	0
1	-1	-1	0
-1	1	-1	0

Table 1: Compatibility condition table $T(i, j, k)$ where i, j denote pairs of intensity spins and k denotes the edge spin inbetween.

The pattern to be processed is coded as a field $F_{i,j} \in \{+1, -1\}$ corresponding to black and white pixels at position (i, j). The interaction between intensity spins and the field is choosen like in the Ising model.

$$E_{i-ifield} = - J_{ifield} \sum_{(i,j)} I_{i,j} \, F_{i,j} \quad (9)$$

The field for edge spins codes intensity changes. The interaction between edge spins and the corresponding fields is

$$E_{h-hfield} = - J_{hfield} \sum_{(i,j)} \delta(H_{i,j}, (F_{i+1,j} - F_{i,j})/2).$$

$$E_{v-vfield} = - J_{vfield} \sum_{(i,j)} \delta(V_{i,j}, (F_{i,j+1} - F_{i,j})/2).$$

(10)

As input for the picture restoration spin system we choose a distorted chessboard pattern, measuring 20×20 pixels. 20 percent of the pixels were randomly reversed from black to white and vice versa. The resulting pattern is presented in Figure 5.a. The aim of the restoration process is to find the chessboard closest to this picture. The interaction paparameters assumed for the restoration are $J_i = 1$, $J_h = J_v = 4.5$, $J_{i-h} = J_{i-v} = 4.5$, $J_{ifield} = 1$, $J_{hfield} = J_{vfield} = 3$. The temperature was lowered in 12 steps from an initial value of $T = 8$ to the final value of $T = 0.05$. Figure 5.b shows the result: the chessboard pattern has been restored to a very large degree.

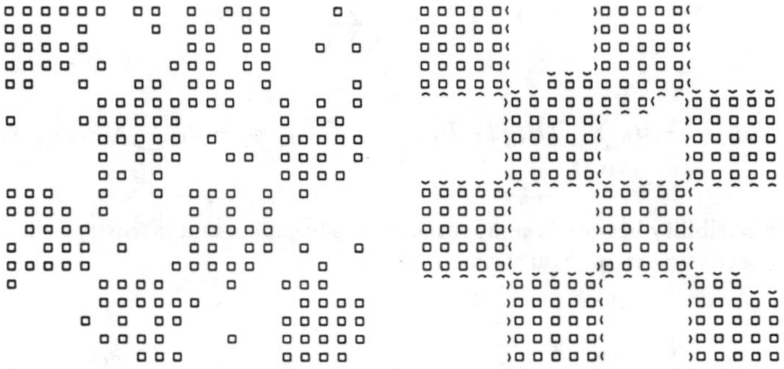

a)   b)

Fig. 5. Input pattern and restored chessboard pattern for the feature spin system described by Eqs. (5)-(10).

REFERENCES

1. J. Kirkpatrick, C.D. Gelatt, M.P. Vecchi, Science, 220, 671 (1983).
2. S. Geman, D. Geman, IEEE Tran. Pattern Anal. Machine Intell., vol. PAMI-6, 721, (1984).
3. P. Kienker, T.J. Sejnowski, G.E. Hinton, L.E. Schuhmacher, Separating Figure from Ground with a Parallel Network, preprint (1986).
4. D. Marr, Vision (Freeman, 1982).
5. B. Julesz, Foundations of Cyclopean Perception (Univ. Chicago Press, 1971).

THE SIGMOID NONLINEARITY IN NEURAL COMPUTATION.
AN EXPERIMENTAL APPROACH.

F. H. Eeckman and W. J. Freeman
University of California, Berkeley, CA   94720

ABSTRACT

A sigmoid function, relating wave amplitude (EEG) to pulse
density, is derived experimentally in the rat olfactory bulb. The
bulb is the first central structure which is related to odor processing. The curve is asymmetric in that the region of maximal
slope is displaced toward the excitatory side. The slope and the
maximum of the curve vary with the state of arousal of the animal.
This is in agreement with earlier experiments by Freeman in cats
and rabbits. The data are explained in terms of his model of the
olfactory system. The relationship of sigmoid slope to efficiency
of neural computation is examined by comparison to known analysis
of neural networks.

INTRODUCTION

The characteristic feature of the electroencephalogram (EEG) of
the mammalian olfactory bulb (OB) in an awake animal consists of
sequences of bursts in the range of 40-80 Hz riding on top of a
baseline shift in potential at the frequency of respiration. Each
burst begins shortly after inspiration and terminates during expira-
tion.

The mitral cells in the OB are large neurons that are mutually
interconnected via excitatory synapses on their re-entering axon
collaterals. A second and larger population of small neurons,
called granule cells, is connected via reciprocal synapses of opposite sign to the mitral cells.[1] The mitral to granule cell synapse
is excitatory, the granule to mitral cell synapse is inhibitory.
This feedback arrangement forms an oscillatory network. The bulb
consists of approximately 2,300 such coupled oscillators.

MATERIALS AND METHODS

A total of 20 rats were used in this study. In 10 (#1-10) rats
acute experiments were done under anesthesia. These experiments
involved an exposure of the olfactory bulb and the lateral olfactory
tract (LOT). Tungsten micro-electrodes were then inserted into the
mitral cell layer at various locations in the bulb. This was done
under full visual control, and was checked electrophysiologically by
reversal of the LOT evoked potential at that layer. A 250 micron
surface electrode was then positioned on the surface in an area
overlying the inserted micro-electrode. Both single and multi-unit
traces were taken together with EEG. No attempt was made to distinguish between single and multiple units.

In 10 rats (#11-20), two 100 micron stainless steel electrodes with sharpened tips were chronically implanted. One electrode was positioned at the mitral cell layer. A reversal of the LOT evoked potential (LOT stimulating electrode 250 micron bipolar) and a good signal to noise ratio for spikes were again used to assess correct positioning. The other 100 micron electrode was either at the surface or in the granule cell layer for EEG recording. The interelectrode distance was no greater than .8 mm.

These chronically implanted animals were recorded from over a one-month period, while awake and attentive. No stimulation (electrical or olfactory) was used. The background environment for recording was the animal's home cage placed in the same room during all sessions. In four of these animals measurements were also taken after anesthesia was induced.

The EEG signal was filtered between 10 Hz and 300 Hz and digitized at 1 ms intervals. It represents the granule cell activity.[2] The unit signal was filtered at 300 Hz and 3 kHz and fed into a threshold device. For every value above a preset threshold (at 2-3X the background level) a standard 5v, 1 ms square wave pulse was stored concurrently with the EEG. These two time series were stored on disk for off-line processing using a Perkin Elmer 3220 supermini.

PROCESSING

The range of EEG amplitude values is divided into w intervals ΔV amplitude, each assigned a bin number. Every value of V(T) lies within a bin and for each a count is added to the bin. The amplitude histogram is normalized to an empirical probability density function. It is truncated at ±3 sd. A second histogram is constructed at the same time: it has one interval corresponding to each ΔV range. For all occurrences of a value $\Delta V(T)$ we check whether a pulse was recorded at the same time (T=0) or in any of the preceding 25 (T=-25) or following 25(T=+25) timebins. This yields a 2 dimensional table for pulses with EEG amplitude and time as the axes.

Cross sections of the table at $\hat{V} \neq 0$ give $\hat{P}(\hat{T})$. This section is often oscillatory at the dominant frequency of the EEG. The frequency is extracted by fitting a sinewave to the data, using nonlinear regression. The model is:

$$P(T) = po \ (1 + \tilde{P} \cos (\omega T + \phi)) \tag{1}$$

where po is the steady state pulse rate, and $\tilde{P}$ is the modulation amplitude.

By counting the number of pulses at a fixed time lead, where the EEG is maximal in amplitude, and plotting them versus the normalized EEG amplitudes, one obtains a sigmoidal function: the pulse probability sigmoid curve $\hat{P}(\hat{V})$.[2] This function is normalized by dividing it by the average pulse level of the record. It is fitted by nonlinear regression. The derivative of this curve gives us the gain for wave to pulse conversion. The model is:

$$Q = Q_{max} \ (1 - \exp \ [-(e^V - 1)/Q_{max}]) \tag{2}$$

where $Q = (P-po)/po$ and $Q_{max} = (P_{max}-po)/po$.

This equation relies on one parameter only to fit both the anesthetized and awake normalized animal data. The derivation and justification for this model are discussed in an earlier paper by Freeman.[3]

RESULTS

The mitral cell firing conditional on the granule cell activity (EEG) was measured in various locations of the olfactory bulb. The data are essentially similar all over the bulb. Units can be found throughout, whose probability of firing varies sinusoidally with the dominant frequency of the EEG. However they have approximately a quarter cycle phase lead to the EEG.

First we looked at the frequency, phase and the amplitude modulation of the fitted waves in animals from both groups. The results obtained from fitting Equation 1 to the data are given in Table I. The data are from 9 acute preparations (anesthetized) and 7 chronic animals (awake). There is a significant difference between the groups for both frequency and amplitude modulation ($p<.05$, two sided t test). There is no difference in phase between the groups.

TABLE I   SINE-WAVE FITS

		Anesthetized					Waking		
Subj.	ω	mod	ϕ	$\sqrt{1-r^2}$	Subj.	ω	mod	ϕ	$\sqrt{1-r^2}$
#1	30.6	44.3	1.05	.16	#11	74.9	91.5	1.26	.28
#2	47.7	11.6	.87	.03	#14	72.9	63.6	.79	.26
#3	38.0	27.5	1.11	.12	#15	65.5	88.2	1.75	.22
#4	56.3	19.6	1.41	.04	#16	82.4	91.0	1.05	.26
#5	43.9	30.3	1.29	.12	#17	81.1	76.3	1.45	.29
#6	48.4	17.0	1.14	.17	#18	75.3	86.2	1.35	.22
#7	56.2	15.0	2.29	.06	#20	73.8	79.0	1.45	.17
#8	39.8	10.0	1.37	.06					
#9	18.0	20.3	2.09	.11					

ω: x=42.1   s=12.3 ω: x=75.1   s=5.6
mod: x=21.7   s=10.8   mod: x=82.3   s=10.1
ϕ: x=1.4   s=0.5 ϕ: x=1.3   s=0.3

Next we examined $\hat{P}(\hat{V})$. In 4 subjects we derived the sigmoid in both waking conditions and under anesthesia. The data were fitted with Equation 2. The differences between the Q_{max} values in both conditions was significant at the .05 level (2 sided, paired t test).

Sigmoids were also derived for all other chronic animals. We compared these to 4 subjects in the acute group, where a $\hat{P}(\hat{V})$ curve was available. The difference in Q_m between both groups is again significant at the .05 level (2 sided, t test for ungrouped data). The results are summarized in Table II.

The overall mean for Q_m in the anesthetized group was 2.36, with a standard deviation of .89. In the awake animals the mean value for

Q_m was 6.34 and the standard deviation 1.46. The ranges were 1.15-
3.62 (anesthetized) and 4.41-9.53 (awake). The values for slope ($\dot{Q}$)
and Q_m are interdependent. The data presented here compare well with
previously published results in rabbits and cats. Freeman[3] found a
Q_m value of 2.64±1.22 for anesthetized animals, and a Q_m of 5.31±2.77
in awake animals. The ranges were 1.69-14.86 (waking) and 1.00-3.29
(anesthetized).

TABLE II SIGMOID FITS

Subj.	Anesthesia Q_m	$\dot{Q}$	Waking Q_m	$\dot{Q}$
#16	3.62	1.76	5.18	2.31
#17	1.15	1.01	5.94	2.58
#19	2.55	1.38	6.70	2.86
#20	3.60	1.50	5.70	2.50

Subj.	Waking Q_m	$\dot{Q}$	Subj.	Anesthesia Q_m	$\dot{Q}$
#11	4.41	2.04	#6	1.87	1.17
#12	6.74	2.88	#7	1.63	1.11
#13	9.53	3.89	#8	2.35	1.32
#14	5.04	2.27	#10	2.08	1.24
#15	6.79	2.89			
#18	7.39	3.11			

Fig. 1. EXAMPLE OF SIGMOID FITS

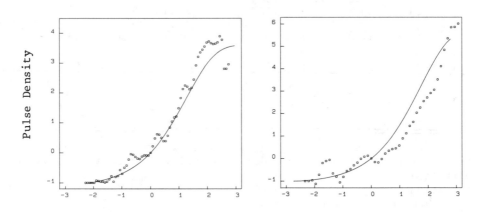

Normalized EEG amplitude (in standard deviations)

#20 Anesthetized
(Q_m = 3.60)

#20 Awake
(Q_m = 5.70)

Note different vertical scales.

DISCUSSION

A model has been proposed by Freeman[2,3] for the generation of the burst. It consists of a feedback mechanism between two sets of densely interconnected neurons: the mitral and the granule cells. The key to the performance of this model is the nonlinear gain curve that relates wave amplitude (EEG) to pulse density (units). This curve was experimentally derived in the manner described above. The maximal slope is displaced to the excitatory side, so that when the input rises on the mitral cells (odor) the system can be driven into an oscillatory burst.[4] This bifurcation into oscillation is essential to information processing in the bulb.[5]

The sigmoid slope decreases under anesthesia and increases with arousal. We are presently investigating short-time changes in slope with varying levels of attention and motivation in animals trained to respond to a stimulus. This short-term modulation (under centrifugal control?) could change the computational capabilities and the level of detail in pattern formation and recognition in the olfactory bulb.

Hopfield[6] has shown how a varying sigmoid slope affects the number of local minima in a system with fixed point attractors. Odor specific EEG spatial patterns emerge and stabilize with learning in rabbits, trained under an odor discrimination paradigm [Grajski and Freeman, this volume]. We can thus postulate that a Hopfield-like network, but one with periodic attractors, exists in the bulb. Both theoretical studies and simulations are being undertaken to investigate this possibility [W. Baird, this volume].

Grossberg[7] has extensively studied the many desirable properties of sigmoid interaction rules. For a comparison and analysis of different models using a sigmoid function, see Grossberg.[8]

Further study of the sigmoid nonlinearity under a variety of experimental conditions is likely to yield other interesting results that will help us to better understand neural computation.

REFERENCES

[1] W. Rall and G. M. Shepherd, J. Neurophysiol. 31: 884 (1968).
[2] W. J. Freeman, Mass Action in the Nervous System (Academic Press, N.Y., 1975), Chapter 3.
[3] W. J. Freeman, Biol. Cybernetics 33: 237 (1979).
[4] D. M. Sunday and W. J. Freeman, Physiologist 18: 42 (1975).
[5] W. J. Freeman, Biol. Cybernetics (1986, submitted).
[6] J. J. Hopfield, Proc. Natl. Acad. Sci. USA 81: 3088 (1984) Biophysics.
[7] S. Grossberg, Studies in Applied Mathematics, Vol LII, 3, (M.I.T. Press, 1973) p. 213.
[8] S. Grossberg, SIAM-AMS Proceedings, Vol. 13, (Math., Psychol. and Psychophysiol., Providence, RI, 1981), p. 107.

Supported by Grant MH 06686 from the National Institute of Mental Health.

CONTEXT-FREE PARSING WITH CONNECTIONIST NETWORKS

M. A. Fanty*

University of Rochester, Rochester, NY 14627

ABSTRACT

This paper presents a simple algorithm which converts any context-free grammar into a connectionist network which parses strings (of arbitrary but fixed maximum length) in the language defined by that grammar. The network is fast, O(n), and deterministic. It consists of binary units which compute a simple function of their input. When the grammar is put in Chomsky normal form, $O(n^3)$ units needed to parse inputs of length up to n.

INTRODUCTION

My goal in designing a connectionist[1] parser was to be able to build, in a systematic way, for any context-free grammar, a network which will parse strings in that grammar (within a length restriction). The network must represent the parse tree for the input when finished, and must clearly indicate when there is no parse. Most important, I wanted the network to be both deterministic and completely general, which is how it differs from previous connectionist parsers.[2,3] My goal is not cognitive modeling per se, but to provide a powerful technique which could prove useful for natural language understanding and other connectionist applications. Context-free grammars have been widely used for linguistics,[4] computer science[5] and pattern recognition.[6] The existence of a fast, simple and relatively efficient connectionist parser may well be of some importance to cognitive scientists and others working with connectionist models. This paper reviews the definition of context-free grammars and present the network in detail.

CONTEXT-FREE GRAMMARS

A context-free grammar consists of symbols and rewrite rules which allow a nonterminal symbol to be replaced by a string of symbols. A derivation begins with a distinguished nonterminal, the start symbol. Rewrite rules are applied in any order until only terminal symbols are left. The string of terminal symbols is a member of the language defined by the grammar. Formally, a context-free grammar is a quadruple, (V_N, V_T, B, P), where V_N is a finite set of nonterminal symbols, V_T is a finite set of terminal symbols, disjoint with V_N, $B \in V_N$ is the start symbol, and P is a set of ordered pairs (L,R) such that $L \in V_N$ and $R \in (V_N \cup V_T)^*$, i.e. R is a possibly empty string composed of terminals and/or nonterminals. These rewrite rules, or productions, are usually written with an arrow separating the right and left-hand sides. The fact that productions have a single nonterminal on the left makes them context-free: the possible rewrites of a symbol are the same no matter where it occurs. The following grammar is meant to capture a subset of syntactically correct English.

$$V_N = \{S, NP, NP2, VP, PP\}$$
$$V_T = \{ verb, noun, determiner, adjective, preposition\}$$
$$B = S$$

*This work was supported by Office of Naval Research grant no. N00014-84-K-0655

$P = \{S \rightarrow NP\,VP$ $S \rightarrow VP$ $VP \rightarrow verb$

$VP \rightarrow verb\,NP$ $VP \rightarrow VP\,PP$ $PP \rightarrow preposition\,NP$

$NP \rightarrow determiner\,NP2$ $NP \rightarrow NP2$ $NP \rightarrow NP\,PP$

$NP2 \rightarrow noun$ $NP2 \rightarrow adjective\,NP2\}$

   The derivation of a string is often written as a tree where the root is the start symbol, the internal nodes are nonterminals, the children of a node correspond to one of its productions and the yield of the tree is the string of terminals whose derivation the tree represents. The derivation of the string "determiner adjective noun verb determiner noun" is depicted in figure 1.

   Parsing is derivation in reverse. Given a string of terminals, a parser must determine if the string is derivable by the grammar and, if so, find the corresponding parse tree.

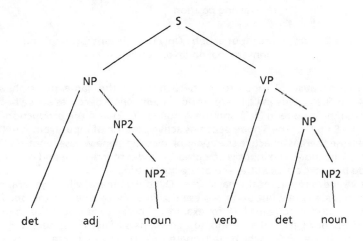

Figure 1. Sample parse tree.

THE PARSING NETWORK

   The strategy used is that of the CYK parser.[7] The network contains nodes representing the terminals and nonterminals of the grammar (several for each, in fact) as well as match nodes which will be explained below. The nodes are best thought of as organized into a table, with columns representing starting positions in the input string, and rows representing lengths. For example, the VP nonterminal in figure 1 has starting position four and length three. Organizing the nonterminals in this way allows all possible parses to proceed in parallel without confusion. There is a node for each terminal symbol in each position of row one. There is a node for each nonterminal at every position in the table (some may be left out because they are impossible). Terminal nodes are activated by some outside source and represent the input to the parser. A nonterminal node will become active if other nodes representing the right-hand side of one of its productions, and having appropriate starting positions and lengths, are active. The parse proceeds in a bottom-up fashion. Figure 2 illustrates the parsing of the string aabbb for the grammar shown (the "|" separates multiple rewrites for the same nonterminal). The terminal symbols are activated as input. The A node in the first row

142

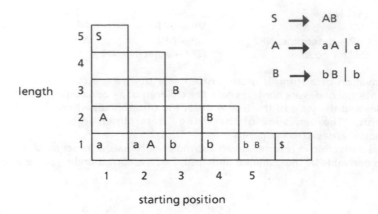

length

S → AB

A → aA | a

B → bB | b

starting position

Figure 2. Parsing the input aabbb. Only the relevant terminal and nonterminal nodes are shown.

becomes active because of input from the a node in the same position and the production A→a. The A node in (2,1) – row two, column one – becomes active because of input from both the a node in (1,1) and the A node in (1,2) and the production A→a A. Similarly for the B nodes. The S node becomes active because of input from A in (2,1) and B in (3,3). Because there is an active start symbol in column one whose length is the same as the length of the input, the input is accepted. In order to mark the end of the input, a special $ node becomes active at the end of the input string.

The active nodes represent the parse tree. Of course, there will in general be many nodes which become active but don't represent a final parse tree. In the above parse, there will be an active A node in (1,1), for example. This will not affect the ability of the network to accept or reject, but a second, top-down, pass of activity is necessary in order to pick out only those nodes which participate in a complete parse; if the string is ambiguous, more than one parse will stay active. The end marker begins the top-down pass by activating the root node of the parse tree. This activity is passed down only to those symbols which satisfied a production on the bottom-up pass. This continues until activity reaches the terminal symbols. Only nodes which receive this top-down input represent the parse tree.

A more detailed account of the network follows. In order to distinguish the two passes, each node mentioned above consists of two units (see figure 3). The left-most unit represents bottom-up activity; the right-most unit represents top-down activity. Each nonterminal node must distinguish activity due to separate productions because in general there will be many ways to realize each production of a nonterminal. If the total bottom-up activity were merely summed, partial realization of several productions would exceed the threshold of any one complete realization and cause the bottom-up unit to fire falsely. In order to achieve this separation, every nonterminal node must be accompanied by a match node for every realization of every production, i.e., every possible combination of component lengths for every production. Figure 3 illustrates one of the nonterminal nodes for the grammar in figure 2. S.1.4 is the S nonterminal in column 1, row 4. It has only one production which can be realized by A.1.1, B.2.3 or A.1.2, B.3.2 or A.1.3, B.4.1.

Match nodes also consist of a bottom-up and top-down unit. The bottom-up match unit receives input from all bottom-up units which represent some realization of a

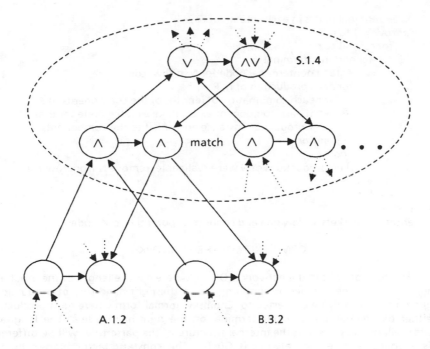

Figure 3. The node for S starting in position 1 and having length 4. One of its match nodes is shown in detail.

production. All the inputs must be on before the match unit comes on, so it computes a logical AND of its inputs. The nonterminal bottom-up unit receives input from all its bottom-up match units. Since only one of a nonterminal's productions need be realized, this unit computes a logical OR of its inputs. Given these rules and an input of length n, bottom-up activation will reach the S symbol in column one, row n if and only if the input is in the language defined by the grammar.

The top-down nonterminal units compute a slightly more complicated function of their inputs. To turn on, they must be receiving input from their bottom-up counterpart and top-down input from at least one external source. For the root symbol the source is the end marker. When a top-down nonterminal unit turns on, it provides top-down activation to all its match top-down units. These units also require their bottom-up counterpart to be on, so only those representing some valid production realization will turn on. They pass top-down activity on to the nodes which activated them bottom-up. The parse tree is represented by activity on the top-down nonterminal nodes.

In the case of ambiguous input, i.e., input with more than one correct parse tree, all parse trees will be on simultaneously. The network can be forced to isolate a single parse by putting the match nodes of a nonterminal node in a winner-take-all[1] network so that at most one can be on. The preference order can be predetermined by the network links or could also depend on which match unit activates first or outside input. The network has been simulated using the Rochester Connectionist Simulator.[8] A program reads a grammar and constructs the network using the following algorithm:

```
enter terminal nodes in each column of row 1
for row = 1 to n
    for col = 1 to n
        for each nonterminal, N
            enter a nonterminal node for N in row, col
            for each production of N, P
                for each combination of lengths of the components of P
                    if each component exits, make a match node for that
              ·         production instance and link it to the components
                endfor
            endfor
            if no productions of N were realizable, remove N from row,col
        endfor
    endfor
endfor
enter endmarkers in row one and connect to potential root nodes
```

COMPLEXITY OF THE NETWORK

The number of units in the network is $O(n^m)$, where n is the length of the input and m equals one plus the number of nonterminals on the right-hand side of productions. Any grammar can easily be converted to Chomsky normal form where each production has either two nonterminals or one terminal on the right-hand side.[7] The language accepted will be the same, but the internal structure of the parse trees will be different. If this is done, the size the network is $O(n^3)$. The constant factor hidden here is approximately the number of productions in the grammar. In any case, the parse completes in $O(n)$ steps, or, more precisely, 4·depth-of-parse-tree steps. The sequential algorithm upon which the network is based will take $O(n^3)$ steps to complete if the grammar is in Chomsky normal form.

The simplicity of the units in the networks would be a real advantage in a VLSI implementation. However the connections would be a real problem. This is another way in which Chomsky normal form helps. If the grammar is in Chomsky normal form, all connections are along the columns or diagonals of the table depicted in figure 2. If wires corresponding to these paths are multiplexed so that the row from which the signal is originating varies with the time step, the total number of wires will be n·c vertically and diagonally, where c is the number of symbols in the grammar and n is the length of the largest input. This could be reduced to n if the lines were multiplexed between the symbols, but this would slow down the parse time.

EXTENSIONS

Various extensions to the basic parsing network have been made using a more complicated model of network nodes,[9] including the parsing of near-miss inputs. These are inputs which are almost but not quite correct. A parse tree of the portion of the input which can be parsed is constructed. The anomalous input is marked by its lack of top-down input. The activation level of the nodes represents the extent to which a production was realized.

The capability to learn new productions has been built on top of this, so that a near-miss parse which would have completed if there were just one more rewrite rule will learn that rule. This does not constitute an adequate learning algorithm for context-free languages, but it does solve the technical problem of distributing a production learned locally to all nodes for the same nonterminal.

DISCUSSION

Any given parsing network is, strictly speaking, a finite-state machine because of the fixed-length restriction. The power of a network can be increased only with a specifically wired extension. Still, when the input is under the limit, the behavior of a network is best characterized as that of a context-free parser. There are applications for which a strict length restriction is acceptable. Another disadvantage of the network, compared to other "neural" networks, is its brittleness. Remove any one node and there will be input for which the network fails. The extensions discussed in the previous section would make the network more robust. The learning mechanism can even be used to re-learn locally-lost rules during periods of network inactivity. Multiple copies of each unit would also increase robustness.

It may be of interest to investigate parallel, connectionist implementations of other dynamic programming algorithms.

REFERENCES

1. J. A. Feldman and D. H. Ballard, Connectionist Models and their Properties, Cognitive Science 6, p. 205-254 (1982).
2. G. W. Cottrell, A Connectionist Approach to Word Sense Disambiguation, Doctoral dissertation, Computer Science Department, University of Rochester (1985).
3. B. Selman and G. Hirst, A Rule-Based Connectionist Parsing System, Proceedings of the seventh Annual Conference of the Cognitive Science Society, Irvine, Cal., p. 212-221 (1985).
4. G. Gazdar, E. Klein, G. Pullum, and I. Sag, Generalized Phrase Structure Grammar (Basil Blackwell Publisher Ltd, Oxford, 1985).
5. A. V. Aho, J. E. Hopcroft and J. D. Ullman, The Design and Analysis of Computer Algorithms (Addison-Wesley Publishing Company, Reading, Mass, 1974).
6. K. S. Fu, Syntactic Methods in Pattern Recognition (Academic Press, N.Y., 1974).
7. J.E. Hopcroft and J.D. Ullman, Introduction to Automata Theory, Languages, and Computation (Addison-Wesley, Reading, Mass. 1979).
8. M. A. Fanty and N. Goddard, The Rochester Connectionist Simulator, forthcoming, Computer Science Department, University of Rochester.
9. M. A. Fanty, Context-Free Parsing in Connectionist Networks, TR 174, Computer Science Department, University of Rochester (1985).

OPTICAL ANALOG OF TWO-DIMENSIONAL NEURAL
NETWORKS AND THEIR APPLICATION IN RECOGNITION OF RADAR TARGETS

N.H. Farhat, S. Miyahara and K.S. Lee
University of Pennsylvania
The Electro-Optics and Microwave-Optics Laboratory
200 S. 33rd Street
Philadelphia, PA 19104-6390

ABSTRACT

Optical analogs of 2-D distribution of idealized neurons (2-D neural net) based on partitioning of the resulting 4-D connectivity matrix are discussed. These are desirable because of compatibility with 2-D feature spaces and ability to realize denser networks. An example of their use with sinogram classifiers derived from realistic radar data of scale models of three aerospace objects as learning set is given. Super-resolved recognition from partial information that can be as low as 20% of the sinogram data is demonstrated together with a capacity for error correction and generalization.

INTRODUCTION

Neural net models and their analogs furnish a new approach to signal processing that is collective, robust, and fault tolerant. Optical implementations of neural nets[1,2] are attractive because of the inherent parallelism and massive interconnection capabilities provided by optics and because of emergent optical technologies that promise high resolution and high speed programmable spatial light modulators (SLMs) and arrays of optical bistability devices (optical decision making elements) that can facilitate the implementation and study of large networks. Optical implementation of a one-dimensional network of 32 neurons exhibiting robust content-addressability and associative recall has already been demonstrated to illustrate the above advantages.[3] Extension to two-dimensional arrangements are of interest because these are suitable for processing of 2-D image data or image classifiers directly and offer a way for optical implementation of large networks.[4]

In this paper we will discuss content addressable memory (CAM) architectures based on partitioning of the four dimensional $T_{ijk\ell}$ memory or interconnection matrix encountered in the storage of 2-D entities. A specific architecture and implementation based on the use of partitioned unipolar binary (u.b.) memory matrix and the use of adaptive thresholding in the feedback loop are described. The use of u.b. memory masks greatly simplifies optical implementations and facilitates the realization of larger networks $\sim (10^3\text{-}10^4$ neurons). Numerical simulations showing the use of such 2-D networks in the recognition of dilute point-like objects that arise in radar and other similar remote sensing imaging applications are described. Dilute objects pose a problem for CAM storage because of the small

Hamming distance between them. Here we show that coding in the form
of a *sinogram classifier* of the dilute object can remove this limitation permitting recognition from partial versions of the stored
entities. The advantage of this capability in super-resolved recognition of radar targets is discussed in the context of a new type of
radar diversity imaging, studied extensively in our laboratory, that
is capable of providing sinogram information compatible with 2-D CAM
storage and interrogation. Super-resolved automated recognition of
scale models of three aero-space objects from partial information as
low as 20% of a learned entity is shown employing hetero-associative
storage where the outcome is a word label describing the recognized object. Capacity for error correction and generalization were
also observed.

TWO-DIMENSIONAL NEURAL NETS

Storage and readout of 2-D entities in a content addressable or
associative memory is described next. Given a set of M 2-D bipolar
binary patterns or entities $v_{ij}^{(m)}$ m=1,2...M each of NxN elements represented by a matrix of rank N, these can be stored in a manner that
is a direct extension of the 1-D case as follows: For each element of
a matrix a new NxN matrix is formed by multiplying the value of the
element by all elements of the matrix including itself taking the
self product as zero. The outcome is a new set of N^2 binary bipolar
matrices each of rank N. A formal description of this operation is,

$$T_{ijk\ell}^{(m)} = \begin{cases} v_{ij}^{(m)} \, v_{k\ell}^{(m)} \\ 0 \qquad\quad i=k, \; j=\ell \end{cases} \qquad (1)$$

which is a four dimensional matrix. An overall or composite synaptic
or connectivity memory matrix is formed then by adding all 4-D
matrices $T_{ijk\ell}^{(m)}$ i.e.,

$$T_{ijk\ell} = \sum_m T_{ijk\ell}^{(m)} \qquad (2)$$

This symmetric 4-D matrix has elements that vary in value between -M
to M also in steps of two as for the 1-D neural net case and which
assume values of +1 and -1 (and zeros for the self product elements)
when the matrix is clipped or binarized as is usually preferable for
optical implementations. Two dimensional unipolar binary entities
$b_{ij}^{(m)}$ are frequently of practical importance. These can be transformed
in the usual way into bipolar binary matrices through $v_{ij}^{(m)} = (2b_{ij}^{(m)} - 1)$
which are then used to form the 4-D connectivity matrix or memory as
described. Also, as in the 1-D neural net case, the prompting entity
can be unipolar binary $b_{ij}^{(m)}$, which would simplify further optical

implementations in incoherent light.

Architectures for optical implementation of 2-D neural nets must contend with the task of realizing a 4-D memory matrix. Here a scheme is presented that is based on the partitioning of the 4-D memory matrix into an array of 2-D matrices of rank N.

Nearest neighbor search of the memory matrix for a given entity $b_{ij}^{(mo)}$ is done by forming the estimate,

$$\hat{b}_{ij}^{(mo)} = \sum_{k,\ell}^{N} T_{ijk\ell} \, b_{k\ell}^{(mo)} \quad \dots \quad i,j,k,\ell = 1,2,\dots N \tag{3}$$

followed by thresholding to obtain a new u.b. matrix which is used to replace $b_{k\ell}^{(mo)}$ in eq. (3) and the procedure is repeated until the resulting matrix converges to the stored entity closest to the initiating matrix $b_{ij}^{(mo)}$. The operation in eq. (3) can be interpreted as first partitioning of the 4-D $T_{ijk\ell}$ matrix into an array of 2-D submatrices of rank N: $T_{11k\ell}$, $T_{12k\ell}$,...,$T_{1Nk\ell}$; $T_{21k\ell}$, $T_{22k\ell}$, ..., $T_{2Nk\ell}$; ..., $T_{N1k\ell}$, $T_{N2k\ell}$,..., $T_{NNk\ell}$ as depicted schematically in Fig. 1(a) where the partition submatrices are arranged in a 2-D array. This first step is followed by multiplication of $b_{k\ell}^{(mo)}$ by each of the partition submatrices, on an element by element basis, and summing the products for each submatrix to obtain the first estimate $\hat{b}_{ij}^{(mo)}$. The tensor multiplications and summation operations called for in eq. (3) are carried out in Fig. 1(a) by placing a spatially integrating photodetector (PD) behind each submatrix of the partitioned

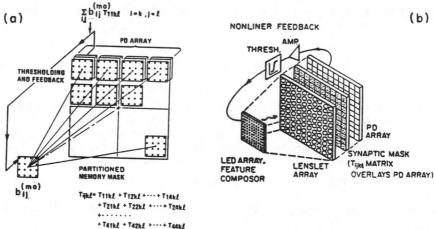

Fig. 1. Optical analog of 2-D neural net. (a) Architecture based on partitioning of connectivity matrix, (b) Opto-electronic embodiment.

memory mask which is assumed for the time being to be realized by
pixel transmittance modulation in an ideal transparency capable of
assuming negative transmittance values. The input entity $b_{ij}^{(mo)}$ is
assumed to be displayed on a suitable LED array. The LED display of
$b_{ij}^{(mo)}$ is multiplied by the ideal transmittance of each of the partition submatrices by imaging the display on each of these with exact
registration of pixels by means of a lenslet array as depicted in
Fig. 1(b). The output of each PD, proportional to one of the components of eq. 3, is thresholded, amplified, and fed back to drive
an associated LED. The (i,j)-th LED is paired with the (i,j)-th PD.
This completes the interconnection of the 2-D array of NxN neurons
in the above architecture where each neuron communicates its state
to all other neurons through a prescribed four dimensional synaptic
or memory matrix in which information about M 2-D binary matrices of
rank N (entities) have been stored distributively. The number of
2-D entities that can be stored in this fashion is $M \simeq N^2/8\ell nN$, which
follows directly from the storage capacity formula for the 1-D neural
net case by replacing N by N^2.

The added complexity associated with having to realize a bipolar transmittance in the partitioned $T_{ijk\ell}$ memory mask of Fig. 1
can be avoided by using unipolar transmittance. This can lead however to some degradation in performance. A systematic numerical
simulation study[5] of a neural net CAM in which statistical evaluation of the performance of the CAM for various types of memory masks
(multivalued, clipped ternary, clipped u.b.) and thresholding schemes
(zero threshold, adaptive threshold where energy of input vector is
used as threshold, adaptive thresholding and relaxation) was carried
out. The results indicate that a u.b. memory mask can be used with
virtually no sacrifice in CAM performance when the adaptive thresholding and relaxation scheme is applied. The scheme assumes an
adaptive threshold is used that is proportional to the energy (total
light intensity) of the input entity displayed by the LED array at
any time. In the scheme of Fig. 1(b) this can be realized by pro-
jecting an image of the input pattern directly onto an additional PD
element. The PD output being proportional to the total intensity of
the input display is used as a variable or adaptive threshold in a
comparator against which the outputs of the PD elements positioned
behind the partitioned components of the $T_{ijk\ell}$ memory mask are compared. The outcomes, now bipolar, are attenuated and each is fed
into a limiting amplifier with delayed feedback (relaxation). Each
limiter/amplifier output is used to drive the LED that each photodetector is paired with. It was found[5] that this scheme yields performance equivalent to that of an ideal CAM with multivalued connectivity matrix and zero thresholding. Note that although the ini-
tializaing 2-D entity $b_{ij}^{(mo)}$ is unipolar binary, the entities fed
back after adaptive thresholding and limited amplification to drive
the LED array would initially be analog resulting in multivalued
iterates and intensity displays. However, after few iterations the

outputs become binary assuming the extreme values of the limiter. The ability to use u.b. memory matrices in the fashion described means that simple black and white photographic transparencies or binary SLMs can be used respectively as stationary or programmable synaptic or programmable synaptic connectivity masks as suggested by Fig. 1.

SINOGRAM CLASSIFIERS AND HETEROASSOCIATIVE STORAGE

Sinograms are object representations encountered in tomography[6,7]. They are also useful as object classifiers specially when the objects are point-like and dilute[8]. Given a set of 2-D dilute objects the Hamming distances between their sinogram classifiers will be greater than the Hamming distances between the objects themselves, with both sets digitized to the same number of pixels, making it easier for an associative memory to distinguish between the sinograms[8]. Sinogram classifiers have additional advantages that enable scale, rotation, and shift invariant recognition of radar target which can not be detailed here because of limited space. A sinogram is a cartesian plot of the polar projections of object detail. For example referring to Fig. 2(a) which represents a dilute object consisting of 16 points on a 32x32 pixel grid, the distance that the projection of each point makes on the y axis as measured from the origin when the object is rotated about the origin traces a sinusoidal pattern when plotted against rotation angle as shown in Fig. 2(b). Figure 2(c) is a

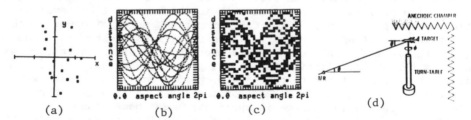

(a)   (b)   (c)   (d)

Fig. 2. Sinogram generation. (a) Sparse object, (b) Sinogram, (c) Digitized sinogram, (d) Experimental sinogram generation in radar by range-profile measurement.

digitized version of the sinogram of Fig. 2(b) plotted on a 32x32 pixel grid. The sinogram of a radar target is produced by measuring the differential range or range-profile of the target employing the arrangement of Fig. 2(d). The system basically measures, with high resolution, the differential distance (differential range or range-profile) from the rotation center of the projections of the scattering centers of the object (here scale models of aerospace targets) on the line-of-sight of the radar system. Cartesian plots of the differential distance or range-profile versus azimuthal angle of rotation ϕ results in a sinogram classifier or feature space of the target which characterizes it at any fixed elevation angle θ. The top row of Fig. 3 shows three digitized sinogram classifiers of scale models of three aerospace targets plotted on a 32x32 pixel

grid. These are treated as a learning set and stored hetero-associa-
tively rather than autoassociatively be replacing $v_{k\ell}^{(m)}$ in eq. (1) by
$r_{k\ell}^{(m)}$ $k,\ell=1,2...32$; $m=1,2,3$ where $r_{k\ell}^{(m)}$ represents abbreviated word
labels shown in the bottom row of Fig. 3 with which the three test
objects are to be associated.

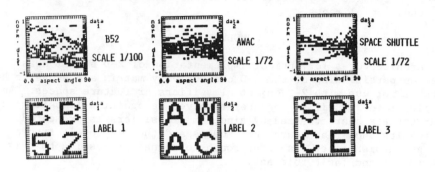

Fig. 3.   Hetero-associative storage. Digitized sinograms (top) and
          associated word labels (bottom).

RESULTS

Representative results of numerical simulation of exercising the
heteroassociative memory matrix with complete and partial versions of
one of the stored entities in which the fraction η of correct bits or
pixels in the partial versions ranged between 1 and .1 are presented
in Fig. 4. Reliable recognition was found to occur for all entities
stored down to η = .2. For η = .1 or less successful recall of
correct labels was found to depend on the angular location of the
partial data the memory is presented with as illustrated in the two

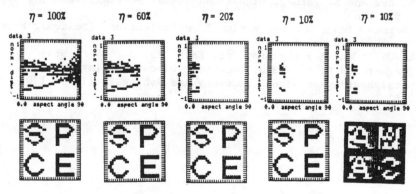

Fig. 4.   Example of recognition from partial information. Complete
          and partial sinograms of data set 3 used as input (top), and
          final memory state-recognized label (bottom).

right-most examples in Fig. 4. Here the memory could not label the partial input correctly but converged instead onto a label that it did not learn before. This appears to be a generalization (mixture) of the three entities stored earlier. This is quite analogous to the generalization capability of the brain. Note the generalization is contrast reversed as we know that stable states of a memory with symmetric connectivity matrix are not only the entities stored but also their compliments.

CONCLUSIONS

Architectures for optical implementation of 2-D neural nets based on partitioning of the 4-D connectivity matrix are shown to be suitable for use with 2-D object classifiers or feature spaces. An example of their utility in super-resolved recognition (labeling) of radar targets characterized by sinogram classifiers is presented. The results show that neural net models and their opto-electronic analogs furnish a new viable approach to signal processing that is both robust and fault tolerant.

ACKNOWLEDGEMENT

The work described was carried out under a grant from DARPA/NRL, and partial support from AFOSR, ARO, and RCA (GE) Corporation.

REFERENCES

1. D. Psaltis and N. Farhat, Opt. Lett., $\underline{10}$, 98, (1985).
2. A.D. Fisher, et. al., SPIE, $\underline{625}$, 28 (1986).
3. N. Farhat, D. Psaltis, A. Prata and E. Paek, App. Optics, $\underline{24}$, 1469, (1985).
4. N. Farhat and D. Psaltis, Digest OSA Annual Meeting, Wash., D.C., p. 58, (1985).
5. K.S. Lee and N. Farhat, Digest OSA Annual Meeting, Wash., D.C., p. 48, (1985).
6. G. Herman, Image Reconstruction From Projections (Academic Press, N.Y., 1980), p. 11.
7. G.R. Gindi and A.F. Gmitro, Opt. Eng., $\underline{23}$, p. 499, (1984).
8. N. Farhat and S. Miyahara, Technical Digest, Spring 86 OSA Topical Meeting on Signal Recovery and Synthesis II, p. 120, (1986).

Application of Neural Network Algorithms and Architectures to Correlation/Tracking and Identification

Sheldon B. Gardner

Naval Research Laboratory, Washington, D.C. 20375

Abstract

Neural network architectures and algorithms provide an anthropomorphic framework for analysis and synthesis of learning networks for correlation, tracking and identification applications. Many researchers in neuroscience believe that through evolution nature has developed efficient structures for multi-sensor integration and data fusion. Consequently, innovations in electronic surveillance and advanced computing may result from current interdisiplinary research in neural networks and natural intelligence (NI). In this paper we propose a network learning paradigm, called entropy learning, based upon the Principle of Maximum Entropy (PME). The close relationship between entropy learning and simulated annealing in network solutions to combinatorial optimization problems is discussed.

I. Introduction

The field of neural networks for computing has recently emerged as an interdisiplinary science [1-3]. Neural networks fall within a broader class of networks which we term learning networks. In general, learning networks can be modeled as dynamic systems using the mathematical tools of non-linear systems theory [4-7]. For the most part, deterministic methods have been used. Learning networks can also be represented by probabilistic models based upon information theoretic concepts. In keeping with the latter approach, we characterize a learning network functionally as a *conditional probability computer* and consider network state variables as multi-dimensional random processes [8-10].

II. Entropy, Uncertainty and Learning

A close relationship exists between the Shannon concept of information [11,12] and the *Principal of Maximum Entropy* (PME) as developed by E. Jaynes [13,14]. PME is a method of statistical inference which sets up a least biased estimate based upon given information. In developing PME, Jaynes has combined established methods of statistical physics (Maxwell, Boltzmann, Gibbs) with methods of statistical inference (Bernoulli, Laplace, Jeffreys, Cox) and information theory (Shannon, Von Neumann). PME has been applied by Jaynes and others to classical statistical mechanics, quantum mechanics and other fields [15-17].

Jaynes was responsible for a resurrection of the important realization that probabilities are best identified with 'states of knowledge' rather than frequencies in a random experiment. Since entropy, uncertainty and learning are closely related, we propose a generalized information theoretic measure for learning *under equilibrium conditions* based upon a definition of information entropy S_1 as a multidimensional integral [18] :

$$S_I = - \int p(x) \log [p(x) / m(x)] \, dx \qquad (1)$$

where $m(x)$ is a measure function associated with prior knowledge, x is assumed to be a vector random variable and we assume that $p(x)$ is a probability density function. Eq.(1) describes the information entropy for equilibrium conditions (i.e. the learning network has reached a steady state where $p(x)$, $m(x)$ are both constant).

S_I can be regarded as a measure of the degree of uncertainty associated with $p(x)$. In this context, maximum entropy learning can be regarded as a minimization of uncertainty subject to constraints. We illustrate maximum entropy learning in the scalar case. Assume constraints fixing the mean values F_k of m different functions $f_k(x)$:

$$F_k = \int p_t(x) f_k(x) \, dx, \qquad k=1, 2, \ldots, m \qquad (2)$$

A solution for the maximization of S_I , Eq.(1), subject to the constraints (2) is given by:

$$p(x) = Z^{-1} m(x) \exp[\lambda_1 f_1(x) + \cdots + \lambda_m f_m(x)] \qquad (3)$$

with the partition function

$$Z = Z(\lambda_1, \cdots, \lambda_m) = \int m(x) \exp[\lambda_1 f_1(x) + \cdots + \lambda_m f_m(x)] \, dx \qquad (4)$$

The Lagrangian multipliers λ_k are determined by

$$F_k = \frac{\partial}{\partial \lambda_k} \log Z(\lambda_1, \cdots, \lambda_m) \qquad (5)$$

Eqs. (3) and (4) have the form of the Gibbs distribution of statistical mechanics[19]. Consequently, maximum entropy inference based upon Eqs. (3) and (4) is identical in mathematical form with the laws of equilibrium statistical mechanics.

III. Entropy Learning using a Gibbs/Markov Model

Consider a network consisting of N interconnected units. Each unit of the network consists of a generalized processing element. The network topology can be regarded as as a finite graph G whose vertices Z represent sites, and whose edges represent mutual interactions. Each site can assume a range of possible (elementary) states. An assignment of values to all of the sites in $z_n \in Z$, $n=1,2,\cdots N$, represents a state of the overall system. We assume that if a site z_1 interacts with a site z_2, and site z_2 interacts with site z_3, it is *not* true that z_1 influences z_3, unless these sites are linked directly[20]. This assumption generalizes the definition of a Markov chain to that of a Markov random field (MRF)[21]. The MRF assumption and the equivalance between a MRF and a Gibbs random field (GRF), as defined by Spitzer[22], determines the form of the Gibbs energy, potential and partition functions (e.g. see Geman and Geman[23] for a comprehensive mathematical treatment on a 2-D lattice).

S_I can be interpreted as an information-theoretic measure of the discrepancy between a learning network's internal model and the environment. Entropy learning is then equivalent to building an internal model of the network weights most likely to have been generated by a set of environmental vectors. Ackley, Hinton and Sejnowski [24,25] (AHS) have developed a parallel constraint satisfaction network, called the Boltzmann Machine. We discuss the Boltzmann Machine as an example of entropy learning. For this case, Eq.(1) can be restated as an entropy measure which AHS call a G metric:

$$G = \sum_a P(V_a) \ln \frac{P(V_a)}{P'(V_a)} \tag{6}$$

where $P(V_a)$ is the probability of the ath state of the visible units when their states are determined by the environment, and $P'(V_a)$ is the corresponding probability when the network is running freely with no environmental input. AHS then proceed to develop an algorithm for weight updating based upon gradient descent in G :

$$\frac{\partial G}{\partial w_{ij}} = - \frac{1}{T} (p_{ij} - p'_{ij}) \tag{7}$$

where T is a parameter that acts like temperature, p_{ij} is the average probability of units i, j both being in the *on* state when the environment is clamping the states of the visible units, and p'_{ij} is the corresponding probability when the environment is not present and the system is at equilibrium. To minimize G, AHS show that it is sufficient to observe p_{ij} and p'_{ij} when the network is at *thermal equilibrium* and to change each weight by an amount proportional to the difference between these two probabilities:

$$\Delta w_{ij} = \epsilon \, (p_{ij} - p'_{ij}) \tag{8}$$

Because of the assumptions inherent in the MRF/GRF formulation, the weight update rule, Eq. (8), uses only locally available information although the change optimizes a global measure.

IV. Entropy Learning and Simulated Annealing

Entropy learning using the Boltzmann distribution (e.g. the Boltzmann Machine) is closely related to a numerical optimization method, called simulated annealing (SA), introduced by Kirkpatrick [26]. SA is based upon the Metropolis Algorithm [27], and has been used to apply numerical simulation tools used in statistical mechanics to combinatorial optimization problems [28,29]. In the field of 2-D processing SA has been used, by Geman and Geman [23] and others [30], to perform Bayesian image reconstruction.

The SA process consists of first 'melting' the system being optimized at a high effective temperature, then lowering the temperature by stages until the system 'freezes' and no further changes occur. At each temperature, the simulation proceeds long enough for the system to reach a steady state. In fact, SA using a Boltzmann machine requires a cooling schedule inversely proportional to log time and is computationally slow.

Another approach to simulated annealing called the Cauchy Machine has been recently developed by Szu [31]. The Cauchy Machine uses the properties of the Cauchy probability distribution to perform fast simulated annealing (FSA) with a cooling schedule inversely proportional to time. Entropy learning as applied to fast simulated annealing (e.g. the Cauchy Machine) requires nonergodic theory. Consequently, the definition of entropy learning based upon equilibrium assumptions (e.g. Eq.(1)) is no longer valid. The extension of entropy learning concepts to fast simulated annealing is a subject of current research.

References

1. J. J. Hopfield, Proc. Nat. Acad. Sci. **79**, 2554-2558,(82).
2. S. Grossberg, Studies of Mind and Brain (D. Reidel Publishing Co.,1982)
3. T. Kohonen,Self-Organization and Associative Memory (Springer-Verlag, 1984)
4. S. I. Amari, IEEE Trans. SMC **13**, 741-748 (1983).
5. M. Cohen, S. Grossberg, IEEE Trans. SMC **13**, 815-826 (1983).
6. E. Goles, SIAM J. Alg. Disc. Meth. **6**, 749-754 (1985).
7. P. Peretto, J. Niez, IEEE Trans. SMC **16**, 73-83 (1986).
8. J. L. Doob, Classical Potential Theory and Its Probabilistic Counterpart (Springer-Verlag, 1984)
9. G. Kallianpur, Stochastic Filtering Theory (Springer-Verlag, 1980).
10. V. Krishnan, Nonlinear Filtering and Smoothing (J. Wiley & Sons, 1984).
11. C. E. Shannon, Bell System Tech. J. **27**, 379, 623 (1948)

12. C. E. Shannon, W. Weaver, The Mathematical Theory of Communication (University of Illinois Press, Urbana, 1949).
13. E. Jaynes, Physical Rev. **106**, 620-630 (1957).
14. E. Jaynes, Physical Rev. **108**, 171-190 (1957).
15. R. D. Rosenkrantz (Ed.), E. T. Jaynes: Papers on Probability, Statistics and Statistical Physics (D. Reidel, 1983).
16. W. T. Grandy, Jr., Physics Reports **62**, 175-266 (1980).
17. J. Shore, R. Johnson, IEEE Trans. IT **26**, 26-37 (1980).
18. E. Jaynes, IEEE Trans. SSC **4**, 227-241 (1968).
19. J. W. Gibbs, Elementary Principles in Statistical Mechanics, Vol. II of collected works (Longmans Green and Company, New York, 1928).
20. J. Moussouris, J. Stat. Phys. **10**, p.11-33 (1974).
21. R. L. Dobrushin, Th. Prob. & Appl. **13**, p. 197-224 (1968).
22. F. Spitzer, Amer. Math. Mon. **78**, 142-154 (1971).
23. S. Geman, D. Geman, IEEE PAMI 6, 721-741 (1984).
24. D. Ackley, G. Hinton, T. Sejnowski, Cog. Sci. 9, 147-169 (1985).
25. G. Hinton, T. Sejnowski, Proc. of the IEEE Computer Society Conf. on Comp.Vision and Pattern Recog.Wash. D. C., 448-453 (1983).
26. S. Kirkpatrick, C. Gellatt, Jr., M. P. Vecchi, Science **220**, 671-680 (1983).
27. N. Metropolis, A. Rosenbluth, M. Rosenbluth, Λ. Teller, E. Teller, J. Chem. Phys. **21**, 1087-1091 (1953).
28. M. R. Garey, D.S. Johnson, Computers and Intractability (Freeman, 1979).
29. J. J. Hopfield, D. W. Tank, Biol. Cybern. **52**, 141-152 (1985).
30. P.Carnevali, L.Coletti, S.Patarnello, IBM J. Res Develop. **29** 569-579(1985).
31. H. Szu, R. Hartley, Cauchy Machines for Fast Simulated Annealing, Neural Networks for Computing, Snowbird, Utah, April 13-16, 1986.

HOPFIELD MODEL APPLIED TO VOWEL AND CONSONANT DISCRIMINATION

Bernard Gold

Lincoln Laboratory, Massachusetts Institute of Technology, Lexington, MA 02173-0073

ABSTRACT

A Hopfield model of 120 "neurons" has been simulated on an LDSP (Lincoln Digital Signal Processor). The model has been applied to the study of problems in automatic discrimination of vowels and consonants. In the first problem, a spectral cross section was extracted by performing a fast Fourier transform on a 20 ms segment from the steady state portion of the vowel in a single syllable word. The spectrum was then smoothed and a one-bit gradient measure applied at 120 frequency values, thus obtaining an assigned state for that vowel. This procedure was repeated until eight such assigned states were obtained. From this data, the connection matrix T_{ij} was obtained using the associative equation,

$$T_{ij} = \sum_{s=0}^{7} \hat{x}_i^s \, \hat{x}_j^s$$

Each $\hat{x}_i^s$ was the component of one of the assigned states.

A discrimination experiment was performed by specifying, as input, half of the 120 bits for each state. This specification was performed randomly, 8 times for each of the 8 states, resulting in a set of 64 runs. For 51 of these runs, the convergence to the correct state was perfect. Of the 13 spurious states generated, most were closest to the correct assigned states.

Using the same T_{ij} matrix, a second experiment was performed. Each of the words containing a given vowel was scanned and a new input pattern was obtained every 10 ms. Each of these patterns was applied as an initial state to the neural network, which was allowed to iterate until convergence. Many (but not all) of the vowel cross sections were most closely identified by the network as the assigned state associated with that word. Outside of the vowel, results were random.

The final experiment was an initial attempt to automate the Diagnostic Rhyme Test that is widely used to rate speech processing devices such as vocoders. In this test, the listener must choose between two initial consonants (e.g., pool vs. tool) that generally differ in only a single distinctive feature. Several algorithmic additions to the Hopfield model have been tried; to date, the most successful one is a two-state model that has some foreknowledge of the critical intervals of the rhyme pair. Work is continuing on this problem.

1. INTRODUCTION

Although the study of both real and artificial neural networks has been a subject of continuing interest for many years, recent work by Hopfield and associates has catalyzed a new spurt of activity [1]. In particular, Hopfield's proposed model of associative memory has raised questions concerning the applicability of the model to traditional problems such as speech and image recognition. This paper describes some computer simulation results on vowel and consonant discrimination based on this associative memory model.

Section 2 of this report reviews several properties of the associative memory model. Section 3 presents the results of a Hopfield network simulation that performs vowel discrimination among eight vowels when random errors perturb the representation. Section 4 describes further work where the network scans an entire word, searching for vowels that are similar to one of the stored vowels. Section 5 describes an approach to the consonant discrimination problem and some preliminary results.

The algorithm that describes the evolution of the network is given by the equations

$$q_i(n) = \sum_{j=0}^{N-1} T_{ij}\, \hat{x}_j(n-1)$$

$$x_i(n) = \alpha\, x_i(n-1) + q_i(n) \tag{1}$$

$$\hat{x}_i(n) = n\ell e\left[x_i(n)\right]$$

In these equations n represents discrete time samples, α is the time constant of the unity gain (at zero frequency) first-order digital filter. (In all our simulations α has been set to .75). $n\ell e$ represents the function of the non-linear element which, in this paper, is taken to be the sign of the argument.

In general, the network can drift to <u>spurious</u> states; that is, stable states that are <u>not</u> one of the assigned states. The precise relationship between a network and its spurious states is not well understood. A useful rule of thumb is this: for m < .15N, there is a high probability that the network will converge to an assigned state. For m > .15N, there is a high probability that the network will converge to a spurious state.

If the initial state of the network is set by some external input that is, in some sense, reasonably close to one of the assigned states, the network has a good chance of converging to that assigned state. Another way to say this is that the network can recognize memorized states that are only partially described. This basic property makes it possible for the network to correctly identify a noisy input and also to continue to operate successfully when individual elements are inoperative.

2. APPLICATION TO SIMPLE VOWEL DISCRIMINATION

In order to apply a Hopfield net to the processing of real data, a suitable representation of that data must first be found. In the case of vowel discrimination, such a representation must be derived from the spectrum. There is much evidence that the peripheral auditory system (8th nerve) of humans performs such an analysis before relaying to higher levels. Our simulation has been restricted to Hopfield net elements with binary outputs; that is, the variables $\hat{x}_i$ are permitted to take on only the values $\pm$ 1. Thus, we need to find a suitable <u>binary</u> representation of the spectral cross section of a vowel. Our procedure is the following:

(a) Perform a high-resolution spectrum analysis on overlapping segments (of 20 ms duration) of a spoken word. This analysis is performed every 10 ms and the signal is first multiplied by a Hamming window before being analyzed. Analysis is accomplished via a 512-point fast Fourier transform, from which a 256-point magnitude function covering the range 0-5 kHz can easily be derived.

(b) This magnitude function is then smoothed in frequency to produce a spectral envelope function.

(c) Finally, a suitable pattern (for entry into the associative memory) is generated in the following way: let $S(f)$ be the magnitude versus frequency function. Let $S(f_i)$, $i = 0,1,...119$ be samples of this function. Then, if $S(f_i) > S(f_{i-1})$, set the initial value of x_i in Eq. (1) to +1. If $S(f_i) < S(F_{i-1})$, set x_i to -1. If $S(f_i) = S(f_{i-1})$, randomly choose +1 or -1.

(It should be noted that a variety of representations can be invented that lead to a binary sequence suitable for input to the Hopfield net. We shall see for instance that a different representation is used for processing of consonants.)

The next task is creation of the network. In our case, the chosen parameters were m=8 and N=120. The parameter m defines the number of stored states. Since m < .15N, our rule of thumb says that convergence to a stored state is to be expected. The waveforms and spectra of eight words were examined and for each word, a steady state region was identified by inspection. Then, a single spectral cross section was subjected to the representation process described above and the T_{ij} matrix was computed and became a fixed part of the Hopfield net.

The main purpose of this experiment was to determine how accurately a given vowel must be represented in order to successfully retrieve its prototype from the network. Another way to ask this question: How much partial information is needed in order to retrieve the total information? (the stored representation of the correct vowel). If we chose as the input or initial state one of the unadulterated stored states, then we would be supplying all, or 100% of the required information. If, on the other hand, we supplied 50% of the required information and allowed the remainder to be supplied randomly, we would, on the average, be supplying 75% of the correct bit pattern. In a formula,

$$PI = 1-2 \ ER \qquad\qquad (2)$$

where PI is the resulting partial information based on a given fractional error rate.

In our experiment, an initial state was derived from one of the stored states by randomly creating errors in the 120-bit representation of that stored state. The system was then subjected to the iterations defined by Eq. (1) until convergence; i.e., the state of the system remained constant despite continued iterations. An example of the evolution of the network is shown in Fig. 1. For convenient viewing, the 120 bits are presented as a 10×12 matrix. Also, Fig. 1 displays the difference between any state and the stored target state. Thus, iteration 0 (upper lefthand corner) of Fig. 1 displays the difference pattern between the perturbed stored state (which is also the initial state of the network) and the unperturbed stored state from which it is derived. Each iteration is numbered and we see that convergence occurs at iteration 7; the network now stays in the original unperturbed stored state despite repeated application of Eq. (1).

Figure 2 shows the evolution for the same initial vowel with a different random error generator. In this case, although the net began with fewer errors than for Fig. 1, the system converged to a spurious state with Hamming distance 4 from the target state. This demonstrates that the behavior of the network is not like that of a correlator or matched filter looking for minimum Hamming distance but a more complicated device. One of the challenges of this work is to understand in finer detail how the system evolves under different conditions.

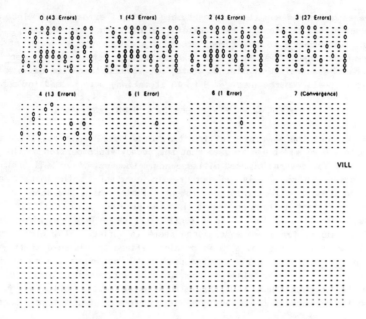

Fig. 1. Evolution of a Hopfield net in vowel discrimination.

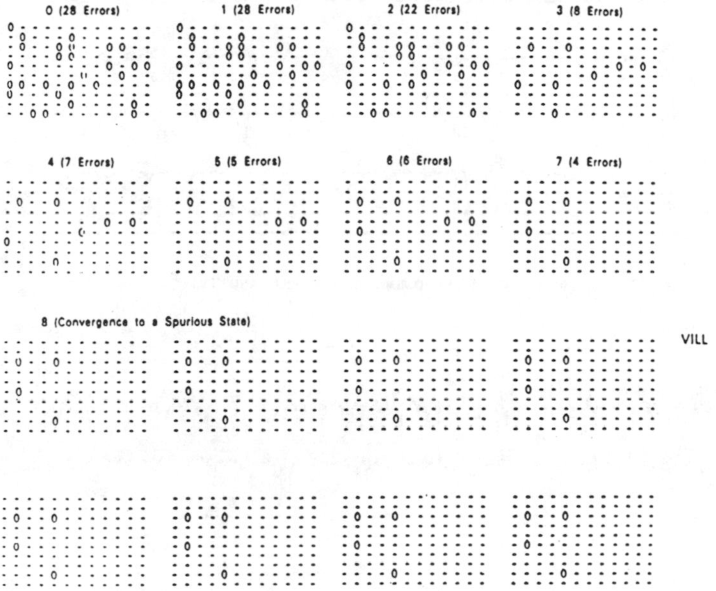

Fig. 2. Evolution of a Hopfield net with different initial conditions.

Each of the eight vowels was randomly perturbed with 25% error rate. Since different sets of random errors create different results, we ran each vowel 8 times, for a total of 64 runs. Convergence to the target stored state took place 51 out of 64 times. Usually, when convergence was to a spurious state, that state was <u>closest</u> (using a Hamming distance measure) to the target state, but several times the network converged to a different vowel. On the average, 8 to 10 iterations were needed for convergence.

3. CONVERGENCE OF NEW VOWELS TO A "FAMILIAR" STATE

Figure 3 shows the waveforms of the two words "vill" and "bill". Each line corresponds to 150 so it can be seen that the steady state portion of the vowels in these words last for several hundred milliseconds. However, it is well known that the measured spectra during vowels do not remain constant; both initial and final consonants can strongly affect this measurement. Furthermore, the vowel in "bill" is modified by a <u>different</u> initial consonant than the same vowel in "vill".

On the fourth line of Fig. 3 is a small asterisk. This signifies that 20 ms portion of the signal for which an assigned state is defined. For each word pair of the list on the right of Fig. 3, a 20 ms portion defines an assigned state. Thus, we construct the <u>same</u> network we constructed previously, but now the task is different. The initial states are now obtained by scanning the entire word pair; every 10 ms a new representation is entered into the network as an initial state and the network iterates until convergence to an assigned or spurious state. At convergence, the

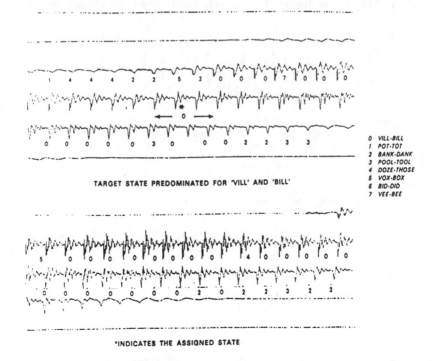

TARGET STATE PREDOMINATED FOR 'VILL' AND 'BILL'

0 VILL-BILL
1 POT-TOT
2 BANK-DANK
3 POOL-TOOL
4 DOZE-THOSE
5 VOX-BOX
6 BID-DID
7 VEE-BEE

*INDICATES THE ASSIGNED STATE

Fig. 3. Convergence of new vowels to a "familiar" state.

final state is compared with all eight assigned states and the assigned state that is minimum Hamming distance is declared to be the "winner". The numbers in Fig. 3 show the sequence of winners every 10 ms. We see that in Fig. 3 the network has found many "familiar" states, that is, states that were identified as the vowel "i" (as in "vill" or "bill"). This "recognition" was based entirely on the single asterisked segment shown in Fig. 3. (It should be pointed out that both "vill" and "bill" and, indeed, all other word pairs processed for this section, were the utterances of a single speaker. Recognition of "familiar" patterns across speakers has not yet been tried.)

All of the 8 word pairs were tested and generally gave results similar to those shown in Fig. 3. Two cases of interest should be mentioned; in the "bank-dank" pair, the assigned state for "bank" was chosen too close to the "b-a" transition, resulting in many errors. This result was rectified by choosing a new assigned state deeper in the steady state vowel portion. The second case of interest was the "bid-did" pair, where the network converged to one or another of these assigned states with comparable frequency. This indicates that the same vowel, but from a different context, can still be classified as "familiar".

4. CONSONANT DISCRIMINATION WITH A HOPFIELD NET

DRT is one of the classical methods used to assess the intelligibility of speech communication systems. The listener is presented with a list of about 200 rhymed word pairs (e.g., pool-tool, bank-dank) and is asked to choose which of the pairs she/he heard. DRT score is defined as,

$$\text{DRT Score} = \frac{\text{correct decisions} - \text{wrong decisions}}{\text{total decisions}} \times 100$$

Thus, a completely random choice would result in a score near zero and a "perfect" score would be 100. There is much data available on the DRT. Typically, high fidelity speech in a quiet acoustic background would yield a score of 98, whereas a 2400 bps vocoder subjected to the DRT words in a noisy environment (0-6 dB signal-to-noise ratio) might yield a score of 60-70.

We asked: Could a structure based on Hopfield nets perform an automatic DRT that would compare favorably with the documented results using human listeners?

There is a relatively small time segment in each of the word-pairs during which a distinction can be made. This segment is either embedded in the consonant or in the transition region between consonant and vowel. We gave our system an a priori determination of these critical regions. The chosen representation involved 4 contiguous spectral envelopes and 30 samples for each 10 ms frame, yielding, 120 bits per word. A total of 16 Hopfield nets were thus defined, each with N=120 and m=2. Given the word pair from which one of two words would be "heard" by the network allowed us to choose which of these 16 nets to employ. Thus, 32 words were processed (1) after passage through a vocoder, and (2) after polluting the speech with additive noise. (Note that the Hopfield nets were derived from the original noiseless data.) Figure 4 shows how well the networks distinguished between the consonants for all 32 words. The curves of Fig. 4 were derived by performing a two-state discrimination every 10 ms, using the Hopfield net corresponding to the input word-pair. Zero on the horizontal axis in all cases corresponds to the critical region; that segment of the signal used to obtain assigned or stored states. The curves show the total number of errors when all 32 words are processed; Figs. 4(a), (b), (c) and (d) are for different degrees of computer-generated white additive noise; Fig. 4(e) is for the output of an 8 kb/s channel vocoder for noiseless input.

164

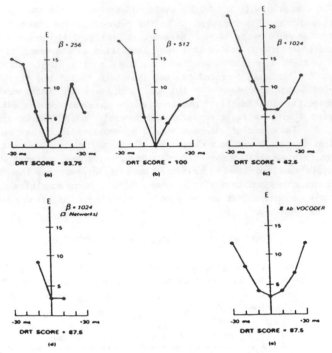

Note: Vertical axes (E) are total number of errors out of 32 tests.

Fig. 4. Results for Automatic Diagnostic Rhyme Test based on 32 words.

The noise variance is proportional to β^2; for Figs. 4(c) and 4(d), $\beta=1024$ corresponds approximately to a 5 dB signal to noise during the consonant. Figure 4(d) is based on a best two out of three decision for 3 nets based on spectra 20 msec apart around the critical region. All other figures are based on a single net at the critical region.

The same set of 32 DRT words with additive noise with β = 1024 was scored using three human listeners; the score was 83.3 which is reasonably close to the score obtained using the 3 network configuration, which yielded the result Fig. 4(d). This very preliminary result encourages us to continue research on an automatic DRT system using Hopfield nets.

REFERENCES

[1]   J.J. Hopfield, "Neural Networks and Physical Systems with Emergent Collective Computational Abilities," Proc. Natl. Acad. Sci. USA, Vol. 79, pp. 2554–2558, April 1982.

This work was sponsored by the Department of the Air Force.

The views expressed are those of the author and do not reflect the official policy or position of the U.S. Government.

LYAPUNOV FUNCTIONS
FOR PARALLEL NEURAL NETWORKS

Eric Goles[1] and Gérard Y. Vichniac

Laboratory for Computer Science
Massachusetts Institute of Technology, Cambridge, Massachusetts 02139

Abstract

We construct additive Lyapunov functions for neural networks defined by threshold transition rules implemented synchronously (à la Little). These functions take the forms $E_P = -\|Tx\|_1$ (Manhattan metric norm in the configuration space) and $E_P = -\sum_i \|\sum_j T_{i,j}\vec{x}_j\|_\infty$ (norm in the space of internal states) if q states per site are present. These functions can be seen as energies for parallel iterations, comparable to Hopfield's energy $E_g - -\langle x, Tx \rangle$ for sequential iterations. Applications to the Marr Poggio cooperative algorithm for stereopsis are presented.

1. Introduction

Neural networks were introduced in 1943 by McCulloch and Pitts[23] to model neural electrical activity. They consist of nets of N highly interconnected elements. Each element i has an internal state (-1 or 1), and interacts at discrete time steps with other elements throught its output lines (identified with its internal state). In turn, it is influenced by other sites through its input lines, and updates its own state x_i according to a *threshold* function—the state will be 1 if a certain linear combination is above a given threshold b_i, and -1 otherwise:

$$x_i^{t+1} = \text{sgn}(\sum_{j=1}^{N} T_{i,j}x_j^t - b_i) \tag{1}$$

The coefficients $T_{i,j}$ of the combination are often called "synaptic weights," or also "Ising bonds," in which case they are denoted $J_{i,j}$.

There are two major ways to implement the updating of the states: synchronously (i.e., in parallel) and sequentially. If time t counts the number of sweeps over the whole network, Equation (1) suggests that the iterations are done in parallel: each site i "computes" its new state x_i^{t+1} as a function of

---

[1]Current address: Dept. Math. Esc. Ing., U. Chile, Casilla 170, Correo 3, Santiago, Chile.

0094-243X/86/1510165-17$3.00 Copyright 1986 American Institute of Physics

the *nonupdated* values x_j^t of other sites. In contrast, in an orderly sequential iteration, where site $i+1$ updates its state immediately after site i does, (1) should be replaced by

$$x_i^{t+1} = \text{sgn}(\sum_{j=1}^{i-1} T_{i,j} x_j^{t+1} + \sum_{j=i}^{N} T_{i,j} x_j^t - b_i) \qquad (2)$$

where again $t+1$ means after an entire sweep over the network. This sequential iteration mixes present and future: site i takes into account the "present" states at i and $i+1$, but the "new" state at $i-1$. Not all sequential iterations are orderly; Hopfield for example uses a dynamics where the sites are updated in random order[16]. That case can also be described by equation (1), but where t counts elementary local steps, not entire sweeps.

Little used parallel iterations in his pioneering work[17,18] that suggested the modeling of the brain's persistent states with ordered phases of Ising spins.

The parallel and sequential iterations present surprising analogies[24]. In particular, as shown by Amit, Gutfreund, and Sompolinski[1], Little's model and the finite temperature version of Hopfield's model have equivalent phase diagrams, i.e., they are essentially identical from the viewpoint of statistical mechanics. Nevertheless, sequential and synchronous dynamics exhibit profound differences. This paper compares these two modes of updating in the deterministic (zero-temperature) case, constructs compact forms of Lyapunov functions for parallel updatings, and study their metric properties.

Lyapunov functions are invaluable tools because they can express complex network dynamics as an optimiztion process. They have been used to an advantage by Hopfield[16] for the *sequential* case, where they take the appealing form of spin-glass energies if the weights $T_{i,j}$ are symmetric. For neural networks in the *continuum* case, Cohen and Grossberg[4] have constructed global Lyapunov functions that apply to a large class of models.

2. Dynamics of symmetric neural networks

Let us consider parallel iterations and assume symmetric interactions (input and output lines are the same, and on them $T_{j,i} = T_{i,j}$). Let us introduce another symmetry, one between the $+1$ and the -1 states, by focusing on the case where all the thesholds b_i are zero.

2.1 Lyapunov function for parallel iterations

We can define the following bilinear operator:

$$E_P(\mathbf{x}^t, \mathbf{x}^{t+1}) = -\sum_{i=1}^{N} x_i^{t+1} \sum_{j=1}^{N} T_{i,j} x_j^t, \qquad (3)$$

which can also be written

$$E_P(\mathbf{x}^t, \mathbf{x}^{t+1}) = -\langle \mathbf{x}^{t+1}, \mathsf{T}\mathbf{x}^t \rangle, \tag{4}$$

where $\mathbf{x}$ is the vector of components x_i, T is the matrix of elements $T_{i,j}$, and where the inner product notation implies a sum over all sites ($\langle u, v \rangle = \sum_i u_i v_i$). The following inequality then holds [12,14]:

$$E_P(\mathbf{x}^t, \mathbf{x}^{t+1}) - E_P(\mathbf{x}^{t-1}, \mathbf{x}^t) \leq 0. \tag{5}$$

The conditions for the strict inequality are known, and it is proven that E_P is constant on an attractor. More importantly for the present discussion, it is also known that attractors are either fixed points or limit-cycles of length two. Replacing x_i^{t+1} by its value from (1), we can write E_P under a form that depends on time t only

$$E_P(\mathbf{x}^t, \mathbf{x}^{t+1}) = -\sum_{i=1}^{N} \left| \sum_{j=1}^{N} T_{i,j} x_j^t \right| = -\|\mathsf{T}\mathbf{x}^t\|_1, \tag{6}$$

where $\| \cdot \|_1$ is the usual 1-norm (in the "Manhattan metric") in R^N. This norm, of course, does not apply on the network itself, but on its configuration space. (It is thus unrelated to the distance used by Little in reference [17], p. 112.) Peretto[24] derived a Hamiltonian of the form (6) by taking the zero-temperature limit of a statistical mechanical description of Little's model.

The nonincreasing quantity $E_P = -\|\mathsf{T}\mathbf{x}^t\|_1$ is a Lyapunov function for the network dynamics with parallel iterations; it depends on one value of time only, and plays a similar role than the dissipative energy $E_S = -\langle \mathbf{x}, \mathsf{T}\mathbf{x} \rangle$ of Hopfield's model, where the updatings are sequential.

2.2 Duality between fixed-points and limit-cycles

The vanishing of all the thresholds b_i immediately yields a ± 1 symmetry: if the configuration $\mathbf{x} = (x_1, \ldots, x_N)$ is a fixed point, its binary complement $-\mathbf{x}$ is also a fixed point. Likewise, if $\mathbf{x} \rightleftharpoons \mathbf{y}$ is a two-cycle, so is $-\mathbf{x} \rightleftharpoons -\mathbf{y}$. Moreover, (6) shows that E_P is invariant if T is changed into $-\mathsf{T}$. The optimization process performed by the network leads to the same attractor sets if all the inhibitory synapses ($T_{i,j} < 0$) are changed into excitatory ones ($T_{i,j} > 0$) and vice versa. We can in fact establish a duality between the fixed-points and the limit-cycles that form this common attractor set:

If the synaptic matrix T is changed into $-\mathsf{T}$, fixed points and their complements become two-cycles and complementary configurations that form two-cycles become fixed points.

(i) If the vector $\mathbf{x}$ is a fixed-point:

$$x_i^t = x_i^{t+1} = \mathrm{sgn}\left(\sum_{j=1}^{N} T_{i,j} x_j^t\right), \tag{7}$$

it is mapped to $-x$ if T changes sign. Thus x and its complement are turned into a two-cycle.

(ii) Conversely, if the vectors x and $-x$ form a two-cycle:

$$- x_i^t = x_i^{t+1} = \text{sgn}(\sum_{j=1}^{N} T_{i,j} x_j^t), \tag{8}$$

each vector is mapped into itself if each weight $T_{i,j}$ is replaced by $-T_{i,j}$.

Consider for example a fixed point z and its three immediate antecedents:

$$w \longrightarrow x \longrightarrow y \longrightarrow z . \tag{9}$$

The following also holds

$$- w \longrightarrow -x \longrightarrow -y \longrightarrow -z , \tag{10}$$

and if the synaptic matrix changes sign, we have

$$\begin{matrix} w & & y & & z & \quad (11) \\ & \searrow \ \nearrow & & \searrow \ \nearrow\!\!\!\!\!\!/ & \\ & -x & & -z & \end{matrix}$$

2.3 An illustration with cellular automata

Consider a network defined on a two-dimensional lattice with $T_{i,j} = 0$ except for i and j equal or labeling nearest neighbors, in which cases $T_{i,j} = 1$. The network is then a cellular automaton in the so-called von Neumann neighborhood, which consists of the considered cell itself and its four adjacent neighbors. (For an informal introduction to cellular automata, see [15], for reviews, see [8,5].) The transition rule (1) is thus a simple majority rule: a cell will assume at time $t + 1$ the state which is most represented (i.e., by at least three cells) in the five-cell neighborhood at time t. This rule, that has been called "V345" and "VGE3" in [27,28], can be seen as describing the dynamics at $T = 0$ of an Ising model with self-interaction (we put $T_{i,i} \neq 0$ to eliminate the tie case that can occur with only four "voting" neighbors, as in the standard Ising model). The dynamics admits two fixed points (all sites $+1$ and all sites -1), as well as one two-cycle made of two binary complementary configurations (regular checkerboard pattern). The majority rule enhances the alignment of each site with its neighbors (just as in the zero-temperature ferromagnetic Ising model), which explains both the stability of the homogeneous array and the blinking of the checkerboards—the blinking occurs because each site in a checkerboard-like region flips to agree with the neighbors, but the neighbors themselves flip too if all the sites are updated synchronously.

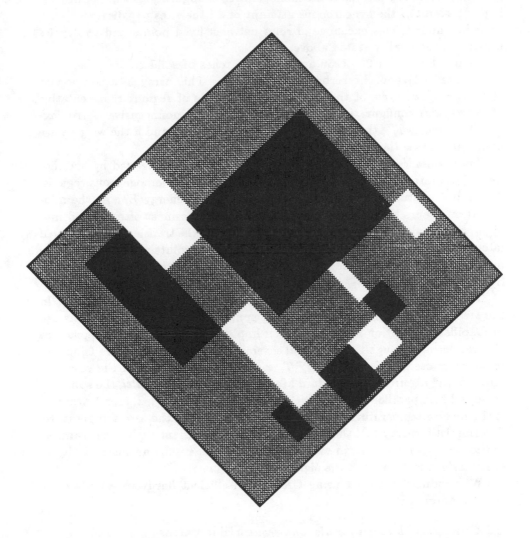

Figure 1. A cellular automaton configuration containing regions characteristic of fixed points and two-cycles for both T and −T.

If the synaptic matrix changes sign, the checkerboard configurations will stabilize (cf. the ground states of the zero-temperature Ising model with *anti-ferromagnetic* couplings), but the homogeneous configurations will start blinking as a result of the synchronous attempt of all the sites to "disagree" with their neighbors! This exchange of roles between fixed points and two-cycles illustrates the duality stated above.

Figure 1 shows a 256×256 array with patches of solid colors (black=+1 and white=−1) and checkerboard-like regions. This array is a component of a typical two-cycle of the dynamics; it is made of regions representative of the regular configurations we just discussed. For nonnegative (ferromagnetic) couplings $T_{i,j}$, the checkerboard regions oscillate, and if the weights are nonpositive, it is the homogeneous domains that blink.

The dynamics of the initial transient period is characterized by the motion of the interface between the checkerboard and the homogeneous regions. This motion actually performs a minimization of the energy E_P. As the minimization occurs by steepest descent, the system gets most often locked in a high-lying local minimum, as in Figure 1. If, during the transient, T is changed into −T, the motion of the interfaces continues, absolutely unaffected. The insensitivity of the interface motion upon sign of the synaptic matrix results from the invariance of E_P under transformation T ↦ −T.

This insensitivity is also observed in synchronous stochastic iterations. In the statistical mechanical study of Ising spins, Monte-Carlo calculations use probabilistic transition rules that, when implemented *sequentially*, reproduce the wandering through the canonical ensemble of finite-temperature Ising systems, characterized by $E_S = -\sum T_{i,j} x_i x_j$. If one implements the so-called "heat-bath" algorithm (a standard Monte-Carlo rule) *in parallel*, the spins are governed by a parallel energy $E_P = -\frac{1}{\beta} \sum_{i=1}^{N} \log \cosh(\beta \sum_{j=1}^{N} T_{i,j} x_i x_j)$, where β is the inverse temperature[24,3]. Though different from the zero-temperature, deterministic energy (6), this form of E_P is also invariant under the transformation T ↦ −T (cosh is an even function!). This entails an insensitivity of the interface motion upon the sign of synaptic matrix.

We produced Figure 1 using CAM-6, an efficient hardware simulator of cellular automata[27,19].

2.4 Comparison between parallel and sequential iterations

The fixed points in parallel and sequential iterations can readily be seen to be the same, and so are the stored patterns (i.e., the memories). Synchronous dynamics also lead to two-cycles that do not occur in sequential iterations. These two-cycles are often unexpected; in some case they lead to undesired effects, in other cases they turn out to be quite useful.

The worst occurence of two-cycles happens if one attempts to reach the minimum of the sequential energy E_S by implementing fully parallel updatings. In that case the system can get locked in a two-cycle whose components correspond to a local minimum of the parallel energy E_P but at the same time to the absolute maximum of the sequential energy E_S! This is the "blinking" effect discussed in the last section and also observed in the standard Ising model at finite temperature[15,28,3].

In numerous instances, however, the two-cycles that are characteristic to parallel iterations are instrumental to describe *bistable* states, such as ambiguities occurring in visual perception (see Figure 2) and in the interpretation of natural language with semantic networks (cf. Pollack and Waltz's example: *"The astronomer married a star"*[26]).

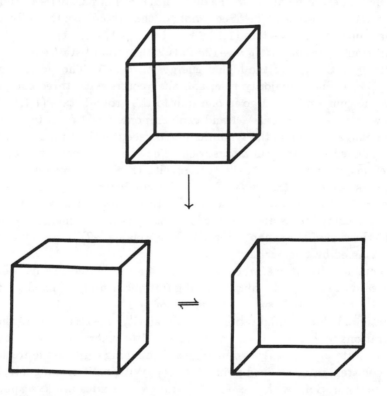

Figure 2. This ambiguity in visual perception can be detected and described by a two-cycle in a synchronous neural network where sites correspond to corners and synaptic links to edges. The synaptic weights determine which corner is in the foreground and which is in the background. Particular values of the weights can lead to a bistable state.

2.5 A very simple example

The duality between fixed points and two-cycles as well as the difference between synchronous and sequential iterations are illustrated by the following very simple example: a zero-temperature Ising model with four spins and no self-interaction. In other words, we consider a network containing four sites, where each site assumes one of the two states 1 or −1 and interacts with the three other sites via a positive (ferromagnetic) unit weight:

$$T_{i,j} = 1 - \delta_{i,j}. \tag{12}$$

An attractive aspect of this model is that transients, when they exist, have length one for both types of iterations (parallel and sequential).

Let us first examine first how the dynamics looks in parallel iterations (see the left part of Figure 3). The fixed points of the system are the fully aligned configurations i.e., the vectors $(1,1,1,1)$ and $(-1,-1,-1,-1)$. By (6), these configurations have energy $E_P = -12$. This is the ground-state energy; all the other configurations have the higher energy $E_P = -6$. The (ferromagnetic) weights (12) define a majority rule, and the vectors with three components $+1$ and one components -1 go in one step to the ground state $(1,1,1,1)$. The vectors with two components $+1$ and two components -1 form two-cycles. If we now change the sign of the synaptic matrix (see the right part of Figure 3), the two aligned vectors form a two-cycle. The (antiferromagnetic) couplings $-T_{i,j}$ define a "minority rule": the vectors with three components $+1$ and one components -1 enter the two-cycle via the configuration $(-1,-1,-1,-1)$. Also, in accord to the duality of section 2.2, the vectors with two components $+1$ and two components are now fixed points. Again, the energy E_P is indifferent to the sign of the synaptic matrix; with weights $-T_{i,j}$, the lowest energy -12 is attained by two-cycles.

In sequential iterations, where sites 1,2,3,4 are updated in this order, the two-cycles disappear. The aligned configurations are again fixed points for positive (ferromagnetic) couplings; the vector $(-1,-1,-1,-1)$ admit a basin of attraction made of configurations with a majority of -1's as well as the three configurations with two $+1$ sites and two -1 sites where site $i = 1$ (updated first) is in state $x_1 = +1$. By symmetry, the remaining eight configurations go, also in one step, into the fixed point $(1,1,1,1)$. With antiferromagnetic bonds, the six configurations with two $+1$ sites and two -1 sites are fixed points.

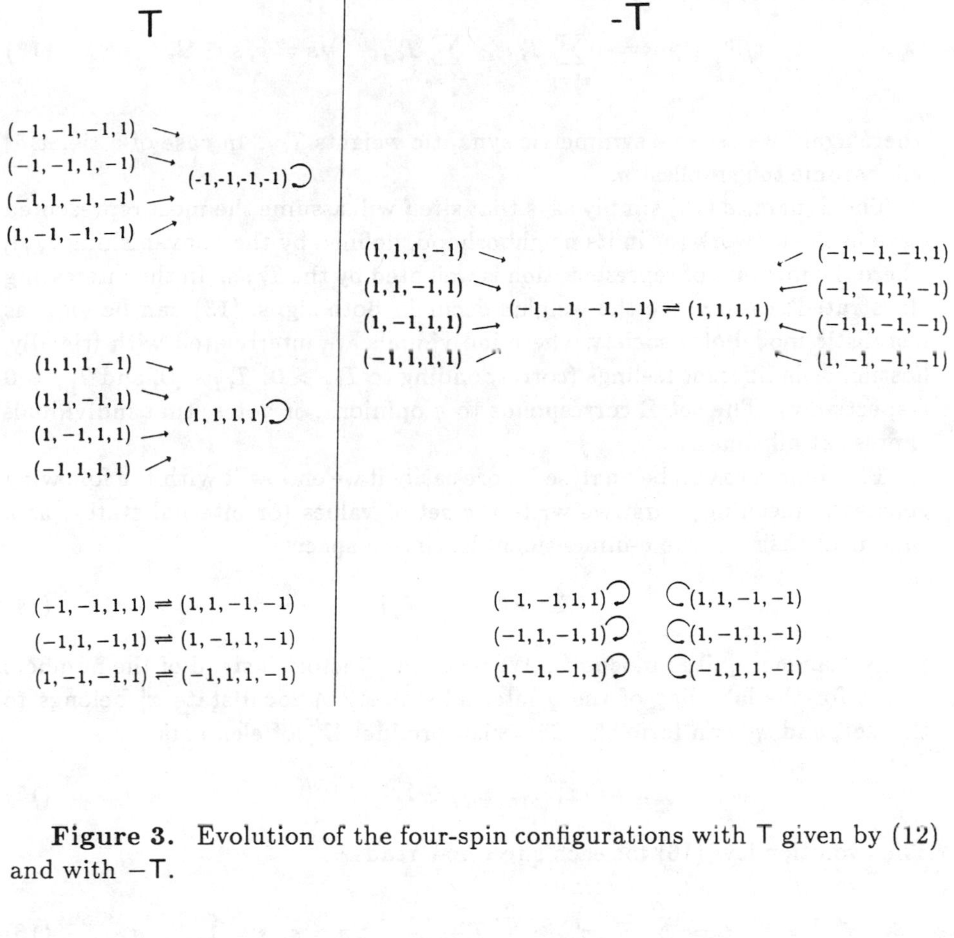

T   -T

$(-1,-1,-1,1)$ ↘

$(-1,-1,1,-1)$ ↗

$(-1,1,-1,-1)$ → $(-1,-1,-1,-1)$ ↻

$(1,-1,-1,-1)$ ↗

$(1,1,1,-1)$ ↘   ↗ $(-1,-1,-1,1)$

$(1,1,-1,1)$ ↘   ← $(-1,-1,1,-1)$

$(1,-1,1,1)$ → $(-1,-1,-1,-1) \rightleftharpoons (1,1,1,1)$ ← $(-1,1,-1,-1)$

$(-1,1,1,1)$ ↗   ← $(1,-1,-1,-1)$

$(1,1,1,-1)$ ↘

$(1,1,-1,1)$ ↘

$(1,-1,1,1)$ → $(1,1,1,1)$ ↻

$(-1,1,1,1)$ ↗

$(-1,-1,1,1) \rightleftharpoons (1,1,-1,-1)$ $(-1,-1,1,1)$ ↻   ↺ $(1,1,-1,-1)$

$(-1,1,-1,1) \rightleftharpoons (1,-1,1,-1)$ $(-1,1,-1,1)$ ↻   ↺ $(1,-1,1,-1)$

$(1,-1,-1,1) \rightleftharpoons (-1,1,1,-1)$ $(1,-1,-1,1)$ ↻   ↺ $(-1,1,1,-1)$

Figure 3. Evolution of the four-spin configurations with T given by (12) and with $-$T.

3. Lyapunov function in generalized majority networks

3.1 Seminorm

We now extend the binary state set to an arbitrary large (but finite) set $\Sigma = \{1, 2, \ldots, q\}$, and a parallel dynamics defined by

$$x_i^{t+1} = p \iff \sum_{x_j^t = p} T_{i,j} \geq \sum_{x_j^t = s} T_{i,j}, \quad \forall s \neq p, s \in \Sigma, \tag{13}$$

where again we assume symmetric synaptic weights $T_{i,j}$. In case of a tie, x_i^{t+1} will assume the smallest p.

The dynamics (13) simply says that site i will assume the most represented state in the network (or in its neighborhood, defined by the nonvanishing $T_{i,j}$), where the measure of representation is weighted by the $T_{i,j}$'s. In the interesting "frustrated" case where the weights occur in both signs, (13) can be view as a stylistic model of a society where individuals are interrelated with friendly, hostile, or indifferent feelings (corresponding to $T_{i,j} > 0$, $T_{i,j} < 0$, and $T_{i,j} = 0$ respectively). The set Σ corresponds to q opinions, or votes, that individuals express at all times.

The dynamics can be analysed more easily if we endow it with the following geometric meaning. First we write the set of values (or internal states) as a canonical basis for the q-dimensional Euclidian space:

$$\Sigma = \{\vec{e}_1, \ldots, \vec{e}_q\}. \tag{14}$$

(This change is quite innocuous. We just use q vectors instead of the numbers 1 to q for the labelling of the q internal states.) A local state x_i^t belongs to this set, and we can form the Cartesian product Σ^N of elements

$$\mathbf{x}^t = (\vec{x}_1^t, \ldots, \vec{x}_N^t) \in \Sigma^N \subset \mathsf{R}^{qN}. \tag{15}$$

The evolution law (16) for each site i now reads

$$\vec{x}_i^{t+1} = \vec{e}_p \iff \sum_{j=1}^N T_{i,j} x_{jp}^t \geq \sum_{j=1}^N T_{i,j} x_{js}^t, \quad \forall p \neq s, s = 1, \ldots, q. \tag{16}$$

With this notation,

$$x_{jp}^t = 1 \iff \vec{x}_j^t = \vec{e}_p. \tag{17}$$

As above, boldface vectors (e.g., $\mathbf{x}^t$) are in the space of sites, which are labeled by the indices $i, j = 1, \ldots, N$. In contrast, we reserve the indices $p, s = 1, \ldots, q$ for internal values at a site, and identify vectors in this value space with an arrow (e.g., $\vec{x}_i^t$).

As the case with two internal values, we construct a bilinear form on vectors $\mathbf{x}^t$ and $\mathbf{x}^{t+1}$, then show that it actually involves only time t and that it is a nonincreasing "energy" E_P (i.e., a Lyapunov function), and finally give it a metric interpretation. Let us write

$$E_P(\mathbf{x}^t,\mathbf{x}^{t+1}) = -\sum_{i=1}^{N}\langle \vec{x}_i^{t+1}, \sum_{j=1}^{N} T_{i,j}\vec{x}_j^t\rangle = -\sum_{i=1}^{N}\langle \vec{x}_i^{t+1}, \vec{y}_i^t\rangle, \tag{18}$$

where $\vec{y}_i^t$ is a vector in the space of values:

$$\vec{y}_i^t = (\sum_{\vec{x}_j^t=\vec{e}_1} T_{i,j}, \ldots, \sum_{\vec{x}_j^t=\vec{e}_q} T_{i,j}). \tag{19}$$

Notice that now the inner product notation involves a sum over q *values* (and not sites, as before). Since the synaptic matrix T is symmetric, we have

$$E_P(\mathbf{x}^t,\mathbf{x}^{t+1}) - E_P(\mathbf{x}^{t-1},\mathbf{x}^t) = -\sum_{i}\langle \vec{x}_i^{t+1} - \vec{x}_i^{t-1}, \sum_{j} T_{i,j}\vec{x}_j^t\rangle. \tag{20}$$

Each term i of the sum satisfies

$$\langle \vec{x}_i^{t+1} - \vec{x}_i^{t-1}, \sum_{j} T_{i,j}\vec{x}_j^t\rangle = \max\{\sum_{\vec{x}_j^t=\vec{e}_1} T_{i,j}, \ldots, \sum_{\vec{x}_j^t=\vec{e}_q} T_{i,j}\} - \sum_{\vec{x}_j^t=\vec{x}_i^{t-1}} T_{i,j} \geq 0. \tag{21}$$

The quantity E_P is indeed nonincreasing, a fact that implies[9] that the attractor set consists of fixed points and two-cycles.

To study the metric properties of E_P, let us define $\vec{\rho}(\sum_{\vec{x}_j^t=\vec{e}_1} T_{i,j}, \ldots, \sum_{\vec{x}_j^t=\vec{e}_q} T_{i,j})$ as the basis vector $\vec{e}_p$ such that $\sum_{\vec{x}_j^t=\vec{e}_p} T_{i,j}$ is maximum. With this definition, the law of evolution (12) takes the form

$$\vec{x}_i^{t+1} = \vec{\rho}(\sum_{\vec{x}_j^t=\vec{e}_1} T_{i,j}, \ldots, \sum_{\vec{x}_j^t=\vec{e}_q} T_{i,j}) \in \{\vec{e}_1, \ldots, \vec{e}_q\}. \tag{22}$$

We have therefore

$$\langle \vec{x}_i^{t+1}, \sum_{j} T_{i,j}\vec{x}_j^t\rangle = \langle \vec{\rho}(\sum_{\vec{x}_j^t=\vec{e}_1} T_{i,j}, \ldots, \sum_{\vec{x}_j^t=\vec{e}_q} T_{i,j}), \sum_{j} T_{i,j}\vec{x}_j^t\rangle = \max_{1\leq p\leq q}\{\sum_{\vec{x}_j=\vec{e}_p} T_{i,j}\} \tag{23}$$

and E_P now reads

$$E_P = -\sum_{i=1}^{N}\max_{1\leq p\leq q}\{\sum_{\vec{x}_j^t=\vec{e}_p} T_{i,j}\} = -\sum_{i=1}^{N}\max(\sum_{j=1}^{N} T_{i,j}\vec{x}_j^t) \tag{24}$$

where

$$\max(\vec{u}) = \max_{1\leq p\leq q}\{u_p\}. \tag{25}$$

The quantity E_P is indeed a nonincreasing function that depends only on a configuration of the network at a single time. Notice, however, that in contrast with the simpler case treated in section 2, Equation (24) does not define a norm but only a seminorm, because one has for instance

$$\max(0, -1, -2, -3) = 0 \quad \text{while} \quad (0, -1, -2, -3) \neq \vec{0}. \tag{26}$$

3.2 Norm in the $\|\cdot\|_\infty$ metric

In order to construct a genuine norm for E_P, it suffices that Σ contain an even number of values of the form

$$\Sigma = \{-\vec{e}_1, -\vec{e}_2, \ldots, -\vec{e}_q, \vec{e}_1, \ldots, \vec{e}_q\} \subset \mathbb{R}^q. \tag{27}$$

The dynamics is now described as

$$\vec{x}_i^{t+1} = \text{sgn}\Big(\sum_{\vec{x}_j^t = \vec{e}_p} T_{i,j} - \sum_{\vec{x}_j^t = -\vec{e}_p} T_{i,j} \Big) \vec{e}_p$$

$$\iff \Big| \sum_{\vec{x}_j^t = \vec{e}_p} T_{i,j} - \sum_{\vec{x}_j^t = -\vec{e}_p} T_{i,j} \Big| \geq \Big| \sum_{\vec{x}_j^t = \vec{e}_s} T_{i,j} - \sum_{\vec{x}_j^t = -\vec{e}_s} T_{i,j} \Big|, \forall s \neq p \tag{28}$$

(in case of a tie, the smallest p is assumed). We have again

$$E_P(t, t+1) = -\sum_{i=1}^N \langle \vec{x}_i^{t+1}, \sum_{j=1}^N T_{i,j}\vec{x}_j^t \rangle \tag{29}$$

and

$$E_P(t, t+1) - E_P(t-1, t) = -\sum_{i=1}^N \langle \vec{x}_i^{t+1} - \vec{x}_i^{t-1}, \sum_{j=1}^N T_{i,j}\vec{x}_j^t \rangle. \tag{30}$$

The ith term A_i of the sum takes the form $A_i = \langle \vec{x}_i^{t+1} - \vec{x}_i^{t-1}, \vec{y}_i \rangle$ where $\vec{y}_i$ is the vector

$$\vec{y}_i = \Big(\sum_{\vec{x}_j^t = \vec{e}_1} T_{i,j} - \sum_{\vec{x}_j^t = -\vec{e}_1} T_{i,j}, \ldots, \sum_{\vec{x}_j^t = \vec{e}_q} T_{i,j} - \sum_{\vec{x}_j^t = -\vec{e}_q} T_{i,j} \Big). \tag{31}$$

Let p and s be the value labels corresponding to times $t+1$ and $t-1$, respectively:

$$\vec{x}_i^{t+1} = \text{sgn}\Big(\sum_{\vec{x}_j^t = \vec{e}_p} T_{i,j} - \sum_{\vec{x}_j^t = -\vec{e}_p} T_{i,j} \Big) \vec{e}_p \tag{32}$$

and

$$\vec{x}_i^{t-1} = \text{sgn}\Big(\sum_{\vec{x}_j^t = \vec{e}_s} T_{i,j} - \sum_{\vec{x}_j^t = -\vec{e}_s} T_{i,j} \Big) \vec{e}_s. \tag{33}$$

Reporting these expressions in the definition of A_i, and using the fact that $|a| = a \times \text{sgn}(a)$, we get the inequality

$$A_i = |\sum_{\bar{x}_j^t = \bar{e}_p} T_{i,j} - \sum_{\bar{x}_j^t = -\bar{e}_p} T_{i,j}| - |\sum_{\bar{x}_j^{t-1} = \bar{e}_s} T_{i,j} - \sum_{\bar{x}_j^{t-1} = -\bar{e}_s} T_{i,j}| \geq 0. \qquad (34)$$

Hence E_P is a Lyapunov function, and reads

$$E_P = \sum_{i=1}^{N} \|\sum_{j=1}^{N} T_{i,j} \bar{x}_j^t\|_{\infty}, \qquad (35)$$

where $\| \cdot \|_{\infty}$ is the usual ∞-norm, but defined in the space of values:

$$\|\vec{u}\| = \max_{1 \leq p \leq q} \{|u_p|\}. \qquad (36)$$

In a general way, other "positive" functions (see [11,13,14]) lend themselves to a similar treatment.

4. Application to vision: stereopsis cooperative algorithms

The results of the last section find an application to Marr and Poggio's model of stereopsis[20,21,25]. This is a fully discrete neural network aimed at reconstructing depth perception out of binocular disparities. The network reconstructs the three-dimensional visual world by seeking matches between corresponding dots in two random-dot stereograms. The model implies two major constraints:

(i) *the uniqueness (or opacity) constraint* implies that only one match can occur at a site (along a sight line), and

(ii) *the continuity constraint* implies that binocular disparities vary smoothly between neighboring sites (perpendicularly to the sight line).

4.1 Lyapunov function for the Marr-Poggio algorithm

Roughly speaking, the Marr-Poggio algorithm involves a neural network on a three-dimensional lattice of nodes that assume values 1 or 0, according respectively to the presence or absence of matches between the two stereograms. For the purpose of the present discussion, the 3-D lattice points can be viewed as a stack of q planes of N nodes each. A site is labeled by a pair of indices ip, $0 \leq i \leq N$, $0 \leq p \leq q$, the p index for the plane and the i index for the site's "address" in that plane. The planes are oriented perpendicularly to the sight line; i thus labels the latter and p labels the depth. An essentially identical representation, but phrased in the language of the Potts model, has been presented by Schulten at this meeting[7].

The dynamics involve threshold functions that can be implemented synchronously:

$$x_{ip}^{t+1} = \Theta\left(\sum_{j,s} T_{ip,js} x_{js} - b_{ip}\right),\tag{37}$$

where the Heavyside Θ step function replaces the sgn function of (1) because the states are now 0 and 1 instead of ± 1.

The weights $T_{ip,js}$ are chosen to implement the constitutive constraints of the model. The uniqueness constraint inhibits the occurrence of several 1's along the sight line: $x_{ip}x_{is} = 0$ for all p and s. This condition is favored if $T_{ip,is} < 0$. The continuity constraint encourages the gathering of 1's in a plane: $x_{ip}x_{jp} = 1$. This relation is enhanced by the choice $T_{ip,jp} > 0$.

It is known[11,12] that the bilinear form

$$E_P(\mathbf{x}^t, \mathbf{x}^{t+1}) = -\sum_{i,p} x_{ip}^{t+1} \sum_{j,s} T_{ip,js} x_{js}^t + \sum_{i,p} b_{ip}(x_{ip}^{t+1} + x_{ip}^t)\tag{38}$$

is nonincreasing for parallel dynamics with threshold functions. In other words, E_P is a Lyapunov function for the network. With parallel updatings, the Marr-Poggio algorithm for stereopsis thus perfoms a minimization of (38), and its only steady states (i.e., local minima) are two-cycles and fixed points.

4.2 An alternative dynamics

The following algorithm automatically satisfies the uniqueness constraint. A unique 1 is deposited on each site line i at depth p such that

$$x_{ip}^{t+1} = 1 \iff \sum_{j,s} T_{ip,js} x_{js} = \max_{1 \leq p \leq q}\left(\sum_{j,s} T_{ip,js} x_{js}\right).\tag{39}$$

If the maximum is obtained for several planes, one choses the plane of smallest depth p. (The next section presents another way to deal with such a tie case.) This algorithm is a slightly simplified version of powerful algorithms recently devised by Drumheller and Poggio[6], and by Marroquin[22]. The weights $T_{ip,js}$ form a $Nq \times Nq$ symmetric matrix. Here again we build the bilinear form

$$E_P(\mathbf{x}^t, \mathbf{x}^{t+1}) = -\mathbf{x}^t \cdot \mathsf{T}\mathbf{x}^{t+1},\tag{40}$$

where $\mathbf{x}^t$ is a vector with Nq components x_{ip} $(i = 1, \ldots, N; p = 1, \ldots, q)$, T is a $Nq \times Nq$ symmetric matrix of elements $T_{ip,js}$, and $\cdot$ denotes the standard inner product in R^{Nq}.

We obtain the difference

$$E_P(t, t+1) - E_P(t-1, t) = -\sum_{i=1}^{N} \langle \vec{x}_i^{t+1} - \vec{x}_i^{t-1}, (T_{i1,\bullet} \cdot \mathbf{x}^t, \ldots, T_{iq,\bullet} \cdot \mathbf{x}^t) \rangle,\tag{41}$$

where $T_{ip,\bullet}$ is the row ip of the matrix T, and, in analogy with the previous section, $\vec{x}_i^t$ is a vector in the q-dimensional space of planes, and $\langle\,,\,\rangle$ denotes the inner product in that space. Each term i of the sum vanishes if $\vec{x}_i^{t+1} = \vec{x}_i^{t-1}$. By construction, $\vec{x}_i^t$ and $\vec{x}_i^{t+1}$ are basis vectors, and thus have one component 1, all the other components being 0. This implies that, if these two vectors are *not* equal, each term i of the sum satisfies the inequality

$$\max\{T_{i1,\bullet}\cdot\mathbf{x}^t,\dots,T_{iq,\bullet}\cdot\mathbf{x}^t\} - T_{ip,\bullet}\cdot\mathbf{x}^t \geq 0 \tag{42}$$

for some plane p, and thus the difference (41) is nonnegative. Just as the previous cases E_P can be written under a form that depends on one time step only:

$$E_P(t) = -\sum_{i=1}^{N}\max \vec{v}_i^t, \tag{43}$$

where $\vec{v}_i^t$ is a vector in the space of planes

$$\vec{v}_i^t = (T_{i1,\bullet}\cdot\mathbf{x}^t,\dots,T_{iq,\bullet}\cdot\mathbf{x}^t). \tag{44}$$

Hence E_P, defined by (43) is a Lyapunov function, or an additive nonincreasing "energy" that the cooperative algorithm attempts to minimize. As above, the algorithm admits a set of attractors simply made of fixed points and two-cycles. Also, a slight modification of the weights $T_{ip,js}$ can eliminate the two-cycles[9,13].

4.3 An alternative treatment of the tie cases

Let us associate at each sight line i the set C_i^t of plane labels p such that $T_{ip,\bullet}\cdot\mathbf{x}^t$ is maximum, and call $\mathcal{N}_i$ the cardinality of this set. For example, $\mathcal{N}_8 = 3$ means that $T_{8p,\bullet}\cdot\mathbf{x}^t$ reaches its largest value at three different planes. One readily sees that with internal states in $\{0, 1, \frac{1}{2}, \dots, \frac{1}{q}\}$, the law of evolution

$$x_{ip}^{t+1} = \frac{1}{\mathcal{N}_i}, \quad \forall p \in C_i^t \tag{45}$$

defines a dynamics with a Lyapunov function of the form (40).

Acknowledgements

This work was supported in parts by grants from DARPA (N0014-83-K-0125), NSF (8214312-IST), and DOE (DE-AC02-83ER13082). One of us (E. G.) wishes to acknowledge the partial support of the FNC 85—86, Chile, and of the PNUD 1986. We are grateful to H. Atlan, N. Margolus, T. Poggio, and T. Toffoli for fruitful discussions and to D. Zaig for his help in the preparation of the figures.

References

[1] D. J. Amit, H. Gutfreund, and H. Sompolinsky, *Phys. Rev.* A 32 (1985) 1007–1018.

[2] E. Bienenstock, F. Fogelman Soulié, and G. Weisbuch (eds), *Disordered Systems and Biological Organization,* Les Houches Spring 1985, Proceedings, NATO ASI Series F: Computer and Systems Sciences, Vol. 20 (Springer-Verlag, 1986).

[3] R. Brower, R. Giles, and G. Vichniac, *to appear.*

[4] M. A. Cohen and S. Grossberg, *Trans. IEEE,* SMC-13 (1983) 815–826.

[5] J. Demongeot, E. Goles, and M. Tchuente (eds), *Dynamical Systems and Cellular Automata,* (Academic Press, 1985).

[6] M. Drumheller and T. Poggio, "On Parallel Stereo," *Proc. 1986 IEEE Int. Conf. Robotics and Automation, San Francisco, CA,* 1439–1448 (1986).

[7] R. Divko and K. Schulten, "Feature Spins: Stochastic Spin Models for Pattern Recognition Problems," *these Proceedings.*

[8] D. Farmer, T. Toffoli, and S. Wolfram (eds), *Cellular Automata* (North-Holland, 1984).

[9] E. Goles, *SIAM J. Disc. Alg. Meth.* 3 (1982) 529–531.

[10] E. Goles, *SIAM J. Disc. Alg. Meth.* 6 (1985) 749–754.

[11] E. Goles and S. Martinez, *R. R.,* Math. Dept., U. Chile (1985).

[12] E. Goles, *Comportement dynamique de réseaux d'automates,* Thèse, Grenoble, 1985.

[13] E. Goles, "Positive Automata Networks," ref. [2], 101–112.

[14] E. Goles, F. Fogelman, and D. Pellegrin, *Disc. Appl. Math.* 12 (1985) 261–277.

[15] B. Hayes, *Scientific American,* 250:3 (March 1984) 12–21.

[16] J. J. Hopfield, *Proc. Natl. Acad. Sci. USA,* 79 (1979) 2554–2558.

[17] W. A. Little, *Math. Biosci.* 19 (1974) 101–120.

[18] W. A. Little and G. L. Shaw, *Math. Biosci.* 39 (1978) 281–290.

[19] N. Margolus, T. Toffoli, and G. Vichniac, *Phys. Rev. Lett.* 56 (1986) 1694–1696.

[20] D. Marr and T. Poggio, *Science* 194 (1976) 283–287.

[21] D. Marr, G. Palm, and T. Poggio, *Biol. Cybernetics* 28 (1978) 223–239.

[22] J. L. Marroquin, *Probalistic Solutions of Inverse Problems,* Ph.D. Thesis, Massachusetts Institute of Technology (1985).

[23] W. McCulloch and W. Pitts, *Bull. Math. Biophys.* 5 (1943).

[24] P. Peretto, *Biol. Cybern.* 50 (1984) 51–62.

[25] T. Poggio, *Scientific American,* 250:4 (April 1984) 106–116.

[26] J. Pollack and D. L. Waltz, *BYTE,* 11:2 (February 1986) 189–198.

[27] T. Toffoli, *Physica* 10D (1984) 195–204; reprinted in ref. [8].

[28] G. Y. Vichniac, *Physica* 10D (1984) 96–115; reprinted in ref. [8].

[29] G. Y. Vichniac, "Cellular Automata Models of Disorder and Organization," ref. [2], 3–20.

VLSI implementation of a neural network memory
with several hundreds of neurons

H.P. Graf, L.D. Jackel, R.E. Howard, B. Straughn,
J.S. Denker, W. Hubbard, D.M. Tennant, and D. Schwartz
AT&T Bell Laboratories, Holmdel, NJ 07733, USA

ABSTRACT

We designed an Electronic Neural Network (ENN) memory with 256 neurons on a single chip using a combination of analog and digital VLSI technology plus a custom microfabrication process. Amplifiers with inverting and noninverting outputs are used for the neurons to make inhibitory and excitatory connections. The connections between the individual neurons are provided by amorphous-silicon resistors which are placed on the CMOS chip in the last fabrication step. This technique allows a very dense packing of the neurons. Electron-beam direct-writing is used to pattern the resistors making it easy to change the information stored in the network from one chip to the next.

INTRODUCTION

Associative ENN memories[1,2,3] combine both data storage and data processing functions making them attractive modules for parallel and distributed data processing. An input vector is compared in parallel to all the vectors stored in the memory and the closest match is determined. Since the content of the memory does not have to be transferred to a processor to perform this task, data buses and processors in a system are free to do other operations.

The ENN consists of a fully interconnected array of amplifiers, i.e. each amplifier has to be connected to every other amplifier with the proper connection strength. This is very difficult to realize when the number of amplifiers is large due to the large number of wires and interconnections needed. An electronic implementation with discrete devices becomes bulky even with a modest number of amplifiers. Only with VLSI technology can one hope to make an ENN feasible for applications. In the design of our chip emphasis was placed upon putting a large number of neurons on a chip and therefore all the components were reduced in area as much as possible. The purpose of the chip is to evaluate the behavior and the capabilities of a large ENN, especially when it is integrated into a system with other processors. Electronic implementations of ENN memories reported so far [4] have had at most a few dozen neurons.

We combine analog and digital technology to get a dense packing of the ENN on the chip. The connections between the neurons are made through resistors and current summing is used to add together all the contributions to the input of an amplifier. In this way the computationally intensive task of forming the weighted sums of all outputs can be done very effectively in a small area. Special attention was paid to reduce the size of the resistors since there are over a hundred thousand of them in the circuit. For a circuit of this size, resistor values of a few megohms are necessary to keep the power consumption of the chip low. A standard CMOS fabrication process does not provide resistors in this range with a size small enough to be useful. We therefore developed a microfabrication process to add high value amorphous-silicon resistors to an otherwise finished CMOS chip. This approach does not allow changing the resistors once the fabrication is finished, however, for the present purpose this is of minor importance. Electron-beam direct-writing used to pattern the resistors makes it easy to change the distribution of the resistors from one chip to the next. All resistors are made to have the same value, i.e., there is only one connection strength between the neurons.

DESIGN OF THE CHIP

The chip is designed in CMOS technology with 2.5μm design rules and contains the amplifier units, the wiring to interconnect the amplifiers plus the I/O control (see Fig. 1). In the center of the chip the output lines and the input lines of the amplifiers form a matrix providing the sites where the resistors can be placed. Modules containing the amplifiers are arranged adjacent to the matrix on all four sides. Each amplifier consists of two simple inverters connected in series plus switches to turn them on or off. Two inverters are used per amplifier to make inhibitory and excitatory connections between the amplifiers. There are 512 of these amplifier units placed on the chip.

The circuit handles several hundred bits in parallel during a run, a number impractically large to bring on and off the chip in parallel. Fig. 2 shows a part of the circuit used to multiplex/demultiplex the data for communication with the outside world. The data are brought in over a 16-bit wide bus and stored in a buffer. When the buffer is filled with the new input data a signal is given that enables the gates, initializing the circuit. All the amplifiers are turned off and their input lines are charged to one of three voltage levels. Once the circuit is initialized, the amplifiers are turned on and the whole circuit settles down freely to a stable state. Then the output voltage of each amplifier

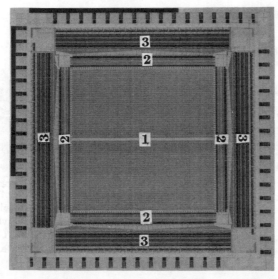

Fig 1. Photograph of the chip.
The different modules are:
1 = Interconnection matrix
2 = Amplifiers
3 = I/O Control

184

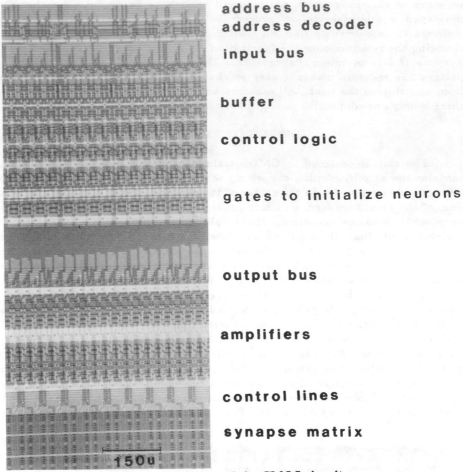

address bus
address decoder

input bus

buffer

control logic

gates to initialize neurons

output bus

amplifiers

control lines

synapse matrix

150u

Fig 2. A magnified view of the CMOS circuit.
This section contains twenty amplifiers plus 17 I/O units.

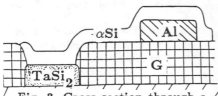

αSi   Al

TaSi₂   G

Fig. 3. Cross section through a
resistor showing the tantalum
silicide line (TaSi$_2$), the
aluminum line (Al), the
insulating glass layer (G), and
the amorphous-silicon resistor
(αSi).

Fig. 4. Electron microscope image of a
section of the interconnection matrix.
Running horizontally are the
aluminum lines and vertically the
silicide lines. The dark rectangles are
the resistors.

is read out over a 16-bit wide output bus. Each data bus actually consists of four individual buses running along only one side of the matrix. This reduces their width to four wires and minimizes propagation delays.

The design is optimized for easy testing of all active devices independent of whether there are resistors in the matrix or not. The different modules are arranged on the chip to minimize the wiring length. This is a crucial aspect in a system consisting of so many individual units. The chip contains roughly 25,000 transistors and has over 130,000 sites for resistors in the matrix. It occupies an area of 5.7mm x 5.7mm.

There are no resistors on the chips when they are received from the silicon foundry. Open via holes in the matrix provide access to the silicide lines where resistors may be placed to connect between the crossing aluminum and tantalum silicide lines (see Fig. 3). To finish the chip, it is covered with amorphous Silicon and the resistors are patterned by electron-beam lithography and reactive-ion-etching. Fig. 4 shows an electron micrograph of the feedback matrix with a few resistors in place.

The CMOS portion of the chip has been tested and determined to function properly. Fabrication tests for the resistors were made and first patterns of resistors fabricated. The amorphous-silicon resistors show good linearity and the variation of their value across a chip is less than 5%.

Although there are 512 amplifiers on the chip the minimum line width available from the silicon foundry is too large to interconnect more than 256. We have developed a fabrication process to make ultra-high density connection matrices where the resistors are stacked vertically between the crossing wires[6]. This technique allows us, to connect all 512 amplifiers. The chips used with this technology are identical to the ones described above except that in the center an open space is left so that the matrix may be added later. The 512 neurons on this chip are by no means the limit for our technology. Connection matrices with as many as 4 resistors per square micron have already been made[6]. Using 1.25μm design rules it will be possible to pack more than a thousand neurons on a single chip of the the same size with essentially the same design as described here.

DYNAMICS OF THE CIRCUIT

To study the dynamics of the circuit a series of simulations were done with a circuit simulator (ADVICE) and a timing simulator (EMU). Most of the simulations used a circuit with only 22 neurons but taking the parameters of the circuit with 256 neurons (e.g., line capacitances). This reduces the simulation time drastically compared to that of a circuit with 256 neurons, which is on the order of a day for a single run. Moreover, the first resistor pattern produced in the matrix consists of a series of 22x22 matrices in order that the results obtained from the chip may be compared with those from our test station [5]. Fig.5 shows an example of a simulation for a circuit programmed to have 4 stable states. Changes of the amplifier output voltages are observed for about 300 ns. After this period no further changes of logic levels occur and therefore it can be defined as the convergence or relaxation time. This convergence time depends on the input vector as well as on several circuit parameters. For 100

# of neuron

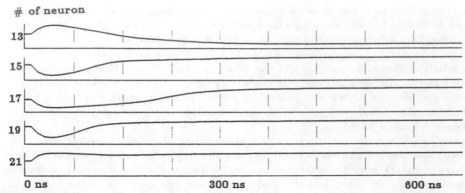

Fig. 5. Result of a simulation of a circuit with 22 neurons. Output voltages of 5 amplifiers as a function of time are shown. The values of the resistors are 200 kΩ. The outputs vary between 0 and 4 volts.

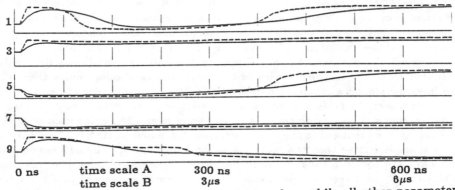

Fig. 6. Two simulations with different resistor values while all other parameters are the same. Solid lines: R = 200 kΩ, (time scale A); Dashed lines: R = 2 MΩ, (time scale B).

random input vectors the observed convergence times were between 200ns and 1400ns, with most results around 400ns. The most critical circuit parameters are the values of the resistors and the capacitances of the wires. Simulations of circuits with 200kΩ and 2MΩ resistors show qualitatively the same voltage curves, but the circuit with the 2MΩ resistors takes about 10 times longer to converge than the one with the 200kΩ resistors (Fig.6). A few simulations of the full circuit with 256 neurons were made with resistor values of 2MΩ. These resulted in convergence times around 700ns.

DISCUSSION

The chip is intended to be used as a specialized processor in a system such as an image processor where the ENN is responsible for extracting features from a picture. The ENN does processing in a totally asynchronous and analog way and special care has to be taken for its behavior to be compatible with a clocked, digital environment. For an application like image processing, it is desirable that the computation the ENN does is completed within a well defined time. The simulations indicate that this might be a problem since the convergence times of the network show quite a large scatter. Applying a time varying bias to the inputs of the amplifiers could improve this. Cycling the bias over a small voltage range prevents the system from remaining stuck on a saddle point of the energy function.

An overall speed-up of the ENN can be achieved in several ways. To a certain limit the resistor values can be decreased to decrease the convergence time. Moreover, it is not necessary to wait until the ENN has fully converged, the result can often be recognized long before the circuit has settled down. This procedure can be improved if a key is connected to every stored vector[2].

Since the ENN handles large bit strings in parallel it is important that the I/O is optimized for specific applications. Otherwise the performance of the ENN will be limited by the time it takes to load and unload the vectors. When a continuous stream of data is to be analyzed, the use of a shift register which transfers the data along the neurons may be the method of choice. In this way, a new input vector can be ready at every clock cycle.

The circuit implemented has a fixed pattern of resistors and therefore the stable states are frozen in once the fabrication is done. However, some variations of the stable states are possible by adding biases. Adding a series of biases with controllable strengths is much easier than making all the resistors switchable. This gives at least some of the flexibility one has with programmable synapses without losing the high packing density of the fixed resistors.

These remarks offer a few ideas for "tuning" the ENN for applications. First however, more knowledge about their behavior is needed before we can clearly optimize them. To explore the possibilities of an ENN with programmable synapses, we recently designed a circuit with 54 neurons where every resistor can be turned on or off by software control. This increases the flexibility of the circuit considerably at the price of a drastically reduced number of neurons.

REFERENCES

1. J.J. Hopfield, Proc. Natl. Acad. Sci. USA, *81* , 3088 (1984).
   Our circuit corresponds to the one described in this reference. Associative Memories have also been discussed extensively by the authors of Ref.2 and Ref.3; see references there.
2. T. Kohonen, Self-Organization and Associative Memory (Springer, New York, 1984).
3. J.A. Anderson, in "Disordered Systems and Biological Organization", E. Bienenstock, F. Fogelman, and G. Weisbuch (Eds.), Springer Verlag, 1986.
4. M. Silviotti, M. Emerling, and C. Mead, Proc. Chapel Hill Conf. VLSI, 1985.
5. W. Hubbard, D. Schwartz, J.S. Denker, H.P. Graf, R.E. Howard, L.D. Jackel, B. Straughn, and D.M. Tennant, this volume.
6. L.D. Jackel, R.E. Howard, H.P. Graf, B. Straughn, and J.S. Denker, J. Vac. Sci.Technol. *B4* , 61 (1986).

A DYNAMIC MODEL OF OLFACTORY DISCRIMINATION

K. A. Grajski* and W. J. Freeman**
Group in Biophysics and Department of Physiology
University of California, Berkeley, CA 94720

ABSTRACT

The organization of the olfactory system represents a biologically adapted solution to a formidable problem. On repeated exposures to a significant odorant, differing subsets of sensitive receptors are stimulated, yet the odor is reliably identified. A hypothesis emphasizing collective neural dynamics is described which predicts that such odor recognition is based on learning dependent changes in olfactory neural networks. Experimental data in the form of EEG spatial patterns obtained from behaving rabbits provides evidence in support of the hypothesis.

INTRODUCTION

The vertebrate olfactory epithelium (100–200 μm thick) covers roughly 100 cm^2 and contains 10^6–10^8 receptor cells. Each receptor cell has a broad response characteristic to odorants.[1] Odorant information is conveyed topographically to the main olfactory bulb (OB) via the Primary Olfactory Nerve (PON) as spatial patterns of spike trains generated by the noninteracting receptor cells. The PON fibers converge onto 10^3 glomerulii in the OB Glomerular layer. Within each glomerulus (80–150 μm in diameter) 10^3–10^5 PON fibers synapse with the apical dendrites of 10^2–10^3 mitral and tufted cells. The mitral and tufted cells form reciprocal synapses with 10^3–10^4 deep lying granule cells. Mitral-granule cell interactions are mediated by a sigmoidal gain function (Eeckman, this volume). Under synaptic driving, the non-spiking granule cells generate electric dipoles with common instantaneous orientation perpendicular to the bulbar surface. These field potentials form the predominant component of the EEG activity recorded at the bulbar surface.[2] Mitral cell axons converge to form the Lateral Olfactory Tract which projects topographically to the Anterior Olfactory Nucleus, but nontopographically to the Pre-Piriform Cortex. These structures project back to the bulb in a similar fashion.

The glomerular layer spatially coarse grains PON input into 10^3 columns, separated by 0.25 mm. We model a bulbar column as the complex of cells driven synaptically under one glomerulus. Lateral excitatory interactions support co-active cell populations across columns. Under a nor-adrenergic process, co-active synapses are strengthened thus forming a neural assembly. This assembly provides a basis for generalization over receptor input, odor discrimination and behavioral responding. The model has been presented in detail

* KAG supported by NIH Pre-Doctoral Training Grant.
** WJF supported by NIMH Contract No. MH06686.

elsewhere,[4,5] (Baird, this volume.)

We test our model by sampling collective neural dynamics of the olfactory bulb. We use an 8 by 8 electrode array (4.0 by 4.0 mm) implanted directly on the lateral surface of the bulb in rabbits. Volume conduction smooths and attenuates (10-fold per 0.5c/mm) high spatial frequencies of EEG at the surface. With a center-center electrode distance of 0.5 mm, each electrode measures the activity of a neighborhood of 4-8 columns. Data consists of spatial patterns of EEG activity from 10-20% of total bulbar surface area. The spatial EEG data transform the rabbits' olfactory discrimination problem into our statistical pattern recognition problem.

STATISTICAL ANALYSIS

We focus on the 40-80 Hz bursting EEG, riding on a "slow-wave" accompanying respiration. Within each 75-100 msec long burst, individual channel series are smoothed, detrended and set to zero mean. Figure 1 shows such a burst sampled on sixty-four channels. Bad channels are replaced with an average of two neighbors. For the ensemble average time series, a Fourier transform yields a peak frequency and phase. These are used as initial guesses in a non-linear fit of an amplitude and frequency modulated cosine wave to the ensemble average. The modulations are set as linear functions about the burst center time. This procedure is repeated recursively for the next four residuals. Thusly, the ensemble average time series is decomposed into five components. The sum of these five components is fitted by linear regression to each of the sixty-four channels to yield five eight by eight amplitude patterns. A spatial filter and deconvolution are applied to each amplitude pattern to compensate for volume conduction. Goodness of fit is measured by the percentage of total spectral energy captured by each component. The component with highest percentage, generally the first, is termed the dominant component. Spatial amplitude patterns of the dominant component form the basis for statistical pattern analysis. Classification is by linear Discriminant Analysis (DA) and Classification Trees[6] (CART) of Factor scores. Utilizing data resampling techniques, CART provides upper and lower bounds for classification performance. Factor scores are obtained from a Factor Analysis (FA) of pooled data which reduces its dimensionality from sixty-four to between six and ten. On average, 90-95% of variance is recovered.

RESULTS

Burst Analysis

Analysis of 10^4 bursts from 30 rabbits provides evidence for the following. Firstly, the burst is a unitary event involving the entire bulb. During a burst, local regions oscillate at different amplitudes, but at the same common frequency. No reproducible phase relationship occurs across bursts.[2] Secondly, across bursts, there is a common amplitude, or "signature" pattern unique to each animal. For robust classification, this pattern is removed by a combination

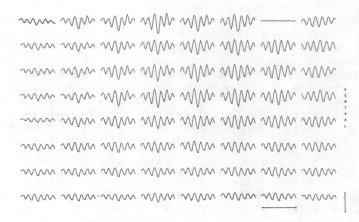

Fig. 1. A typical spatial sample of bursting EEG recorded from the rabbit olfactory bulb (Time is 76 msec. Voltage is 100 μV).

of spatial filtering and normalization by channel to zero mean and unit standard deviation. Thirdly, bursts fall into two classes: orderly and disorderly. The dominant component of orderly bursts contains a sharp spectral peak in the 55-85 Hz range, 50% of total spectral energy, less than 25% frequency modulation and stable spatial pattern, (within group, Fisher Z-transformed correlation is 0.92). Disorderly bursts have a broad distribution of spectral energy across components, frequency modulation in excess of 25% and unstable spatial pattern (within group, Fisher Z-transformed correlation is 0.12).

For burst amplitude pattern classification, the orderly bursts are selected. The disorderly bursts, however, are not merely statistical outliers. It has been suggested that they manifest a failure of convergence to a stable spatial mode of oscillation in the bulb[4,5] (Baird, this volume).

Behavioral Experiments

Novel Odors

Our model predicts that no novel odor-specific spatial patterns should exist in the bulbar EEG signal. Six rabbits were presented with ten trials each of three novel odors and ten air trials, randomly interspersed at one to two minute intervals. On average, the rabbits sniffed on nine of ten odor and on three of ten air trials. There was no evidence for Habituation to any of these odors during the session. Up to twenty-four bursts per six-second trial were selected for analysis to yield a data base of 2700 bursts. If novel odor-specific patterns emerge in the bulb, then it should be possible to classify each burst as belonging to an air class, or one of the three odor classes.

In this data base, commonality of burst waveform, presence of a

signature pattern and orderly and disorderly bursts are observed.
Disorderly bursts occur on average for 25% of air and 33% of odor
bursts. Table I lists the results of DA of Factor scores for the
three odor groups. The average percent correct classification
across subjects is 42.7%; random classification gives 33%. Odor
bursts are also pooled into one class and discriminated against air
bursts. Across subjects, percent correct classification for air and
odor bursts is 67.1% and 47.1%, respectively. Thus, while there is
some differentiation of air from pooled odor bursts, no odor-
specific burst amplitude patterns are found.

Table I Novel Odor Burst Classification (% Correct)

Subj.	Odor A	Odor B	Odor C	N	Air	Pooled Odor	N
1	31.	20.	50.	254	46.	72.	525
2	18.	58.	66.	62	65.	56.	133
3	25.	48.	78.	60	91.	22.	164
4	21.	45.	55.	105	74.	36.	221
5	36.	34.	67.	176	79.	31.	365
6	38.	43.	32.	241	48.	65.	497
Mean	28.	41.	48.		67.	47.	
S.E.	3.	5.	7.		7.	8.	

Differential Conditioning

Our model further predicts that with learning, odor-specific
patterns will emerge and stabilize in the spatial patterns of
bursting EEG. Five thirsty rabbits were appetitively conditioned to
respond by licking to a CS+ odor; a CS- odor served as a discrimina-
tive control. Three CS+, CS- combinations were studied, each such
stage spanning six sessions. (Stage I: A+,B-; Stage II: C+,B-;
Stage III: C+,A-). Using a standard six-second trial, air, CS+, CS-
odor bursts were recorded concurrently with behavioral responding.[7]
For classification, bursts from the last three sessions within
each stage are pooled. Typical results are those from Stage I
sessions four through six. Only the dominant components of orderly
bursts from trials with "correct" responding are included. Burst
dominant components are subjected to dimensionality reduction by a
Factor analysis and expressed as six-dimensional data vectors.
These reduced patterns are classified using the CART methodology.
Table II lists the upper and lower bound estimates of correct
classification for two distinct runs. In the first, classification
of air, CS+ and CS- concurrently, averages across subjects to a
lower bound (Cross-Validated) of 47% and an upper bound (Learning)
of 80% correct; random classification gives 33% correct. In the
second, differentiating air from pooled odor bursts gives a classification range from 65 to 80 percent correct.

Table II Conditioned Odor Bursts
CART Learning and Cross-Validated % Correct
Classification

Subject	(Air, CS+, CS−)		(Air, Pooled Odor)	
	Learning	Cross-V.	Learning	Cross-V.
1	83.	44.	84.	58.
2	75.	47.	87.	65.
3	88.	51.	75.	72.
4	70.	48.	72.	64.
5	67.	60.	80.	65.
Mean	77.	50.	80.	65.
S.E.	4.	3.	3.	2.

To determine how these patterns change with new learning, the correlation between factor loadings in each session with the other seventeen is constructed. The within-stage loading correlations are higher than between stages. Changes in loadings occur between stages, but the amount of difference is small. Within-stage values average 0.703, compared with 0.645 between stages.[5]

DISCUSSION

The olfactory bulb is to first order a massively parallel, interconnected network of identical units, with globally synchronized oscillatory dynamics. A generalization of memory and computational properties[8,9] in such systems provides a model mechanism for olfactory discrimination (Baird, this volume). Through spatial analysis of EEG we have sought experimental evidence for the operations of such a system in the olfactory bulb. Two experiments show that: 1. stable odor-specific patterns do not emerge for novel odors, 2. under associative learning, stable, co-existing, odor-specific burst amplitude patterns emerge with learning, persist for weeks and reorganize with new learning.

The present findings are a necessary step in a program to define how collective neural dynamics underly vertebrate olfactory discrimination. Further studies are needed on the relationship between EEG and unit activity. Further studies are also needed on bulbar projection sites such as the anterior olfactory nucleus and pre-piriform cortex. These structures share bulbar structural and dynamical properties, perform spatio-temporal integration on their input, feedback to one another and undergo changes in EEG activity patterns with learning.[10] Lastly, mathematical analysis is needed to more rigorously define neural dynamics in model systems and state-space reconstructions from experimental data.

REFERENCES

[1]D. Lancet, Ann. Rev. Neurosci. 9, 381 (1986).
[2]W. J. Freeman and W. Baird, Beh. Neurosci. to appear (1986).

[3]C. Gray, et al., Beh. Neurosci. to appear (1986).
[4]W. J. Freeman and C. Skarda, Brain Res. Rev. 10, 147 (1986).
[5]W. J. Freeman and K. A. Grajski, Beh. Neurosci. to appear (1986).
[6]K. A. Grajski, et al., IEEE Trans. Biomed. Eng. to appear (1986).
[7]G. Viana di Prisco and W. J. Freeman, Beh. Neurosci. 99, 964 (1986).
[8]J. J. Hopfield, PNAS. 79, 2554 (1986).
[9]T. Kohonen, Associative Memory (Springer-Verlag, Berlin, 1977).
[10]S. Bressler, Ph.D. Thesis in Physiology, UC Berkeley (1982).

A COMPARISON OF NEURAL NETWORK AND MATCHED FILTER PROCESSING FOR DETECTING LINES IN IMAGES[+]

P. M. Grant[*] and J. P. Sage
Lincoln Laboratory, Massachusetts Institute of Technology
Lexington, MA 02173.

ABSTRACT

This paper compares the performance of matched filters with several versions of nonlinear associative memory for a specific pattern recognition problem, where we attempt to identify the presence and orientation of an angled line. Comparative simulations are presented on the above processors for 5-by-5 and 9-by-9 pixel data fields when the input line patterns are corrupted by Gaussian noise. We concluded that considerable sophistication was required in the nonlinear associative memory processor design to achieve equivalent performance to the matched filter. In all cases the matched filter was shown to exhibit a reduced computational load.

INTRODUCTION

We have been studying a restricted pattern recognition problem where we attempt to detect the presence or absence of line patterns in an optical image which is corrupted by Gaussian additive noise. We have been particularly interested in examining the applicability of the recently reported artificial-neural-network-based information processing techniques. We have carried out computer simulations to compare the probabilities of false alarm and missed detection using conventional two-dimensional matched filtering or template matching techniques [1] with several versions of nonlinear associative memory [2]. For our analysis we assumed that the object giving rise to the observed image is known to be one of a limited set of ideal patterns, which in this example are either line segments oriented at specific angles or a uniform field without the presence of any pattern. The image of the object, however, is corrupted by the addition of Gaussian noise.

ASSOCIATIVE MEMORY FOR PATTERN RECOGNITION

For associative memory processing the two-dimensional image is mapped (for example, using a raster scan) into a one-dimensional "state" vector of pixel intensity values. The most prominent of the artificial neural network techniques in the engineering literature is the linear associative or correlation matrix memory [3,4]. In this memory the analog input state vector is multiplied by a matrix built from information related to the patterns to be recognized. After a single operation, the resulting output is a more accurate version of the true input pattern, i.e., one in which the noise has been reduced.

Hopfield's nonlinear associative memory [2], Fig. 1, differs in several respects. First, the input state vector has binary quantized components; second, the output from the linear multiplication by the memory matrix is binary quantized; and third, the

---

+ This work is supported by the Department of the Air Force.
* Permanent address: Department of Electrical Engineering, University of Edinburgh.

output vector after each step is treated as an input vector and reprocessed by the memory matrix until a stable state is reached (Fig. 1). Ideally this stable output pattern would correspond exactly to the true input pattern with the noise removed. The basic step in this iterative process is represented mathematically by the equation

$$V^{out} = sign \ (\ T \cdot V^{in} \ ) \tag{1}$$

where V^{in} and V^{out} are the input and output state vectors, T is the association matrix, and the sign operator indicates that each component in the output vector is replaced by either +1 or -1 depending on its sign.

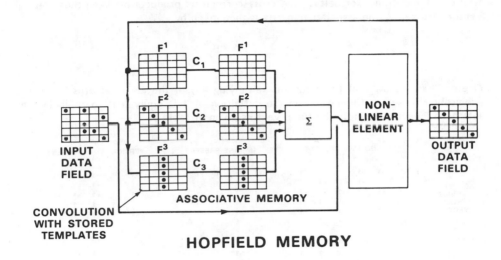

HOPFIELD MEMORY

Figure 1

ASSOCIATIVE MEMORY MATRIX CONSTRUCTION

Our initial simulations used a Hebbian memory matrix constructed as described by Hopfield [2] directly from the M ideal state vectors to be recognized, $F^1...F^M$, each vector having N components. The N-by-N association matrix T has coefficients given by

$$T_{ij} = \sum_k (\ F_i^k \cdot F_j^k \ ) \tag{2}$$

Thus the matrix is the sum of the M outer products of each pattern with itself. Three 25-sample pattern vectors F -- comprising a field with no object, a line oriented at 45 degrees, and a line oriented vertically -- are shown as 5-by-5 arrays in Table 1 row (a). These vectors give rise to a 25-by-25-coefficient T matrix. Note that our T matrix is multivalued and not binary quantized as in some implementations. The simulated pattern recognition error rates using this matrix were found to be very much higher than those obtained from a two-dimensional matched filter bank processor implementing the maximum-likelihood detection criterion [1]. In the latter, the ideal pattern with the largest filter output, i.e., C_k value in Fig. 1, is directly selected without implementing the subsequent processing. Analysis showed that the high error rate of the associative memory results from the significant cross-correlation coefficients between pairs of different ideal patterns. We next investigated a memory matrix built using a modified technique that guarantees perfect recognition of perfect input patterns. Instead of building the association matrix entirely from the ideal state vectors F, we used an additional set of basis vectors B that obey a mutual orthonormality relationship to the F vectors

$$B^p \cdot F^q = \delta_{pq} \tag{3}$$

where δ is the Kronecker delta. The cross-correlation products between these sets of vectors are zero. The association matrix is then given by

$$T_{ij} = \sum_k (F_i^k \cdot B_j^k) \tag{4}$$

One can easily show that the result of multiplying any of the ideal patterns F^k by T is precisely F^k itself. This change in the definition of T is equivalent to replacing the left hand F vectors in Fig. 1 by the B vectors given in Table 1.

	NO LINE PRESENT F^1					45° LINE PRESENT F^2					VERTICAL LINE PRESENT F^3				
(a) INPUT STATE VECTORS	-1	-1	-1	-1	-1	+1	-1	-1	-1	-1	-1	-1	+1	-1	-1
	-1	-1	-1	-1	-1	-1	+1	-1	-1	-1	-1	-1	+1	-1	-1
	-1	-1	-1	-1	-1	-1	-1	+1	-1	-1	-1	-1	+1	-1	-1
	-1	-1	-1	-1	-1	-1	-1	-1	+1	-1	-1	-1	+1	-1	-1
	-1	-1	-1	-1	-1	-1	-1	-1	-1	+1	-1	-1	+1	-1	-1
	B^1					B^2					B^3				
(b) MUTUALLY ORTHOGONAL BASIS VECTORS	0	-1	0	-1	-1	+4	-1	-1	-1	-1	-1	-1	+4	-1	-1
	-1	0	0	-1	-1	-1	+4	-1	-1	-1	-1	-1	+4	-1	-1
	-1	-1	-16	-1	-1	-1	-1	+4	-1	-1	-1	-1	+4	-1	-1
	-1	-1	0	0	-1	-1	-1	-1	+4	-1	-1	-1	+4	-1	-1
	-1	-1	0	-1	0	-1	-1	-1	-1	+4	-1	-1	+4	-1	-1
	B^1					B^2					B^3				
(c) OPTIMUM ORTHOGONAL BASIS VECTORS	-3.4	-0.4	-3.4	-0.4	-0.4	+4	-1	-1	-1	-1	-1	-1	+4	-1	-1
	-0.4	-3.4	-3.4	-0.4	-0.4	-1	+4	-1	-1	-1	-1	-1	+4	-1	-1
	-0.4	-0.4	-6.4	-0.4	-0.4	-1	-1	+4	-1	-1	-1	-1	+4	-1	-1
	-0.4	-0.4	-3.4	-3.4	-0.4	-1	-1	-1	+4	-1	-1	-1	+4	-1	-1
	-0.4	-0.4	-3.4	-0.4	-3.4	-1	-1	-1	-1	+4	-1	-1	+4	-1	-1

5-BY-5 INPUT STATE AND ORTHOGONAL BASIS VECTORS FOR THE NO-LINE FIELD AND TWO-LINE PATTERNS ORIENTED AT 45° AND 0° TO THE VERTICAL.

Table 1

Since the number M of F vectors is less than their dimensionality N, the B vectors are not determined uniquely. We initially chose an arbitrary set of B vectors by inspection (Table 1, row (b)). Although any set of B vectors satisfying Eq. 3 leads to a T matrix that will regenerate an ideal state vector F exactly, T matrices constructed from different sets of B vectors will give different results when the input vectors deviate from an ideal F state. Analysis of the ratio of average rms deviation in the output state to average rms deviation in the input state revealed that the optimum performance with this type of association matrix is achieved with B vectors chosen to minimize the sum of the squares of the elements in the **T** matrix. Although such optimum B vectors were found initially by symmetry, we later derived the following general expression using methods of linear algebra:

$$\mathbf{B} = \mathbf{F} \cdot \mathbf{C}^{-1} \qquad (5)$$

In this expression $\mathbf{C} = \mathbf{F}^T \cdot \mathbf{F}$ is the M-by-M correlation matrix for the F vectors, where the M ideal patterns are represented as a single two-dimensional matrix **F** comprising M columns and N rows. The inverse of the **C** matrix will exist provided the F vectors are linearly independent. This solution turns out to be the optimal linear association matrix [3]. A more general solution using Moore-Penrose matrix inversion treats the case where the F vectors are linearly dependent. This optimum set of B vectors (multiplied by a scale factor) is shown in Table 1 row (c). With this choice of association matrix, significantly better results were obtained.

COMPARATIVE SIMULATIONS

Simulations were run at first on 5-by-5 pixel fields and later on 9-by-9 fields using ideal patterns represented by values of -1 and +1. The initial input states for the simulations were generated by taking one of these binary ideal patterns and adding analog, Gaussian-distributed noise to each pixel value. In some of the simulations this analog state vector was quantized to values of -1 and +1 before submitting it to the processor; in other cases the analog vectors were used. The baseline simulation used matched-filter maximum-likelihood detection on full analog data. This technique provides the theoretically most accurate decision.

Simulations were performed on 9-by-9 pixel fields at an input per-pixel signal-to-noise ratio (SNR) of 3.5 dB with up to eight different line orientations in addition to the no-line state. Figure 2 summarizes the results by plotting the probability of detection against the number of ideal patterns for several types of processor. This shows separately the results for a noisy no-line input pattern and a noisy line input pattern. The types of processor are: (a) the Hopfield processor, with the **T** matrix of Eq. 2; (b) the Hopfield processor, with the **T** matrix of Eqs. 4 and 5, and using fully quantized data throughout; (c) the processor of (b) but with analog processing on the first pass through the associative memory; and (d) the matched-filter processor. These simulations clearly show that the Hopfield associative memory is inadequate for our problem. The performance with orthonormal basis vectors (curve (b)) is better but only becomes acceptable when the analog data is preserved for the first pass through the processor (curve (c)).

When the line patterns all overlap on the center pixel, the associative memory was found to settle often into a state where the center pixel was in error, significantly reducing the detection probability. One way to overcome this problem, we thought, might be to build the basis vectors from 8-pixel lines where the center pixel was omitted and quantized as -1. Despite the fact that the sum of the squares of T_{ij} was 170% larger with these basis vectors, the resulting detection performance was superior. It is these results that are shown in curves (b) and (c) of Fig. 2.

With the matched-filter processor the only type of error which can occur is recognition of an incorrect pattern. The probability of this happening is a monotonically decreasing function of the Hamming distance between the states. When a line is present, the only likely error with the SNR used in our simulations is misidentification as the no-line pattern, since the distance to the wrong line pattern is much farther. When no line is present, the matched filter error rate is significantly larger than when the line was present, as there is an equal probability of falsely identifying the noisy pattern as any one of the possible line patterns.

The associative memory calculation was found to exhibit a second type of error. Besides settling into wrong states, it can also settle into a spurious state, one that is not any of the ideal states but is a superposition of two of them. In the matched filter processor such ambiguities are resolved by comparing the relative amplitudes of the individual filter outputs ($C_1..C_M$ in Fig. 1). In the associative memory access to this information is not available, and the nonlinear processing can force weak lines to be quantized as valid patterns. This kind of error becomes increasingly significant with the larger number of stored patterns.

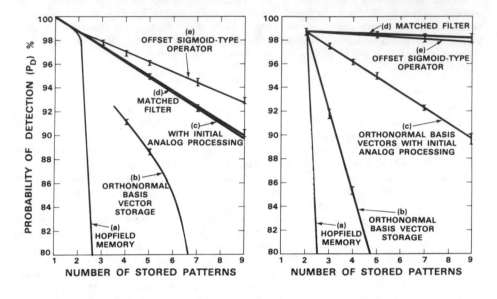

RECOGNITION OF NO LINE STATE
IN 9-BY-9 IMAGES

RECOGNITION OF LINE PATTERNS
IN 9-BY-9 IMAGES

Figure 2

The performance in Fig. 2 was considered to be acceptable for the no-line case, but when detecting noise-corrupted line patterns the simulations showed a progressive increase in the incidence of double-line detection when a larger number of line patterns was stored. We thus concluded that some method was required to reduce the occurrence of dual pattern states in the associative memory processor.

The most elaborate processor we examined in our attempt to raise the performance level of an associative-memory-type processor toward the level of the matched filter was one using a continuous sigmoid-type nonlinear operator whose steepness was increased with successive iterations. For programming convenience we chose an arctan function scaled by $2/\pi$ to provide an output range from -1 to +1. In the 9-by-9 field analysis an offset sigmoid-type nonlinear operator achieved the performance which came closest to the matched filter. The results with sigmoids crossing the abscissa at an offset of +0.05 are shown in Fig. 2 curve (e). The incidence of spurious states was eliminated, but detection was always biased in favor of no-line identification, no matter what threshold offset was chosen. Consequently, the apparent probability of detection for no-line patterns was very high, but was balanced by a precisely compensating loss of detection probability for the line states (since there are more of these states, the loss is smaller for each one than the gain for the no-line state). The results are identical to those that would be obtained using matched filtering with offset thresholds.

COMPUTATIONAL CONSIDERATIONS

The associative memory requires storage of N squared T matrix coefficients rather than the M-by-N matched filter coefficients. It is normally accepted that the associative memory can accommodate a maximum of approximately 0.15N distinct patterns [2]. Thus the required associative memory storage is approximately an order of magnitude larger than the matched filter, and this comparison is worse when only a small number of patterns are stored. The associative memory requires N squared multiplies per pass or iteration to process an N pixel or sample vector where N = 25 or 81 in the above examples. The matched filter requires N multiplies per filter for each of the M patterns to be recognized, which is again an order of magnitude less computation than with the associative memory. When multiple passes are required to resolve the expected 1-5% of difficult cases, the computational load of the associative memory further increases. However, the matched-filter processor does require a comparison operation to identify the state with the highest matched-filter response.

CONCLUSIONS

The associative memory is very sensitive to the choice of pattern data stored in the association matrix. The direct storage of the patterns themselves, as proposed by Hopfield, has been shown to be inadequate for our problem. The storage of orthonormal basis vectors overcomes these problems to some extent, but the associative memory still generates spurious states comprising superimposed patterns, which results in a considerable degradation compared to the template-matching or matched-filter approach. The associative memory performance can be made to approach that of matched-filter detection with sufficient optimization of the stored patterns and sophistication of the nonlinear operator. However, the matched filter techniques, besides offering greater computational efficiency, can handle analog pattern data and filter coefficient values, which are required to process the output from practical optical imagers, where the line pattern intensity values inevitably spread into neighboring pixels. It should be noted, furthermore, that these simulations only calculate the performance for one specific spatial position, assuming that the data are correctly registered with respect to the stored patterns. In a practical system all possible registrations of the data must be examined.

REFERENCES

1. R. O. Duda and P. E. Hart, Pattern Classification and Scene Analysis (Wiley, New York, 1973).
2. J. J. Hopfield, Proc. Natl. Acad. Sci. USA, 79 (Biophysics), 2554 (1982).
3. T. Kohonen, Self-Organization and Associative Memory (Springer-Verlag, Berlin, 1984).
4. T. Kohonen, IEEE Trans. Computer C-21, 353 (1972).

Motion Correspondence
and
Analog Networks

Norberto M. Grzywacz and Alan L. Yuille

Center for Biological Information Processing
and Artificial Intelligence Laboratory
Massachusetts Institute of Technology

Abstract

We model the process of visual motion correspondence by minimizing a cost function using an analog Hopfield-style network. We compare results of our model with psychophysical data.

1. Introduction

This paper has two main themes. The first is to analyze processes in early vision, in this case visual motion, and formulate them mathematically. The second is to define efficient methods to solve these problems.

In recent years many vision problems have been formulated in terms of minimizing cost functions. A classic example is the work on shape from motion by Ullman[1]. For this process the data alone is not sufficient to determine a unique solution. To solve this problem Ullman proposes that the human visual system uses assumptions about the world, such as rigidity of objects, to constraint the solution. These assumptions can be elegantly incorporated into a cost function and the resulting theory accounts for a range of psychophysical experiments. Ullman used methods from linear programming to minimize the cost function.

Ullman[1] used psychophysical data to argue that the process is divided into two stages. The first consists of matching tokens, such as points or straight lines, between different image frames; solving the so called correspondence problem. He proposes a minimal mapping scheme, implemented by minimizing a cost function, to find the correspondence. Once this matching is done the second stage assumes rigidity to recover the structure of the object. Ullman[2] later suggested an alternative method of finding the structure (see also Grzywacz and Hildreth[3]) capable of dealing with non-rigid motion. He again used psychophysics to argue that the three dimensional structure of an object was not perceived immeadiately but developed gradually over time. He proposed that the visual system constructed an internal model of the object, initially flat, which was updated over time by assuming the minimal change of rigidity between sucessive image frames, the so called *incremental rigidity scheme.*

It is natural to ask whether errors are caused by dividing the process into two stages. Both are solved using different assumptions and it is possible that these conflict for some stimuli. It is also interesting to see if rigidity alone is sufficient to solve the correspondence problem. To investigate this we define a cost function that minimizes incremental rigidity and solves the correspondence problem simultaneously. We show that further constraints are usually needed to get the correct answers.

Finding the correct cost function, however, is only half the problem. We need a quick and reliable method to minimize it. If the cost function is convex there exist many fast and reliable methods for finding the global minimum. For non-convex cost functions strategies like simulated annealing will generally find the global minimum, but takes a long time to do so. Ullman[1] used a linear programming method to solve the correspondence problem, but although this always converged correctly it did so very slowly (Ullman, pers comm). Instead of a slow algorithm that always converges it may often be a better strategy to use a fast algorithm that converges most of the time. This suggests trying to implement the problem in terms of neural networks.

Neural, or analog neural-like, networks consist of simple units connected to each other by links consisting of resistors, capacitors and inductances. It seems technically possible to build such networks in the near future. If so they will perform calculations extremely quickly because of their parallel, analog nature. They are also of interest because of their possible biological plausibility. Hopfield and Tank[4] have shown that these networks are capable of calculating good approximate solutions to complex minimization problems, such as the as the Travelling Salesman Problem. Koch, Marroquin and Yuille[5] successfully applied them to the surface interpolation problem of early vision. Our present work (Grzywacz and Yuille[6]) attempts to use such networks to do motion correspondence. We compare different types of constraints for motion matching and propose psychophysical experiments.

2. Theory Implementation

We first investigate using rigidity to solve the correspondence problem and determine the structure simultaneously. In the incremental rigidity scheme

(Ullman[2]) an object with N points is described by a model $(x_i(t), y_i(t), z_i(t))$, for $i = 1, ..., N$. The x, y components are directly observable (assuming orthographic projection) and the z components are deduced. Initially the z components are set to zero. Then for each time frame define $L_{ij}(t)$ by

$$L_{ij}(t) = (x_i(t) - x_j(t))^2 + (y_i(t) - y_j(t))^2 + (z_i(t) - z_j(t))^2. \qquad (2.1)$$

The $z_i(t + \delta t)$ are defined to minimize the change in rigidity ΔR between frames

$$\Delta R = \sum_{i,j}^{N} (L_{ij}(t) - L_{ij}(t + \delta t))^2. \qquad (2.2)$$

We now define a set of binary *correspondence* variables $[V_{ia}]$. If point i in the first frame goes to point a in the second frame then $V_{ia} = 1$, otherwise $V_{ia} = 0$. We can define a matching cost E_R by

$$E_R = \sum_{i,j,a,b}^{N} (L_{ij}(t) - L_{ab}(t + \delta t))^2 V_{ia} V_{jb}. \qquad (2.3)$$

To find the correspondence and structure we minimize E_R with respect to $[z_a(t + \delta t)]$ and $[V_{ia}]$ requiring that all points in the first frame are matched to exactly one point in the second.

In this paper we use a method proposed by Hopfield and Tank[4]. We first define a new array of variables, $[U_{ia}]$. These, are internal variables of the new problem and have a monotonically increasing relationship to V_{ia}:

$$V_{ia} = \frac{1}{1 + e^{-2\lambda U_{ia}}}, \qquad (2.4)$$

where λ is a positive parameter of the problem. Although $-\infty < U_{ia} < \infty$, one can see from Eq. 2.4 that V_{ia} is still bound between 0 and 1. We next define the full energy function to be:

$$
\begin{aligned}
E = \ &\frac{A}{2} \sum_{a=1}^{N} \sum_{i=1}^{N} \sum_{\substack{j=1 \\ j \neq i}}^{N} V_{ia} V_{ja} + \frac{B}{2} \sum_{i=1}^{N} \sum_{a=1}^{N} \sum_{\substack{b=1 \\ b \neq a}}^{N} V_{ia} V_{ib} \\
&+ \frac{C}{2} \left(\sum_{i=1}^{N} \sum_{a=1}^{N} V_{ia} - N \right)^2 + \frac{D}{2} E_R \\
&+ \frac{F}{2\lambda} \sum_{i=1}^{N} \sum_{a=1}^{N} (V_{ia} \log(V_{ia}) + (1 - V_{ia}) \log(1 - V_{ia})),
\end{aligned}
\qquad (2.5)
$$

where A, B, C, D, F are positive parameters of the problem. (We will informally identify each of the terms of the right hand side of Eq. 2.5 by the parameter leading it.). Minimization of the A term forces each feature in the second frame to maintain correspondence with as few features as possible in the first frame, (and vice versa for the B term). Minimization of the C term, forces the total amount of correspondeces to be N. Thus the terms A and C will force N correspondences of strength 1 to be established in such a way that each feature of the first frame will tend to have a correspondence with a feature in the second frame and so that the correspondences will be evenly distributed among the features of the second frame. It follows that the process will tend not to leave any feature unmatched.

The F term is necessary to give a time constant for convergence of the network. Observe that if the U_{ia} variables are updated by the differential equations:

$$\frac{dU_{ia}}{dt} = -\frac{\partial E}{\partial V_{ia}}, \qquad 1 < i < N, \quad 1 < a \le N, \qquad (2.6)$$

then the system will stop in a point of the solution space in which the function E is at one of its minima. To see this observe that because of the monotonicity between U_{ia} and V_{ia} expressed in Eq. 2.4 the update rule, Eq. 2.6, will tend to force V_{ia} to descend down the gradient of E. Note that if λ is large enough the variables V_{ia} will tend to be either 0 or 1 and thus, in spite of the fact the the searching process is in a continuous space, it will tend to force a binary decision to determine whether a correspondence is to be established or not.

We simulated this network on a Symbolics 3600 LISP machine. The results are described in detail in Grzywacz and Yuille[4]. To summarize them: despite extensive experimentation with the parameters the system rarely converged to the correct answer unless given a hint of the correct matches. The system made some interesting mistakes, it would sometimes choose matches which were almost rigid but which corresponded to complicated motion of the object between different frames. This suggested that rigidity alone was not a strong enough constraint and we should introduce another term in the energy function corresponding to smoothness of motion between frames. After some experimentation we fell back on the energy function E_{MM} used by Ullman[1] to solve the correspondence problem.

$$E_{MM} = \frac{M}{2} \sum_{i=1}^{N} \sum_{a=1}^{N} V_{ia}^2 d_{ia}^2. \qquad (2.7)$$

When we added the E_{MM} term to the energy E the network gave consistently good results for a wide range of data Grzywacz and Yuille[4]. It gave a high percent of correct matches for systems of up to twenty points.

To see what contribution the rigidity term made to the matching we removed it and ran the system using the minimal mapping term. The system gave identical results suggesting that the rigidity term was usually uneccessary for matching. A possible exception is at the occluding boundary of an object. Here the order of points can reverse between frames and the minimal mapping scheme gave incorrect results. For small angles and for some values of the parameters the rigidity term obtained the correct matching. We plan psychophysical experiments to investigate this case.

In our simulations using the E_{MM} term we did not try to optimize the parameters A, B, C, F, M and λ in any sense. Instead we found that the asymptotic behavior of the system was the same for a large range of parameter values (few orders of magnitude). Typical values used during the course of this research were $A = B = 50000, C = 500000, F = 1, M = 50$ and $\lambda = 1$, where the distances between features in a given frame ranged from 1 to 10. Finally, we used homogeneous initial conditions for our simulations, i.e.:

$$V_{ia}(t = 0) = \frac{1}{N}, \tag{2.8}$$

3. Some Special Cases

It is interesting to run our network for simple situations and compare the predictions with the experimental results. For most structures the correspondence elements, the V_{ai}, converged to zero or one, correctly predicting matches or non-matches. Figure (1) shows several ambiguous situations. In figure (A1) the network no longer converges to binary values, but instead gives a chance of a half for the two possible matches, (A2) and (A3). The psychophysics experiments give the two possible results, (A2) and (A3), equally often, it never predicts dots splitting.

Situations with different number of dots in different frames require special care. One can require the total number of matches to be the number of dots in either the first or second frame. Suppose we set the number of matches in figure (B) to be one. Then the triangle will split into the two nearest triangles with equal strength of near one half. If instead we make the number of matches equal to three the triangle will go to all three squares, consistent with psychophysics. Now consider a hypothetical situation where the squares are lined up as in figure (C). If three points are matched then the triangle will be matched to all three squares, which seems unlikely.

Other ambiguous situations are discussed in Grzywacz and Yuille[2]. An obvious question is how stable the split matches are if noise is added to the system. We plan to do psychophysical experiments to test this and compare the results with the predictions of our network.

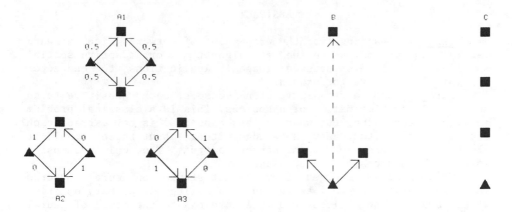

Figure 1. Ambiguous matches. The triangles and squares represent dots in the first and second frames respectively. Figure (A1) shows the output of our network, each point in the first frame has a correspondence of one half to each point in the second frame. Figures (A2) and (A3) shows the two, equally likely, experimental outcomes. In figure (B) the triangle will split to the nearest squares, and may also go to the third square. Figure (c) is a hypothetical situation, will the triangle go to the nearest square, or to all of them?

References

1. S. Ullman, *The Interpretation of Visual Motion* MIT Press, Cambridge. (1979).

2. S. Ullman, Perception. **13**, 255–274. (1984).

3. N. M. Grzywacz and E. C. Hildreth. MIT Artificial Inteligence Memo 845. (1985).

4. J. J. Hopfield and D. W. Tank. Biological Cybernetics., 52, 141. (1985).

5. C. Koch, J. Marroquin and A.L. Yuille. Proceedings of the National Academy of Science. USA. In press. (1986).

6. N. M. Grzywacz and A. L. Yuille. MIT Artificial Intelligence Memo 888. To appear. (1986).

SPACE OR TIME ADAPTIVE SIGNAL PROCESSING BY NEURAL NETWORK MODELS

J.HERAULT, C.JUTTEN

INPG-Lab. TIRF, 46 avenue Félix Viallet   F-38031 GRENOBLE Cedex

ABSTRACT

Part I. Starting from the properties of networks with backward lateral inhibitions, we define an algorithm for adaptive spatial sampling of line-structured images. Applications to character recognition are straightforward.

Part II. Let be an array of n sensors, each sensitive to an unknown linear combination of n sources. This is a classical problem in Signal Processing. But what is less classical is to extract each source signal without any knowledge either about those signals or about their combination in the sensors outputs. The only assumpsion is that the sources are independant.

This problem emerged from recent studies on neural networks where any message appears as an unknown mixing of primary entities which are to be "discovered". According to the model of neural networks, we propose an algorithm based on :

i - a network of fully interconnected processors (like neurons in a small volume of the Central Nervous System).

ii- a law which controls the weights of the interconnections, derived from the Hebb concept for "synaptic plasticity" in Physiology, and very close to the well known "stochastic gradient algorithm" in Signal Processing.

This association results in a permanent selflearning mechanism which leads to a continuously up-dating model of the sensor array information structure.

After convergence, the algorithm provides output signals directly proportionnal to the independent primitive source signals.

PART I. ADAPTIVE SPATIAL SAMPLING BY BACKWARD LATERAL INHIBITION

I.1. PRINCIPLE

Let us consider a 2-dimensional network of neurons with backward lateral inhibitions [1]. We define i/ an input plane corresponding to the sampled input image E(m,n), ii/ a processing plane of neurons facing point to point the input samples, iii/ an output plane corresponding to the outputs S(m,n) of the neurons.

Each input pixel E(m,n) spreads forwards to some neighbour facing neurons with an excitatory weighted function $\alpha(i,j)$. Each output S(m,n) spreads backwards to a greater neighbourhood in the processing plane with an inhibitory weighted function $\beta(k,l)$. Each neuron has a non-linear characteristic as drawn in figure 1.

This architecture is very closed to a 2-D recursive filter of high-order (the number of terms of the function $\beta(k,l)$) according to the difference equation :

$$S(m,n) = F [\sum_{ij} \alpha(i,j).E(m-i,n-j) - \sum_{kl} \beta(k,l).S(m-k,n-1)] \quad (1)$$

with $\qquad F[.] = \text{Min}(S_{max}, \text{Max}([.], 0)).$ (2)

In the linear case, these high-order recursive filters are known to be easely unstable [2]. But, by means of the non-linear function F[.] of the neuron, the oscillations initiated at some pecular pixel of the input image are clipped within the limits of F[.] : only some clusters of neurons give a non-zero response.

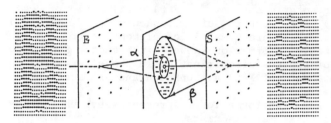

Figure 1 The 2-D network for adaptive character processing. Example of input and output images.

It appears like a new sampling function of the input image with interesting properties :
1. the "samples" are first set at the spatial discontinuities of the input image : ends of lines, sharp angles.
2. The sampling period, roughly determined by the size of the convolution kernel $\beta(k,l)$, adapts to the size of the input shape : for example an increase of 10% of the size leads to the same sampling, but a 15% increase leads to one sample more inside the shape, preserving the place of the samples at the discontinuities.

I.2. APPLICATIONS

A hard-wired processor has been designed on this principles with the following features : 32x128 neurons; a and b kernels are 7x7 pixels wide; F[.] is an unit step function. Thus, by means of hybrid (digital and analog) implementation of the weighted sums, we obtained a very high speed of computation : the more than 400,000 weighted sums are computed within 5ms. The results of the processing performance of this algorithm are shown in figure 2.

Notice the interest of adaptive sampling : i/ the input shape is described with very few samples, ii/ because of property 1, the ambiguities between characters like 5-S, O-D or B-8 are easely resumed. This leads to very simplified structures of classifiers suitable for efficient and fast recognition algorithms.

Figure 2 Examples of the processing capabilities of the network.

PART II. TIME ADAPTIVE SIGNAL PROCESSING BY SYNAPTIC PLASTICITY

II.1 - P R O B L E M

Let be an array of n sensors, each sensitive to an unknown linear combination of n sources $X_j(t)$. The response $E_i(t)$ of sensor i is :
$$E_i(t) = \sum_j a_{ij}.X_j(t) \qquad 1 \le i \le n \qquad (3)$$
with matrix and vector notations : $\quad \underline{E}(t) = [A].\underline{X}(t) \qquad (4)$
where $\underline{E}(t)$ and $\underline{X}(t)$ are vectors of components $E_i(t)$ and $X_i(t)$ respectively, and [A] is a square n-matrix of elements a_{ij}.

At any time t, the only information known is the vector $\underline{E}(t)$. Now we state the problem of sources discrimination in these terms :
Hypothesis 1 :the components of $\underline{X}(t)$ are unknown and are statistically independent sources.
Hypothesis 2 : the linear transformation represented by the matrix [A] is regular : the inverse matrix exists.
Hypothesis 3 : the unknown mixing matrix [A] is supposed to be time independant.
QUESTION : Is it possible, with only the help of observation of $\underline{E}(t)$, i.e. without a priori knowledge either about mixing [A], or about sources $\underline{X}(t)$, to discover the primary signal $\underline{X}(t)$?
ANSWER : There is no classical solution to this problem which in this form is a new problem, more general than the classical sources discrimination problem in Signal Processing [3,4]. We propose an auto-adaptive algorithm to solve it.

II.2 - PRINCIPLES OF SOLUTION

We use a recursive architecture (figure 3) made up of n fully interconnected neuron-like operators : the output $S_i(t)$ of the i th operator is defined by :
$$S_i(t) = E_i(t) - \sum_j C_{ij}.S_j(t) \qquad (5)$$

In vector notation, $\underline{S}(t) = \underline{E}(t) - [C].\underline{S}(t) \qquad (6)$
where [C] is a square n-matrix (C_{ij}) with $C_{ii} = 0$
we deduce : $\qquad [I+C].\underline{S}(t) = \underline{E}(t) , \qquad (7)$
and if we assume : $\qquad \underline{E}(t) = [A].\underline{X}(t) , \qquad (8)$
we get : $\qquad \underline{S}(t) = [I+C]^{-1}.[A].\underline{X}(t) \qquad (9)$
if the matrix $[I+C]^{-1}$ exists.

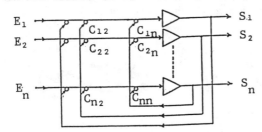

Figure 3. Network of fully interconnected operators. Every triangle N_i is an operator the output of which is S_i . The horizontal line on the left of operators is an addition line, which gives :
$$E_i - \sum_j C_{ij}.S_j$$

To solve the discrimination problem, the outputs $S_i(t)$ must be equal or at most proportional to $X_k(t)$: $\underline{S}(t) = [D].[P].\underline{X}(t) \qquad (10)$
where [D] is a diagonal n-matrix, and [P] is a permutation matrix.

II.3 - SIMULATION RESULTS

II.3.1. <u>2-source problems</u>   We begin this study with the easiest
case, that of discrimination of two sources (figure 4).   In this
case, we consider only two sensors, two operators and two elements
C_{ij} : C_{12} and C_{21}. We present here two examples : in the first one,
we build ourselves the mixing matrix [A], which permits us to verify
convergence and accuracy of the algorithm because we can calculate
the theoretical solution : point P* in figure 5 ; in the second
example we describe an application to extraction of some time
dependant signals.

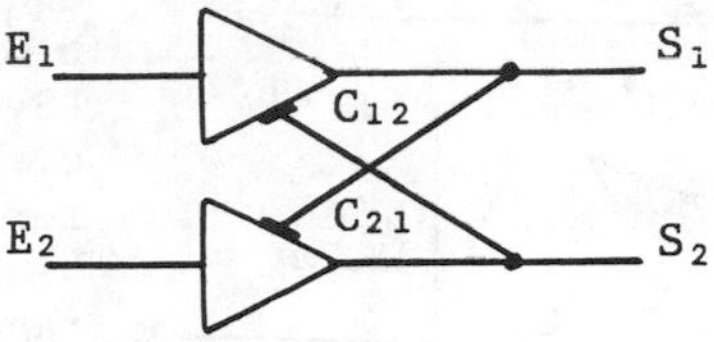

Figure   4.   2-operator network.
In this case, we have only 2
coefficients, C_{12} and C_{21}.

E_1 ——▷ S_1 C_{12}
E_2 ——▷ C_{21} ——— S_2

Example 1 : We choose a regular mixing matrix [A]. Then we consider
two signals $X_1(t)$ and $X_2(t)$, with a range of properties : deterministic or random, with large or narrow bandwidth, etc, but satisfying
the independance assumption (hypothesis 1 in § II.1). At any time t,
we choose $X_1(t)$ and $X_2(t)$,   and we compute the mixed inputs $E_1(t)$
and $E_2(t)$; then we deduce :

$$S_1(t) = E_1(t) - C_{12}(t).S_2(t)$$
$$S_2(t) = E_2(t) - C_{21}(t).S_1(t) \qquad (11)$$

If the operators N_1 and N_2 are linear :

$$S_1(t) = \frac{E_1(t) - C_{12}(t).E_2(t)}{1 - C_{12}(t).C_{21}(t)}$$

$$S_2(t) = \frac{E_2(t) - C_{21}(t).E_1(t)}{1 - C_{12}(t).C_{21}(t)} \qquad (12)$$

Equations (12) are valid if 1 - $C_{12}(t).C_{21}(t)$ ≠ 0, i.e. if
$[I+C]^{-1}$ exists. Then we compute new values of matrix [C] elements
according to the adaptation law, derived from generalization of
Hebb's law [5,6] : $C_{ij}(t+1) = C_{ij}(t) + \alpha.f(S_i(t)).g(S_j(t)-\hat{S}_j(t))$ (13)
where f(.) and g(.) are monotonous odd functions of (.), and $\hat{S}_j(t)$
is the mean value estimation of $S_j(t)$ by a first order low-pass
filter. After 500 to 1000 adaptive modifications, we note that :

$$< \frac{dC_{ij}}{dt} > = 0 \qquad (14)$$

where < > represents time averaging symbol, the algorithm converges.
Let us consider the series expansion of f and g, the zero mean value
of the product f(.).g(.) implies zeroing of all high order
moments $S_i^{2k+1}.S_j^{2l+1}$ which constitutes an approximation of an
independance test more powerfull than a covariance one. Figure 5.c
displays trajectories of points ($C_{12}(t)$), $C_{21}(t)$) in the plane (C_{12},
C_{21}) leading to the theoretical point P*. In this example, the
sources $X_1(t)$ and $X_2(t)$ are random signals with uniform probability
density. At the beginning of learning ($C_{12}(0) = C_{21}(0) = 0$), the
outputs $S_1(t)$ and $S_2(t)$ are proportional to inputs $E_1(t)$ and $E_2(t)$,

i.e. the outputs consist of a linear combination of the primary sources $X_1(t)$ and $X_2(t)$ (figure 5.a). After the algorithm converges, (figure 5.b), $S_1(t)$ and $S_2(t)$ are continuously proportional to $X_1(t)$ and $X_2(t)$ respectively : we see a rectangular distribution, because $X_1(t)$ and $X_2(t)$ are independent signals.

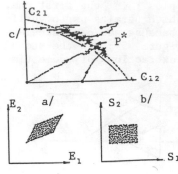

Figure 5. 2-operator problem.
a. Outputs distribution in cartesian plane (S_1, S_2) at time $t=0$. It is equivalent to inputs distribution,
b. Outputs distribution in cartesian plane (S_1, S_2) after convergence,
c. Trajectories of points $(C_{12}(t), C_{2i}(t))$ for various initial conditions.

In this example $X_1(t)$ and $X_2(t)$ are random sequences generated by the computer.

<u>Example 2</u> : Let us consider now two independant signals $X_1(t)$ and $X_2(t)$. Here, $X_1(t)$ is some non-stationnary signal and $X_2(t)$ (non represented) is a wideband noise. On the figure 6, before learning $(C_{ij}=0)$ $S_1(t)$ is a mixing of $X_1(t)$ and $X_2(t)$. After the beginning of learning $(t=0)$, the signal $X_1(t)$ is progressivly extracted by the output $S_1(t)$, while $S_2(t)$ extract the noise $X_2(t)$ (figure 6).

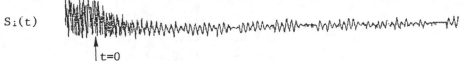

Figure 6. Progressive extraction of some non-stationnary signals.

II.3.2. <u>More general results</u>   Problem generalisation to an n-operator architecture is theorically straightforward. But in practice, simulation is very difficult because of computing duration and of the C_{ij}-trajectories and outputs S_j distributions which are $n(n-1)$ and n-dimensional spaces, respectively. We have simulated a stereoscopic vision model which uses a 4-operator architecture.

<u>Example 3</u> : An object (figure 7) in 3-D cartesian space (u_1, u_2, u_3) is seen by two TV cameras. The object forms on the retina of each receptor two different images (binocular disparity) $I(E_1, E_2)$ and $I'(E_1', E_2')$. At any time t, we illuminate with a laser a random point $M(u_1, u_2, u_3)$ of

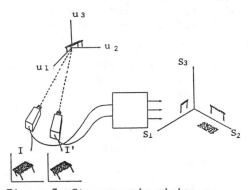

Figure 7. Stereoscopic vision.

this object. We observe luminous points $P(E_1, E_2)$ in image I and point $P'(E_1', E_2')$ in image I'. We associate with point M a vector $E(t)$ of which the 4-component $E_1(t)$, $E_2(t)$, $E_1'(t)$ and $E_2'(t)$ are <u>non</u> <u>linear</u> mixtures of cartesian coordinates. By means of random points illumination, we obtain a random vector series $\underline{E}(t)$, which provides input vectors of a 4-operator autoadaptive network. We observe convergence of the 12 coefficients C_{ij} after about 3000 iterations. In output space we note that two outputs are equal $(S_3(t) = S_4(t))$. The distribution of outputs represented by projections on the planes (S_1, S_2), (S_1, S_3) and (S_2, S_3) is shown in figure 7. We recognize front, side and top wiews of the original object. It is important to note that computing architecture shows the object by a standard representation, unknown to the receptors.

II.3. GENERALIZATION

* we cannot prove the convergence of this algorithm because of the non linearity of the adaptation law and the non stationarity of signals [7,8]. This difficult problem is also encountered in adaptive filtering, where often practical algorithm implementations have been used before deriving proofs of their convergence [9,10].

* we have studied particular cases 11, in which the number p of independent sources was different of number n of sensors :

<u>p < n</u> , after convergence, p operators among the n operators of the network extract the p independent sources : the n-p remaining are equal to zero or constant.

<u>p > n</u> , there is more sources than operators. The problem is undetermined. However, the network provides an optimal solution.

* When the mixing is non-linear, the algorithm provides good results [11] ; example 3 in part II.3.2 demonstrates it perfectly.

* Learning is permanent : adaptation gain remains constant, which involves the drawback of a limited convergence accuracy, but allows to follow possible variations of mixing, the causes of which can be various : ageing of sensors, relative movement between the object and the sensors in stereoscopic vision.

REFERENCES

1. J.HERAULT, G.BOUVIER, A.CHEHIKIAN J. Phys. Let. 41, 75-77, 1980
2. R.M.MERSEREAU, D.E.DUDDGEON Proc. IEEE 63, 610-623, 1975
3. F.CHRISTOPHE, C.MORISSEAU GRETSI, 677-682, may 1985
4. H.W.STRUBE Signal Processing, 3, 355-364, 1981
5. D.O.HEBB Wiley, 1949
6. J.P.RAUSCHEKER, W.SINGER J. Physiol., 310, 215-239, 1981
7. M.L.HONIG IEEE Trans. on ASSP, 31, n°2, 415-425, April 1983
8. A.WEISS, D.MITRA IEEE Trans. Inf. Theo., 25, n°6, 637-652, 1979
9. C.W.K.GRITTON, D.W.LIN IEEE ASSP Magazine, 30-37, April 1984
10. O.MACCHI, L.GUIDOUX Ann. Télécommu., 30, n°9-10, 331-338, 1975
11. J.HERAULT, C.JUTTEN, B.ANS GRETSI, 1017-1022, may 1985

MEMORY NETWORKS WITH ASYMMETRIC BONDS

J. A. Hertz
Nordita, Blegdamsvej 17, 2100 Copenhagen, Denmark

G. Grinstein
IBM Thomas J. Watson Research Center, Yorktown Heights NY 10598

S. A. Solla
AT&T Bell Laboratories, Holmdel NJ 07733

ABSTRACT

We study the way in which the properties of associative memory networks are changed when the interactions between the units are not symmetrical. For a class of analog networks subject to thermal noise (Langevin models), the results indicate that the memory states are not seriously degraded; their critical temperature is simply lowered from its value in the corresponding symmetric model. On the other hand, spin glass states, which occur in the symmetric case when the number of memories becomes a finite fraction of the number of units in the system, are rendered unstable by the introduction of asymmetry. This suggests that asymmetric couplings may make retrieval of the desired memory states faster, since the system will not get trapped in the spin glass states.

PACS numbers: 87.30.Gy, 64.60.Cn, 75.10.Hk, 89.70.+c

INTRODUCTION

Usually, spin models for associative memories[1,2] have been based on an Ising Hamiltonian with symmetric bonds ($J_{ij} = J_{ji}$), which guarantees detailed balance in the statistical dynamics.[3] This allows one to use the standard equilibrium formalism of Gibbsian statistical mechanics.[4,5] In particular, the memories are associated with locally stable configurations of the magnetization, which are identified with minima of the free energy. However, in both biological and device contexts where these ideas might be relevant, the interactions between neurons or circuit elements need not be symmetric. We therefore investigate here the degree to which the properties of symmetric nets carry over to the asymmetric case.

0094-243X/86/1510212-7$3.00 Copyright 1986 American Institute of Physics

The definitive study of the statistical mechanics of symmetric Ising nets was done by Amit et al.[4,5] They found the following: If the number p of stored patterns is finite, and the number, N, of units (spins) is infinite, all these patterns may be stored with negligible error below a certain critical temperature. However, if p is a finite fraction α of N, then in addition to the states corresponding to the stored patterns and combinations of them (or instead of them, for sufficiently large α) one finds a spin glass phase. Its many metastable states are uncorrelated with the memory states. Here we study what happens to both the memory states and the spin glass states for asymmetric networks.

We use a Langevin model to describe the dynamics of the network. The equation of motion is

$$- \frac{1}{\gamma_0} \frac{dS_i}{dt} = r_0 S_i + u S_i^3 - \sum_j J_{ij} S_j - h_i - \eta_i(t)/\gamma_0 \tag{1}$$

The state of unit i is described by the continuous variable $S_i(t)$. It feels a local effective potential $r_0 S_i^2/2 + u S_i^4/4$ which suppresses too-large values of S_i, the internal field from the other units S_j through J_{ij}, an external field h_i, and a noise field $\eta_i(t)$. Ising spins ($S_i = \pm 1$) can be recovered by taking the limit $-r_0 = u \to \infty$. The statistics of the noise are given by

$$< \eta_i(t)\eta_j(t') > = 2T\gamma_0 \delta_{ij} \delta(t - t') \tag{2}$$

where T is a "temperature". For symmetric J_{ij}, T really is a thermodynamic temperature, that is, the parameter which appears in the Gibbs distribution $\exp(-H/T)$. For random symmetric J_{ij}, this model has been used to study the dynamics of spin glasses.[6,7] Here we will use the same methods employed in that work to study an asymmetric case. Space in this volume only allows us to sketch the derivations and to give the results. A more complete account for publication elsewhere is under preparation.

We take our bonds J_{ij} to be generated from the patterns we want the network to remember by a slight variation on the Hebb rule used in refs. 4 and 5. We let

$$J_{ij} = \frac{1}{N} \sum_{\mu=1}^{p} \xi_i^\mu \xi_j^\mu w_{ij} \tag{3}$$

where w_{ij} and w_{ji} are independent and can take on the values 1 or 0 with equal probability. Thus if we think of the symmetric case as having bonds of equal strength running from unit i to unit j and from unit j to unit i, the model described here corresponds to randomly removing half of these directed connections.

Following Amit et al., we take the patterns ξ_i^μ to be independent binary (± 1) variables. We therefore find

$$< J_{ij}^2 >_J = \alpha/2N, \qquad (4)$$

while

$$< J_{ij}J_{ji} >_J = \alpha/4N, \qquad (5a)$$

$$< J_{ij}J_{jk}J_{ki} >_J = \alpha/8N^2, \qquad (5b)$$

and so on for higher order moments. The notation $< >_J$ indicates an average over the bond distribution, i.e. an average over the random patterns ξ_i^μ.

MEMORY STATES: THE LIMIT $\alpha = 0$

In the limit of an infinite system storing a finite number of patterns p, we can immediately see that all the patterns are locally stable at zero temperature. This is because, as in the symmetric model, if the system is in a state with $S_i = \xi_i^o$, the average value of the field

$$H_i = \frac{1}{N}\sum_{\mu, j}\xi_i^\mu\xi_j^\mu\xi_j^o w_{ij} \qquad (6)$$

is just half of ξ_i^o (plus some noise which vanishes like $N^{-1/2}$ in the large N limit). Thus each spin lies parallel to its molecular field and the state is stable.

To study the network at finite T, we follow the conventional[7,8] procedure for iterating the equation of motion (1) and averaging over the noise and the J_{ij}'s. It is convenient to work in a basis of states in which the first p eigenfunctions are just the (normalized) memories. The remaining states may be any orthonormal set orthogonal to these. We then find a susceptibility $G_\lambda = \partial < S_\lambda > /\partial h_\lambda$ of the form

$$G_\lambda(\omega) = [-i\omega/\gamma_0 + r_0 + \Sigma(\omega) - < <\lambda|J|\lambda > >_J]^{-1} \qquad (7)$$

Here Σ is a self energy which comes from expanding in u. The term involving J in (7) is equal to 1/2 for modes λ in the set of memory states and zero otherwise. Note that the fluctuations in J vanish in the thermodynamic limit, so that except for the factor of 1/2, which is simply a measure of the dilution of the bonds (half the directed connections which are present in the symmetric model have been removed), this problem is identical to the symmetric one. While we cannot solve for G_λ (or Σ) in closed form, the important point is that the asymmetry of the bonds is just equivalent to a change in the bond strength in the symmetric model.

Things simplify in the Ising spin limit $S_i = \pm 1$. This is because the average local susceptibility is, quite generally,

$$G(\omega) = N^{-1}\sum_i G_{ii}(\omega) = N^{-1}\sum_\lambda G_\lambda(\omega) = [\,-i\omega/\gamma_0 + r_0 + \Sigma(\omega)]^{-1} \quad (8)$$

(since the memory eigenstates are a negligible fraction of the total basis), and when $S_i^2 = 1$, we have simply $G(0) = 1/T$. Therefore, for this case we have, for the memory states,

$$G_\mu(0) = (T - \tfrac{1}{2})^{-1} \quad (9)$$

i.e. the transition temperature is just half its value in the symmetric case.

FINITE α: ABSENCE OF A SPIN GLASS TRANSITION

We now look at the case where the number of patterns p is a finite fraction of the size of the network. Then the higher moments (5) of J_{ij} are nonvanishing and make contributions to the self energies. These are shown in fig. 1. We evaluate the average local susceptibility $G(\omega)$, working this time in the site representation:

$$G^{-1}(\omega) = -i\omega/\gamma_0 + r_0 + \Sigma(\omega) - \frac{(\tfrac{1}{4})\alpha G(\omega)}{1 - (\tfrac{1}{2})G(\omega)} \quad , \quad (10)$$

where Σ represents the self-energy graphs that occur in the $\alpha = 0$ case discussed above, plus extra contributions which arise from incorporating factors of J_{ij} in these graphs and averaging over them.

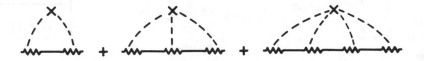

Fig. 1. Self energy graphs proportional to the moments (5) of the bond distribution. Wavy and solid lines represent J's and G's respectively. Dotted lines with crosses denote averages over the J's.

Now in the symmetric case, one way to look for a spin glass transition is to look for slowing down of the dynamical response in $G(\omega)$. That is, we can define an effective kinetic coefficient γ_1 as $(\partial G^{-1}(\omega)/\partial(\,-i\omega))^{-1}$ and ask whether γ_1 goes to zero at some temperature. In the Ising spin limit, this turns out to give a transition at $T_g = 1 + \sqrt{\alpha}$, in agreement with the equilibrium calculation of ref. 5. However, for asymmetric bonds, we can show that the effective kinetic coefficient defined this way from the susceptibility and that defined from the long-time behavior of the correlation function are different, and, furthermore, the one defined in terms of the correlation function is always smaller. This is the way in which the fluctuation-dissipation theorem is violated for this system. Thus if the kinetic coefficients are shrinking as one approaches a hypothetical spin glass transition, it is the behavior of the correlation function that we must examine in order to find the transition. We now do this and show that no transition can occur. In this calculation, we will consider the limit of very small asymmetry, so that eqn. (4) will be replaced by

$$N < J_{ij}^2 >_J = \alpha/4 + \delta , \quad \delta << 1 \tag{11}$$

while eqns. (5) remain the same. We argue that if a very small amount of asymmetry suppresses the transition, then so must a larger amount, such as the original model in which (4) holds.

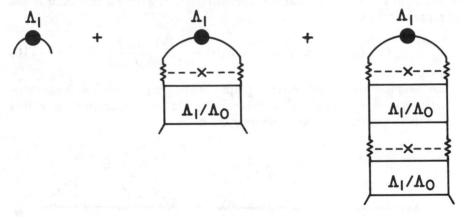

Fig. 2. Dressed noise vertex graphs for the weakly asymmetric case (11). Λ_1 consists of all diagrams obtained by replacing a G line in a self-energy graph (e.g. those in (1)) by C, and the combination of a pair of wavy lines connected by a dotted line stands here for the asymmetry parameter $\delta = (< J_{ij}^2 - J_{ij}J_{ji} >_J)/N$.

The Fourier transform of the correlation function $C(t) = < < S_i(t)S_i(0) > >_J$ is given generally by

$$C(\omega) = G(\omega)\Lambda(\omega)G(-\omega) \tag{12}$$

where we have introduced the noise vertex function Λ. Its zeroth-order value is $\Lambda_o = 2T/\gamma_o$, and the leading diagrams which contribute to it here are shown in fig. 2. We now suppose that the symmetric case $\delta = 0$ gives an effective kinetic coefficient γ_1 at low frequencies, so the susceptibility is $G(\omega) = (-i\omega/\gamma_1 + r)^{-1}$. (Its static limit r^{-1} will be determined later.) Then the noise vertex for this limit must be (by virtue of the fluctuation-dissipation theorem) $\Lambda_1 = 2T/\gamma_1$, and the ladder sum of fig. 2 gives the full vertex in the presence of the asymmetry as

$$\Lambda(\omega) = \Lambda_1\{1 - \delta|G(\omega)|^2\Lambda_1/\Lambda_o\}^{-1} . \tag{13}$$

(Note that the small-ω approximation for $G(\omega)$ breaks down in the critical region, $\gamma_1 \to 0$, of the symmetric problem. It is, however, adequate for arguing, as we now do, that the *asymmetric* problem can never become critical. A more careful, detailed discussion of this point is relegated to a separate publication.)

Combining (12) and (13), we obtain a Lorentzian peak in $C(\omega)$ at small ω, indicating exponential decay in $C(t)$ with a characteristic time $\tau = \gamma_1^{-1}(r^2 - \delta\gamma_o/\gamma_1)^{-\frac{1}{2}}$. For fixed r, τ would diverge before γ_1 went to zero. However, this cannot happen because then the integral of $C(\omega)$ over ω, which gives the mean square spin magnitude $<<S_i^2>>_J$, would diverge. Since this quantity must always be finite, r must increase in such a way as to keep τ from diverging. Consequently, we cannot have a transition signaled by dynamical slowing down in either the response function G or the correlation function C.

In general, r and $<<S_i^2>>_J$ must be calculated self-consistently. However, the case of the Ising limit again affords a simple explicit illustration. In the symmmetric case, we would have $r = T$ fixed by the fluctuation-dissipation theorem. For finite δ, the requirement $<S_i^2> = 1$ fixes

$$r^2 = T^2 + \delta\gamma_o/\gamma_1, \tag{14}$$

showing how the static susceptibility r^{-1} is suppressed by the asymmetry. It is easy to check that this suppression also prevents the system from ever reaching the instability at $r = 1/2$ for a second-order transition to one of the memory states. Thus the paramagnetic state remains locally stable against transitions to either spin glass or memory states at all temperatures.

However, we argue that in a memory state, the new self-energy and vertex terms introduced at finite δ and α are finite and go to zero for small values of these parameters. We therefore expect the memory states to remain dynamically stable for small enough α, just as Amit et al. found for the symmmetric case.

COMMENTS AND CONCLUSIONS

We have argued quite generally above against a second-order transition to a spin glass state. We have not been able to exclude categorically a first-order spin glass transition. However, if the spin glass state is of the conventional sort, which is characterized by divergent relaxation times[6] everywhere below T_g, our arguments also appear to rule out such a phase, since an infinite τ would imply an infinite $<< S_i^2 >>_J$. The fact that the memory states are stable and the spin glass phase is not is thus a consequence of the marginal character of the stability of the latter.

Accepting then, provisionally, the absence of a spin glass phase, we expect to find a line $T_o(\alpha)$ in the $T - \alpha$ plane, below which the memory states are locally stable, just as in the symmetric case. The dependence of this boundary on the degree of asymmetry remains a very interesting question for future study.

Put most simply, the result is that asymmetry does only quantitative damage to the memory states (lowering their transition temperature), while, at least in the cases we can do calculations for, it destroys the spin glass phase. This is potentially relevant to the design of devices along the lines of such models. Spin glass states are undesirable because if the system gets trapped in them, it will slow up the retrieval of the true memories. The asymmetric couplings can eliminate these states and thus, in principle, improve the performance of the device.

REFERENCES

1. W. A. Little, Math. Biosc. <u>19</u>, 101 (1974); W. A. Little and G. L. Shaw, Math. Biosc. <u>39</u>, 281 (1978).

2. J. J. Hopfield, Proc. Nat. Acad. Sci. USA <u>79</u>, 2554 (1982); <u>81</u>, 3088 (1984).

3. P. Peretto, Biol. Cybern. <u>50</u>, 51 (1984).

4. D. J. Amit, H. Gutfreund and H. Sompolinsky, Phys. Rev. A <u>32</u>, 1007 (1985).

5. D. J. Amit, H. Gutfreund and H. Sompolinsky, Phys. Rev. Lett. <u>55</u>, 1530 (1985); and preprint, 1986.

6. H. Sompolinsky and A. Zippelius, Phys. Rev. Lett. <u>47</u>, 359 (1981); Phys. Rev. B <u>25</u>, 6860 (1982).

7. J. A. Hertz, J. Phys. C <u>16</u>, 1219, 1233 (1983).

8. S-k. Ma, *Modern Theory of Critical Phenomena* (W. A. Benjamin, 1976), chapters 11-14.

NEURONS WITH HYSTERESIS FORM A NETWORK THAT CAN LEARN WITHOUT ANY CHANGES IN SYNAPTIC CONNECTION STRENGTHS

Geoffrey W. Hoffmann

Theoretical Biology and Biophysics, Los Alamos National Laboratory,
Los Alamos, New Mexico 87545, U.S.A.,

and

*Departments of Physics and Microbiology, University of British Columbia,
Vancouver, B.C., Canada V6T 2A6

Maurice W. Benson

Center for Nonlinear Studies, Los Alamos National Laboratory,
Los Alamos, New Mexico 87545, U.S.A.,

and

*Department of Mathematics, Lakehead University,
Thunderbay, Ontario, Canada P7B 5E1

ABSTRACT

A neural network concept derived from an analogy between the immune system and the central nervous system is outlined. The theory is based on a neuron that is slightly more complicated than the conventional McCullogh-Pitts type of neuron, in that it exhibits hysteresis at the single cell level. This added complication is compensated by the fact that a network of such neurons is able to learn without the necessity for any changes in synaptic connection strengths. The learning occurs as a natural consequence of interactions between the network and its environment, with environmental stimuli moving the system around in an N-dimensional phase space, until a point in phase space is reached such that the system's responses are appropriate for dealing with the stimuli. Due to the hysteresis associated with each neuron, the system tends to stay in the region of phase space where it is located. The theory includes a role for sleep in learning.

INTRODUCTION

It is difficult to overstate the importance of analogy in the development of theories. An example from physics is the development of the theory of electromagnetism in the last century, which involved the use of complicated mechanical analogies.[1] An example from biology is Burnet's clonal selection theory of the immune response, a theory that is modelled on Darwin's theory of evolution, and describes a survival of the fittest process at the level of single cells within an animal.[2] The precursor of the clonal selection theory was Jerne's natural selection theory of antibody formation[3], which marked the transition from older instructionist ideas to Darwinian selectionist theories. Analogies clearly played a similarly crucial role when Jerne subsequently postulated that the immune system functions as a network; this time he was impressed by analogies between the structures and the functions of the immune system and those of

---

*Permanent address

the central nervous system.[4,5]

In order to explore a variety of ideas on how the brain could work, we need to utilize whatever analogies we can think of. Little has formulated a neural network theory[6] based on the analogy between an Ising spin system and a neural network. Klopf[7] and Barto[8] have described neural network models in which they assume that the selfish behaviour of neural networks might be a reflection of analogous selfishness at the level of the neuronal building blocks. It would be interesting to know what analogies with other physical system, if any, aided Hopfield in the formulation of his model,[9] which has had such a profound influence on this field during the last four years. His treatment of the effect of symmetric synaptic coefficients, T_{ij}, was probably inspired by familiarity with many physical systems in which symmetric couplings lead to interesting properties.

As mentioned above, similarities between the central nervous system and the immune system led to a new way of viewing the immune system. An exciting possibility is that we can now reverse the process, and apply insights gained in the study of the immune system network to the central nervous system problem. This possibility has been considered also by Edelman and Reeke.[10] The immune system is a complex system, but in the last few years we have learned an enormous amount about how it functions. We now have at least the outline of an immune system network theory that seems to work quite well.[11–16] Since many of the similarities between the immune system and the central nervous system are at a high level,[17] we considered the possibility that the same kind of mathematical model could be applicable to both systems. That idea led to a theory that we here review briefly, and that has been developed in more detail elsewhere.[17,18]

At one level our theory is slightly more complex than other theories, in that we invoke hysteresis at the level of single neurons. This added complexity is compensated by a new simplicity at the level of the network; the network can learn without any changes in the synaptic connection strengths. Learned information is associated solely with a state vector; memory is a consequence of the fact that due to the hysteresis associated with each neuron, the system tends to stay in the region of an N-dimensional phase space to which its experiences have taken it. Its stimulus-response behaviour is determined by its location in that space.

A NEURON WITH HYSTERESIS

Neurophysiologists tell us that we cannot, on the basis of presently available data, exclude the possibility that at least some neurons exhibit hysteresis at the single cell level. A simple, plausible biochemical model that could lead to single cell hysteresis is as follows. Let X be a key neural metabolite that controls the rate of firing of the cell, such that the rate of firing is (say) proportional to the concentration of X. The concentration of X is assumed to be affected by a second substance Y that reflects the net level of inhibitory signals the cell receives. We consider the following simple reaction scheme:

$$A \rightarrow X$$

$$X \rightarrow B \text{ (slow)}$$

$$X \rightarrow B \text{ (catalysed by Y)}$$

Excess X inhibits Y

X is produced at a constant rate from A. X is broken down by two processes, one of which is slow and is independent of Y. The second breakdown process involves Y as an enzyme or the activator of an enzyme. High levels of X inhibit the enzymatic breakdown of X by Y. This phenomenon, known to biochemists as substrate inhibition of enzymes, can occur when an enzyme has two binding sites for the substrate. If we denote the concentrations of X and Y in neuron i by x_i and y_i, respectively, an appropriate differential equation for x_i as a function of time can have the form

$$\dot{x}_i = 1 - x_i - \frac{x_i \, y_i}{1 + \alpha x_i^2} \tag{1}$$

where α is a constant. y_i is given by

$$y_i = \sum_{j=1}^{N} \beta_{ij} \, x_j \tag{2}$$

β_{ij} is the synaptic connection strength from neuron j to neuron i. The β_{ij} correspond to the T_{ij} of Hopfield, except that learning can occur with the β_{ij} fixed (see below).

There can be three steady-state values for each x_i, two of which are stable (A and C, Fig. 1), and one of which is unstable (B). With N neurons, each of which can be in either a high x_i or a low x_i steady state, the number of attractors can be almost 2^N. The number of unstable steady states in the system can be close to 3^N.

THE DYNAMICS OF AN N-DIMENSIONAL SYSTEM DISPLAYED ON A PHASE PLANE

It is convenient to consider a simple variant of the system (1) that also has about 2^N attractors:

$$\dot{x}_i = (1 - |x_i|)x_i - y_i \tag{3}$$

When the system (3) is at equilibrium, all the neurons are located on the reverse S-shaped curve

$$y_i = (1 - |x_i|)x_i$$

which is shown in Figure 2. Figure 2a shows trajectories following a small transient stimulus. The same stimulus with a larger amplitude results in a qualitatively identical perturbation of the network (Figure 2b). A still larger stimulus causes the x_i value of one or more neurons to change sign, and the other neurons then relax back to slightly different positions on the equilibrium curve (Figure 2c). The system can make a large number of different responses to a correspondingly large number of different stimuli, and unless the amplitudes of the stimuli exceed a certain threshold level, the

222

Fig. 1. The rate of change in the rate of firing, $\dot{x}_i$, as a function of x_i for three different values of the input y_i: low ($y_i = y_L$), intermediate ($y_i = y_I$), and high ($y_i = y_H$). For $y_i = y_I$ the neuron has two stable steady state rates of firing, $x_i = x_A$ and $x_i = x_C$.

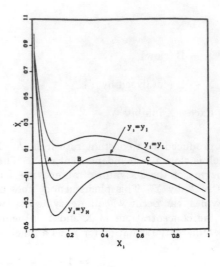

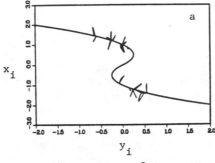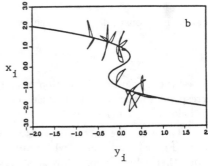

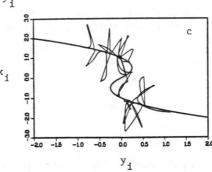

Fig. 2. The dynamics of a network with 20 neurons portrayed in the y_i, x_i phase plane. The system is at equilibrium when the y_i, x_i coordinates of every neuron is on the reverse S shaped curve. (a) A transient small perturbation causes the neurons to leave the curve, and then relax back to their original positions. (b) A larger stimulus causes a qualitatively similar response. (c) A still larger stimulus causes two neurons to switch from one branch of the curve to the other, and the equilibrium values of each of the other neurons changes somewhat. For these plots we used a network with 20 neurons with a random β_{ij} matrix of connectance 0.5. (The matrix was not symmetric, in contrast to the matrix used for similar figures in reference 18.)

system stays in the same region of phase space.

The network has been shown to be capable of producing arbitrary outputs, providing N is sufficiently large.[17]

A particular output is produced if the system has an initial condition that can lead to that output. The hysteresis associated with each neuron tends to keep the system as a whole in a restricted region of the N-dimensional phase space, and this can account for memory. The particular region in that space where the system is located at any instant depends on the set of stimuli to which it has been subjected. Hence, in contrast to conventional neural network models, the formation of memories does not need to involve changes in the synaptic strengths, β_{ij}.

TEACHING

These networks can be taught to exhibit prespecified stimulus-response behavior. We systemmatically perturb the network until it reaches a region of phase space where it gives the desired stimulus-response behavior. Since the system tends to stay in a particular region of the phase space in the absence of systemmatic perturbations, the system is able to retain trained stimulus-response behavior or memories. The idea is that a brain is moved around in the N-dimensional phase space by the stimuli that it receives, until it reaches a point in that phase space such that its responses to the environment are such that it is not strongly perturbed by the environment.

We simply apply stimuli and observe responses. There is no "twiddling" of the synaptic connection strengths. Only the various stimuli themselves are used to move the system around in the N-dimensional phase space until a satisfactory region in that space is found. If at some point a satisfactory response to a stimulus is not obtained, the stimulus is reapplied and can intensify. A wrong response means the system was in an unsatisfactory region of the phase space. Applying the same stimulus again takes the system further away from the region of phase space that did not give the correct response. If a good response is obtained, we pass on to another stimulus. This algorithm presumes to mimic the way real brains learn by experience. In an abstract sense there needs to be some complementarity between the stimulus and the response. If we are exposed to a stimulus and respond to it appropriately, our response can lead to the elimination of, or escape from, the stimulus. If we do not respond appropriately, the stimulus can persist, and possibly intensify, independently of whether the stimulus is being provided by a teacher or aspects of the non-human or inanimate environment. In a recent series of experiments we trained networks consisting of 20 neurons to respond to stimuli in prespecified ways using an algorithm based on these ideas.[18] The correct responses were defined in a binary fashion, so that untrained networks had a fifty percent rate of making correct responses. In one series of experiments, we achieved an average accuracy of 73% when we were training for a single stimulus, 70% when there were two stimuli, 66% with 3 stimuli, and 59% when there were 4 stimulus-response pairs being taught. The teaching procedure has not been fine-tuned, so we might be able to do a little better than this. More recent results indicate that networks with completely random β_{ij} can learn just as well as the networks with symmetric β_{ij} that we used initially.

SLEEP

There is an automatic motion of each neuron from the central region (near $x_i = 0$ in the model (3)) towards one of the attractors (near $x_i = \pm 1$). The system can be strongly influenced by small signals when neurons are in the central region, while much larger stimuli are required when the system is at or near a system attractor. The system might therefore be much more easily taught if there were some means of keeping at least some of the neurons near the central regions. Perhaps during sleep the $\dot{x}_i$ versus x_i characteristic changes to, say,

$$\dot{x}_i = - ux_i$$

where u is a constant. The x_i of the system then relax back towards the origin during sleep, while maintaining their respective relative magnitudes. This part of our theory leads to the prediction that the variance in the rates of firing of neurons involved in memory should increase during waking hours and decrease during sleep.

PROSPECTS

The potential possibilities for systems of the above type has scarcely begun to be explored. In this final section we speculate about the directions that future research might take.

Since the β_{ij} matrix is fixed, special forms of that matrix that permit very rapid computation by digital or optical techniques can be used. As discussed in more detail elsewhere,[18] the use of circulant matrices will permit computations that are up to a factor of $N/\log N$ faster than calculations with random matrices, and such systems are also amenable to the application of extremely rapid analogue optical methods.

The brain is organized largely as groups of neurons that have high internal connectance, with lower connectance between groups. For artificial intelligence applications, it might be profitable to mimic this architecture, and thus obtain reliable behavior from unreliable components. If we couple together groups of our neurons that have been previously trained separately, it should be possible that the larger aggregates will give an improved fraction of "correct" responses than the smaller groups are able to do independently.

A scientific and perhaps philosophical mystery in neural science is the question of the neural bases of pain and joy. The above teaching method is something of a brute force approach, that might be viewed as being based on using only "pain." (If the response is incorrect, we increase the magnitude of the stimulus.) Presumably the efficiency of our teaching could be much greater if we could add something to the algorithm equivalent to joy. A very speculative neural interpretation of pain and joy is in terms of familiar and unfamiliar regions of phase space. Perhaps pain is associated with stimuli that take one much away from familiar regions of phase space (as most completely random strong stimuli would do), and joy results from stimuli that are special in the sense that they move one back towards more familiar regions. When a brain has developed a set of memories, it presumably spends most of its time in a relatively small section of the total phase space. Specific stimuli that it receives periodically (for instance the specific stimuli experienced as a result of consuming food) would be expected to play an important role in keeping it in that limited region of phase space. Stimuli that cause pain are normally not received periodically, and they are likely to be more random in nature. There are many directions in phase space away from any restricted (familiar) region, and only a small number of directions back towards the most familiar region if one is slightly removed from it. Thus this idea would account for the fact that a limited set of specific stimuli can impart joy, while a broad range of relatively nonspecific stimuli can impart pain. New stimuli could move one back towards familiar regions of phase space if they are somehow correlated (in the context of the wiring diagram of our brains) with other stimuli with which we are familiar. If this speculation is correct, we might be able to simulate joy by making a network accustomed to a certain set of stimuli during its "ontogeny." Stimuli that are correlated with the early set might then be used to impart "joy."

The reader might feel that we are being excessively ambitious in trying to account for the phenomena of pain and joy so simplisticly. Such an assessment could well be correct. On the other hand, if we are serious about trying to understand the brain, we need to face up to these more difficult questions, and we have no way of knowing in advance how complex the solutions will be. Furthermore, these ideas can be tested at two levels. We can firstly see whether they do in fact work at the level of our computer model, that is whether they prove useful in the development of a more successful teaching algorithm. Secondly, this concept predicts that the periodic application of a specific stimulus to an animal during its early ontogeny should lead to the animal becoming addicted to that stimulus. That is, if such an animal were then given control over the application of the stimulus, it would be likely to learn to apply the stimulus to itself, and then continue to apply the stimulus with the same periodicity. If both of these predictions can be verified, the hypothesis will become worthy of serious consideration.

ACKNOWLEDGEMENTS

This work was done while we were both on sabbatical leave at the Los Alamos National Laboratory. We thank George Bell and the other members of the Theoretical Biology group and David Campbell of the Center for Nonlinear Studies for their kind hospitality.

This research was financed by the Natural Sciences and Engineering Research Council of Canada, Grant No. A-6729 to G. W. Hoffmann, and the United States Department of Energy.

REFERENCES

1. F. Hund, Geschichte der physikalischen Begriffe (Bibliographisches Institut, Mannheim 1972).
2. F. M. Burnet, The Clonal Selection Theory of Acquired Immunity (Cambridge University Press, Cambridge, 1959).
3. N. K. Jerne, Proc. Nat. Acad. Sci. (USA) 41 849-857 (1955)
4. N. K. Jerne, Sci. Amer. 229 (1), 52-60, (1973).
5. N. K. Jerne, Ann. Immunol. (Inst. Pasteur) 125C 373-389 (1974).
6. W. A. Little, Math. Biosci. 19 101-120 (1974).
7. A. H. Klopf, The Hedonistic Neuron: A Theory of Memory, Learning and Intelligence (Hemisphere, Washington, D.C., 1982).
8. A. G. Barto, Human Neurobiol. 4 229-256 (1985).

9. J. J. Hopfield, Proc. Nat. Acad. Sci. (USA) 79 2554-2558 (1982).

10. G. M. Edelman and G. N. Reeke Jr., Proc. Nat. Acad. Sci., 79 2091-2095 (1982).

11. G. W. Hoffmann, Eur. J. Immunol. 5 638-647 (1975).

12. G. W. Hoffmann, in Theoretical Biology, G. I. Bell, A. S. Perelson and G. H. Pimbley, (eds.), Marcel Dekkar (New York, 1978), pp. 571-602.

13. G. W. Hoffmann, in Contemp. Topics in Immunobiol., vol. XI, 185-226 (1980).

14. N. Gunther and G. W. Hoffmann, J. Theoret. Biol. 94 815-855 (1982).

15. G. W. Hoffmann, in Regulation of Immune Response Dynamics, C. DeLisi and J. Hiernaux, (eds.), (CRC Press, Boca Raton, Florida, 1982), pp. 137-162.

16. G. W. Hoffmann, A. Cooper-Willis and M. Chow, in Lecture Notes in Biomathematics, vol. 36 (Springer-Verlag, 1986), p. 15.

17. G. W. Hoffmann, J. Theor. Biol., in press.

18. G. W. Hoffmann, M. W. Benson, G. M. Bree and P. E. Kinahan, Physica D, in press.

Electronic Neural Networks

W. Hubbard, D. Schwartz, J. Denker, H. P. Graf,
R. Howard, L. Jackel, B. Straughn, D. Tennant
AT&T Bell Laboratories, Holmdel, N. J. 07733

ABSTRACT

We have constructed electronic neural networks with 22 x 22 arrays of microfabricated resistive synapses. The purpose of this work is both to develop a VLSI-compatible process for making large networks, and to extend our understanding of electronic neural network dynamics.

INTRODUCTION

As part of our project to construct practical circuits for massively-parallel analog computation, we have developed a VLSI-compatible process for making resistive "synapses." We have constructed chips containing a 22 x 22 array of microfabricated resistors. These chips are plugged into a circuit containing 22 amplifiers and a computer interface. The resulting network is used to test the synapse chips and to extend our understanding of neural network "learning rules" and dynamics.

For demonstration purposes, we programmed the network to serve in a Content Addressable Memory (CAM) mode[1]. Memories are coded as vectors of 22 bits, one bit per neuron. To implement a CAM, the practical task is to choose the synapse values so that the vectors corresponding to the desired memories are attractive fixed points of the circuit equations. The actual map of resistor placement is derived from the target memories using an adaptive learning rule[2].

The resistor arrays are fabricated on oxidized silicon substrates [fig.1], and wire-bonded into industry-standard 44-pin chip carriers. The resistor matrix itself takes up only a very tiny area (88 x 88 microns), so nearly all of the area of the chip is covered by the leads and pads required for wire bonding [fig.2]. The carrier is plugged into a socket on the prototype board [fig.3]. A minicomputer is used to control experiments and record data.

NETWORK FABRICATION

Traditional VLSI design techniques do not generally include the use of resistors. We have shown not only that resistors can be made, but indeed that they can be made extremely small, reliable, and compatible with other VLSI devices and procedures[3]. Since dense memories require the smallest cell possible, traditional circuits are limited by the size of a transistor. In this CAM architecture, resistors are used as the storage element. Using layered thin films, we have made resistors considerably smaller than the smallest state-of-the-art transistor.

0094-243X/86/1510227-8$3.00 Copyright 1986 American Institute of Physics

The details of the design are constrained by certain fundamental considerations. Resistances greater than hundreds of kilohms per synapse are needed in order to limit the current carried through any individual row or column in the matrix to a reasonable value. Sufficient resistivities cannot be achieved in a metal, so amorphous silicon was chosen as the resistor material. The resistivity of the Si is high enough that a layer 0.1 - 0.3 microns thick between two tungsten conductors provides the desired resistance in an area of a few square microns. Electromigration is minimized by using tungsten for the long, thin "axon" and "dendrite" wires.

The two levels of tungsten metallization are patterned using contact photolithography and reactive-ion etching. After patterning the lower layer of metal, the wafer is spin-coated with a polyimide dielectric. The programming is performed by making holes in the dielectric with electron-beam lithography and oxygen reactive-ion etching. The silicon and second layer of metal are deposited in situ by E-beam evaporation, together with ion-beam bombardment of the wafer both prior to and during the deposition. The use of the ion beam minimizes the contact resistance and makes the conduction ohmic. Each wafer contained four sets of four devices consisting of a diagonal matrix (for diagnostic purposes) and three CAM chips containing two, three and four memories, respectively. The uniformity across the 2" wafer was sufficient to produce resistor tolerances of better than +/- 5 percent.

The size of the network is severely limited by the pin-count of available packages. To maximize the number of neurons in our network, we use a special learning rule that requires only positive resistances and inverting amplifiers. Implementing both signs of synapses would have required twice as many input wires. The 22 amplifiers are made from CMOS inverters, which appear [fig.3] as the four chips surrounding the resistor matrix. The rest of the circuitry implements a simple D/A, gain control, and input and output interfacing.

The input vector (initial condition) is a binary string of 22 bits, which is converted to analog logic levels by the D/A converter. Typically a true level is 2.5 volts above the amplifier threshold, and a false level is 2.5 volts below; the effect of smaller voltages is discussed below. These voltages then precharge the dendrites until the run command is issued.

TESTING

The first step in testing is checking all connections into and out of the resistor array, and verifying resistor placement. To do this, the feedback path is broken by disconnecting all dendrites from their amplifiers by means of FET switches. Using the initial-value circuits, the test program sets the output of all neurons to ground, and then sequentially pulses each neuron in turn with a voltage. A scope probe placed on any dendrite wire will show a waveform that represents resistor placement on that dendrite [fig.4]. Resistor tolerances can be determined by observing the amplitude variations of the pulses. Shorted or open tungsten wires and bad wire bonds can be spotted easily.

ANALYSIS

To characterize the operation of the network, test programs are used to map stable states and their basins of attractions. The procedure is to generate an input vector, submit it to the network (precharge the dendrites), allow the network to settle, and record the stable state. Typical amplifier waveforms are shown in figure [5]. Because the computer interface is not designed for speed, this procedure of setting up the initial conditions for one vector and recording the subsequent stable state requires about 10 ms. To try all possible combinations of 22 ones and zeros (about 4 million) requires just under half a day.

For most of our studies, it was unnecessary to try all possible starting conditions. We used a sampling technique that first searched to identify the stable states [Table I], and then completely mapped out the basin of attraction near (in hamming distance) each of those states [fig.6].

Two special tests were conducted. The first was with low-gain. This was implemented by placing feedback resistors around the CMOS amplifiers. We were able to vary the gain about one order of magnitude, from Av = 40 (open loop) to approximately Av = 4. Voltage gain less than four began to make the network "forget" stable states. Surprisingly, lowering the gain of the neurons made the network converge faster. This results from the feedback resistors lowering the RC time of charging and discharging the dendrites, which may not be observed when using "conventional" (op-amps or models thereof) neurons, or with other ways of lowering gain.

The second special test involved allowing the input vector to have analog depth. This allowed us to experiment with setting some bits in the input vector to the "I don't know" state. We also allowed the input bits to map to voltages closer to the inverters' threshold. This showed that precharging the dendrites with voltages that saturate the amplifiers causes the network to settle slower[4].

In general, we find that the distribution and size of the basins of attraction does not change significantly with amplifier gain, or with magnitude of dendrite precharge, over the ranges tested. Network settling time is proportional to the RC time of the dendrite, and increases with the hamming distance of the input state from the final state. We have never observed a "spurious" state, that is, a stable state that was not programmed into the matrix.

SUMMARY

We have demonstrated micro-fabrication of resistive synapse matrices. The technique is suitable for programming very large pre-fabricated amplifier arrays of the type discussed by Graf et al[5]. We have tested some of the principles needed to construct these larger arrays. We have studied basin distributions and network speed as functions of: learning rules, memory loading, coding, amplifier gain, and initial value technique. These exhaustive studies that we have described would not have been feasible to simulate, but required an actual functioning electronic neural circuit.

REFERENCES

[1]   J. J. Hopfield, Proc. Natl. Acad. Sci. USA, **81** , p. 3088, (1982).

[2]   J. S. Denker, to appear in Physica D, (1986).

[3]   L. D. Jackel, et al., J. Vac. Sci. Technol., **B4** , p. 61 (Jan/Feb 1986).

[4]   J. S. Denker, this volume.

[5]   H. P. Graf, et al., this volume.

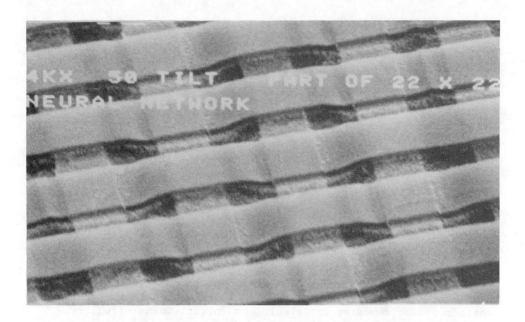

Fig. 1. Electron micrograph showing part of 22 x 22 chip. Tilt is 50 degrees, magnification is 4000 times. Pitch shown is 4 microns.

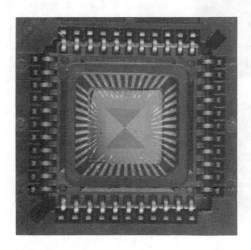

Fig. 2. Sample shown wire bonded into standard 44 pin carrier.

Fig. 3. Carrier shown mounted on prototype board. The two chips on either side of the array contain the amplifiers. All other chips contain either computer interfacing or initial-value circuitry.

This is the programming matrix for four memories

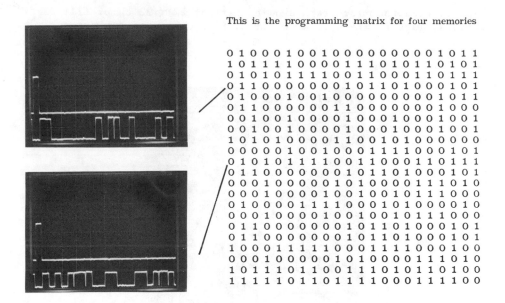

```
0 1 0 0 0 1 0 0 1 0 0 0 0 0 0 0 0 1 0 1 1
1 0 1 1 1 0 0 0 0 1 1 1 0 1 0 1 1 0 1 0 1
0 1 0 1 0 1 1 1 1 0 0 1 1 0 0 0 1 1 0 1 1 1
0 1 1 0 0 0 0 0 0 0 1 0 1 1 0 1 0 0 0 1 0 1
0 1 0 0 0 1 0 0 1 0 0 0 0 0 0 0 0 1 0 1 1
0 1 1 0 0 0 0 0 0 1 1 0 0 0 0 0 0 1 0 0 0
0 0 1 0 0 1 0 0 0 0 1 0 0 0 1 0 0 0 1 0 0 1
0 0 1 0 0 1 0 0 0 0 1 0 0 0 1 0 0 0 1 0 0 1
1 0 1 0 1 0 0 0 0 1 1 0 0 1 0 1 0 0 0 0 0 0
0 0 0 0 0 1 0 0 1 0 0 0 1 1 1 1 0 0 0 1 0 1
0 1 0 1 0 1 1 1 1 0 0 1 1 0 0 0 1 1 0 1 1 1
0 1 1 0 0 0 0 0 0 0 1 0 1 1 0 1 0 0 0 1 0 1
0 0 0 1 0 0 0 0 0 1 0 1 0 0 0 0 1 1 1 0 1 0
0 0 0 1 0 0 0 0 1 0 0 1 0 0 1 0 1 1 1 0 0 0
0 1 0 0 0 0 1 1 1 1 0 0 0 1 0 1 0 1 0 0 0 0 1 0
0 0 0 1 0 0 0 0 1 0 0 1 0 0 1 0 1 1 1 0 0 0
0 1 1 0 0 0 0 0 0 0 1 0 1 1 0 1 0 0 0 1 0 1
0 1 1 0 0 0 0 0 0 1 0 1 1 0 1 0 0 0 1 0 1
1 0 0 0 1 1 1 1 1 0 0 0 1 1 1 0 0 0 1 0 0
0 0 0 1 0 0 0 0 0 1 0 1 0 0 0 0 1 1 1 0 1 0
1 0 1 1 1 0 1 1 0 0 1 1 1 0 1 0 1 1 0 1 0 0
1 1 1 1 1 0 1 1 0 1 1 1 1 0 0 0 1 1 1 1 0 0
```

Fig. 4. Verifying resistor placement. A test program sequentially pulses the axon wires, so that a probe on a dendrite wire will yield a waveform (as shown) indicating resisitor placement. Neuron #1 is used as scope trigger (top trace).

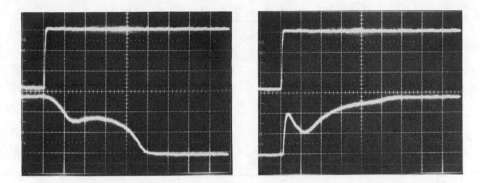

Fig. 5. The bottom traces show typical waveforms as the network settles. The top trace is the run pulse. (2 usec/div 2 v/div)

Table I. State distributions for the 2, 3, and 4 memory chips. The 22 bit stable states are written as octal numbers; numbers followed by the [ ]'s indicate the complement (or mirror image) states.

sample DS3   (2 memory matrix)

state(1) =	1222273		# hits = 25358	% hits = 25.36
state(2) =	4361507		# hits = 25127	% hits = 25.13
state(3) =	16555504	[-1]	# hits = 24925	% hits = 24.92
state(4) =	13416270	[-2]	# hits = 24590	% hits = 24.59

total   100000

sample DS5   (3 memory matrix)

state(1) =	1222273		# hits = 38821	% hits = 27.73
state(2) =	4361507		# hits = 8284	% hits = 5.92
state(3) =	12605705		# hits = 12909	% hits = 9.22
state(4) =	16555504	[-1]	# hits = 47743	% hits = 34.10
state(5) =	13416270	[-2]	# hits = 9068	% hits = 6.48
state(6) =	5172072	[-3]	# hits = 23175	% hits = 16.55

total   140000

sample DS2   (4 memory matrix)

state(1) =	1222273		# hits = 25645	% hits = 12.82
state(2) =	4361507		# hits = 22236	% hits = 11.12
state(3) =	12605705		# hits = 39607	% hits = 19.80
state(4) =	11763264		# hits = 15924	% hits = 7.96
state(5) =	16555504	[-1]	# hits = 25668	% hits = 12.83
state(6) =	13416270	[-2]	# hits = 18611	% hits = 9.31
state(7) =	5172072	[-3]	# hits = 35339	% hits = 17.67
state(8) =	6014513	[-4]	# hits = 16970	% hits = 8.48

total   200000

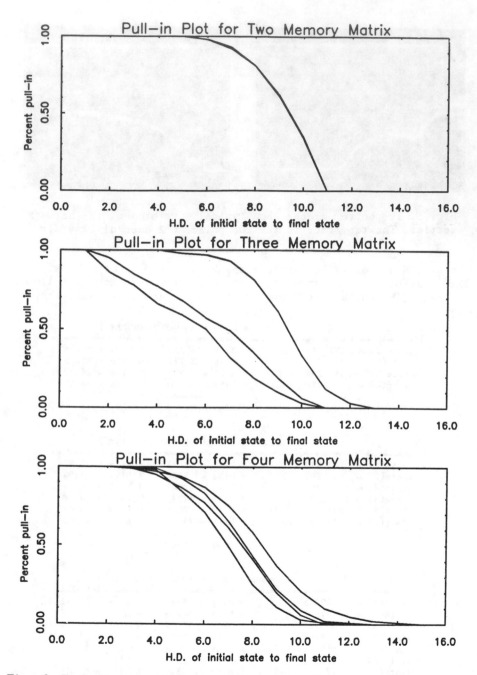

Fig. 6. Each line represents a stable state. Data was accumulated by trying all possible input vectors up to hamming distance 6, then continuing with vectors of random distances.

SIMPLE ANALOG AND HYBRID NETWORKS
FOR SURFACE INTERPOLATION

James M. Hutchinson[1] and Christof Koch[2]
[1]Artificial Intelligence Laboratory
[2]Center for Biological Information Processing
Massachusetts Institute of Technology, Cambridge, MA. 02139

ABSTRACT

Fitting a smooth surface through a set of sparse observations in the presence of both noise and discontinuities can be formulated as an energy minimization problem. In this paper, we will discuss simple networks of analog and mixed analog/digital components for rapidly solving this class of problems.

SMOOTH SURFACE RECONSTRUCTION

Surface reconstruction is a typical problem encountered in computer vision. It occurs in several situations. If a stereo algorithm computes depth values at specific locations in the image the surface must be interpolated between these points in order to yield a dense depth representation. In some other process, it may simply be to expensive to compute or measure a value everywhere or data is given everywhere but is noisy and needs to be smoothed. Grimson[1] studied surface interpolation in the context of stereo matching. He proposed an interpolation scheme to obtain depth values throughout the image, corresponding to fitting a thin flexible plate through the observed data points. Both Grimson's[1] and Terzopoulos's[2] more efficient multigrid interpolation scheme are guaranteed to converge (since the expressions to be minimized are convex).

A different approach to solving this problem is based on the use of analog networks. As shown by Poggio and Koch[3] these and similar variational problems in early vision can be cast into a problem of minimizing the power dissipated in a network of linear resistive elements. In the following, we will use the membrane-type of surface interpolation, which involves minimizing $(\int \int (f_x^2 + f_y^2) dx dx)^{1/2}$. On a 2-D lattice, the appropriate energy function can be expressed as

$$E(f) = \frac{1}{2} \sum_{i,j} (f_i - f_j)^2 + \frac{1}{\lambda} \sum_i (f_i - d_i)^2, \qquad (1)$$

where j ranges over the immediate neighbors of node i (on a square lattice the four nearest neighbors) and the second sum only includes those sites where data is present and measures the deviation of the reconstructed surface f_i from the data d_i ($1/\lambda$ is a measure of the signal-to-noise ratio). If the measured data is very reliable, $1/\lambda$ will be large and deviating too much from the data will cost

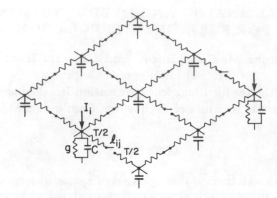

Figure 1: Surface interpolation network. Switches in transversal conductance T implement "line processes" which enable the formation of discontinuities in the computed field (they are not present in the *smooth* surface network).

the amount $(d_i - f_i)^2/\lambda$. If no data is available, $E(f)$ will be minimized if the Laplacian of f is minimal. Eq. (1) can be transformed into

$$L(V) = \frac{1}{2}\sum_{i,j} T_{ij}V_iV_j + \sum_i V_iI_i. \tag{2}$$

In an electric circuit implementation, V_i corresponds to the voltage (relative to ground) at the processing element i and T_{ij} to the conductance of the connection between nodes i and j. If no connection exists between i and j, T_{ij} is set to zero. At those nodes where data is present a current $I_i = d_i/\lambda$ is injected and the node is "grounded" by a resistance $R_i = \lambda$ (with $T_{ii} = \sum_j T_{ij} + 1/R_i$). L now corresponds to the total power dissipated by the resistive circuit as heat[3]. In order to introduce dynamics into the circuit, we associate with every node a capacity C_i. The equation describing the change in potential is then given by: $C_i dV_i/dt = -\partial L/\partial V_i$. Eq. (2) can be interpreted as the Lyapunov function of the circuit; and given this update function (equivalent to a steepest descent rule) the circuit will always converge to the unique global minimum. In order to interpolate an image from sparse data a positive current, proportional to the data, is injected into those nodes where data is given and an appropriate resistance to ground is connected to those nodes. Once the network has settled to its ground state, the voltage at the lattice nodes corresponds to the interpolated and smoothed image (see Fig. 2a). Smoothing of the input data is accomplished by varying the ratio of T to g in the network: the higher the ratio, the more smoothing done. Since $\lambda \propto T/g$, we can vary this ratio by making $g \propto 1/\lambda$. For Gaussian input noise, we choose $\lambda \propto 2\sigma^2$, where σ is the standard deviation of the measurement process[4].

How fast will the network converge? If data is present at every node it

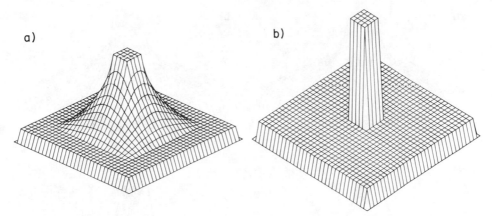

Figure 2: Stable states reached for a simple "tower" image by the surface interpolation network (a) without using line processes and (b) using line processes. In both cases, noiseless data is given along rim of tower and midway around platform.

is possible to prove[5] that the time required for the voltage V_i at the "slowest" node i to undergo some fraction ϵ (e.g. 90%) of its total transition is bounded from above by $-C_i R_i ln(1 - \epsilon)$. But what happens in the more realistic case when the network is only driven at some isolated nodes? If reliable depth data is present along edges in the image the conductance to ground $g = 1/R_i$ will have a large value, decoupling the image along the edge. This situation can be approximated by modeling a single slice of the network which is driven at one end by a current source with a resistance to ground. The typical settling time of this circuit is given by its "Elmore" time constant[6]: $\tau = C(n^2 - n)/2T$, where n the total number of nodes. On a 2-D lattice, the equivalent situation is a square patch of n by n nodes, driven by data along its perimeter. Empirically, the corresponding convergence time is slightly less than the convergence time for a one-dimensional network with $n/2$ nodes. Using state-of-the-art $1.2\mu m$ cMOS VLSI technology, Hutchinson[5] estimates convergence times on the order of $30\mu sec$ for a 1024 by 1024 node resistive network.

How sensitive is the network to noise in the individual circuit elements? Its structure is, fortunately, very simple. The transversal conductance T is constant throughout the lattice. The conductance to ground g, however, is only connected to those nodes where data is available and a current is being injected. For a given noise process, its value is assumed constant throughout the image. Nonetheless, we must worry about the effect of fabrication error on the actual values of network components. Fig. 3a,b shows the effect of noise in the transversal conductances. Each conductance $\overline{T}$ was replaced by $T = \overline{T}(1 + N(0, \sigma_T))$ where $N(0, \sigma)$ is a Gaussian probability distribution with mean 0 and stan-

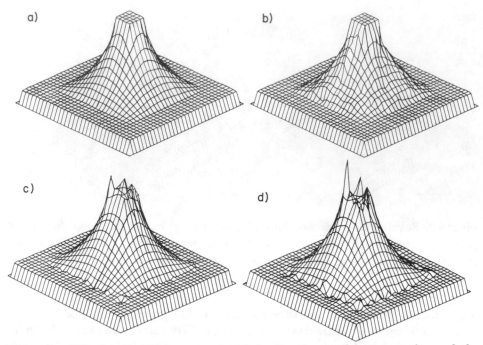

Figure 3: Effects of additive gaussian noise in the conductance values of the "tower" smooth interpolation network. (a) and (b) show noise in the transversal conductance T, with $\sigma_T = 0.2T$ and $0.5T$ respectively. (c) and (d) show noise in the conductance to ground g, with $\sigma_g = 0.1g$ and $0.2g$ respectively.

dard deviation σ_T. The network is robust even for widely varying conductances ($\sigma_T = 0.5$). Intuitively, noise in T corresponds to distorting the sampling lattice. Fig. 3c,d demonstrates a similar point for noise in the conductance to ground g. Varying g corresponds to placing more or less confidence in the data at that node. If a node where no data is present becomes accidentally grounded, it will shunt all the current to ground, "pulling" the surface to the ground. Conversely, if at a node at which data is present no path to ground exists, the node will be pinned to some fixed value.

LINE PROCESSES

However, quadratic variational principles have limitations. The main problem is the degree of smoothness required for the unknown function that is to be recovered. The surface interpolation scheme outlined above smoothes over edges and thus fails to detect discontinuities. Marroquin[4,7] has proposed a scheme to overcome this difficulty. Following[8], he models the behavior of a piecewise smooth surface by two coupled Markov random fields: a continuous-valued one that corresponds to the depth f_i and a binary one l_i, whose variables are located at sites between the depth lattice. The function of this unobservable "line process" is to indicate the presence or absence of a discontinuity between two adjacent depth elements. Using Bayes theory, it is found[4] that the maximum *a posteriori* estimate of the surface corresponds to the global minimization of the "energy" function:

$$E(f,l) = \sum_{i,j}(f_i - f_j)^2(1 - l_{ij}) + \frac{1}{\lambda}\sum_i(f_i - d_i)^2 + \sum_i V_C(l), \qquad (3)$$

where the term $V_C(l)$ measures the cost that has to be paid for the introduction of a specific configuration of lines. This term embodies the *a priori* knowledge about the geometry of the discontinuities, for instance that they occur along lines that rarely intersect. If the depth gradient $(f_i - f_j)^2$ becomes too large, it becomes cheaper to break the surface, that is $l_{ij} = 1$, paying the price $V_C(l)$ rather than to interpolate smoothly. To find an optimal solution to this non-quadratic minimization problem, we follow Koch, Marroquin and Yuille[9] and Marroquin, Mitter and Poggio[7] and use a mixed analog/digital network. Onto the mesh-like architecture of the resistive network we impose a grid of digital processors, each able to break one of the conductances $T_{i,j}$ between nodes, thus mimicking the line process l_{ij}. The hybrid machine has two cycles. For a given distribution of line processes, the analog network will find the state of lowest "energy". Subsequently, the steady-state voltages are read out from the nodes of the resistive network and digitized (A/D conversion). The line processes are now updated in an asynchronous and purely deterministic manner. l_{ij} will be turned on if $E(f, l_{ij} = 1) < E(f, l_{ij} = 0)$, otherwise it will be turned off. After each l_{ij} has been updated, the appropriate resistances are broken and the machine switches into its analog cycle. Although there is no guaranty that the system will converge to the global minimum, is seems to find next-to-optimal solutions (Fig. 2b). Fig. 4 shows a "Tower of Babylon" image, sampled along its edges, where the A/D conversion is limited to 8 and 6 bits respectively. The network converged within about 15 digital cycles. The effects of fabrication errors in T, g or in the digital processors will be similar to those seen in the resistive network and will always be of a local nature.

Koch *et al.*[9] propose and simulate a purely analog implementation of the piecewise smooth surface interpolation algorithm, based on Hopfield's and Tank's[10] mapping of a binary variable into a continuous one. Thus, the binary line process is replaced by a smoothly varying one with $0 < l_{ij} < 1$. Their network converges within a few time-constants of the system with a performance similar to our hybrid machine, implying that the smoothing of the solution space via a continuous line process is not essential for convergence, although in a few cases we observed local oscillation (eg. if the A/D samples too few bits). Our mixed analog/digital network offers substantially greater versatility and

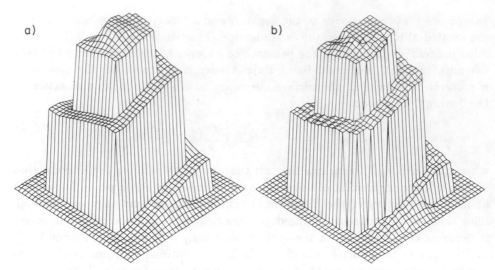

Figure 4: A hybrid network implementation of the surface interpolation problem with line processes necessitates A/D conversions between analog and digital cycles. Computed stable states are shown for (a) 8 bit and (b) 6 bit conversions.

programmability (eg. implementing a stochastic update rule would be a simple software change at the digital nodes) than purely analog networks. Also, analog network convergence times may not be negligible compared to digital network cycle times, while state of the art A/D conversions will be[5]. Nonetheless, both types of implementations offer the promise of cheap but real-time artificial vision systems.

We greatfully acknowledge many helpful discussions with John Wyatt, Tomaso Poggio and John Harris.

REFERENCES

1. W.E.L. Grimson, Phil. Trans. R. Soc. Lond. B, **298**, 395 (1982).
2. D. Terzopoulos, Comp. Vision Graphics Image Proc., **24**, 52 (1983).
3. T. Poggio and C. Koch, Proc. R. Soc. Lond., B, **226**, 303 (1985).
4. J. Marroquin, AI Lab Memo No. **792**, MIT, Cambridge MA (1984).
5. J.M. Hutchinson, MS Thesis, MIT, Dept. of EECS, Cambridge MA (1986).
6. W.C. Elmore, J. Appl. Physics, **19**, 55 (1948).
7. J. Marroquin, S. Mitter and T. Poggio, Proc. Image Understanding Workshop, Florida, (1985).
8. S. Geman and D. Geman, IEEE Trans. PAMI, **6**, 721 (1984).
9. C. Koch, J. Marroquin and Y. Yuille, Proc. Natl. Acad. Sci. USA, in press.
10. J.J. Hopfield and D.W. Tank, Biol. Cybern., **52**, 141 (1985).

NEURAL NETWORK PROCESSING AS A TOOL FOR FUNCTION OPTIMIZATION

W. Jeffrey and R. Rosner

Harvard-Smithsonian Center for Astrophysics

ABSTRACT

We summarize our recent work on the development of "neural network"-like processing for function optimization, and demonstrate how this method can be designed so as to avoid trapping in local extrema. We illustrate the application of our algorithm by considering the inversion of severely ill-posed remote sensing data and the solution of variational problems. The algorithm described here has been implemented on a serial processor, but is cast in a form which is ideally suited for parallel processing.

INTRODUCTION: SOLVING OPTIMIZATION PROBLEMS VIA NEURAL NETWORKS

Use of neural networks for combinatorial optimization problems (such as the Traveling Salesman problem) has been extensively discussed[1], here we summarize an extension of this earlier work to continuous optimization problems, and describe several applications of this extention we have recently carried out. A more detailed presentation will be published in the _Astrophysical Journal_.

Consider the standard form for the objective function[2]

$$H = -0.5 \sum_i \sum_j T_{ij} q_i q_j - \sum_i I_i q_i \qquad (1)$$

To determine how H evolves in time we have

$$dH/dt = \sum_i \{\partial H/\partial q_i\}\{dq_i/dt\} = -\sum_i \{dq_i/dt\} [\sum_j T_{ij} q_j + I_i], \qquad (2)$$

where dq_i/dt is given by the update equation

$$dq_i/dt = \lambda_i [\sum_j T_{ij} q_j + I_i]. \qquad (3)$$

Notice that for λ positive, this update equation insures that dH/dt as given by equation (2) is negative semi-definite; hence the objective function can only decrease in energy. Furthermore, for the combinatorial optimization problem, the components of **q** are often bounded (i.e., $0 \le q_k \le 1$ for all k), and the optimal solution lies along the _surface_ of the energy hypercube (i.e., $q_k = \{0,1\}$ at optimum). Thus the gain term, λ, is often allowed to go to infinity in order to force the solution to the boundary. For typical continuous optimization problems, this is rarely, if ever, true: instead, the optimal solution generally lies within the _interior_ of the hypercube. Consider then the discretized version of equation (2),

242

$$\Delta H_k = - [\sum_j T_{kj} q_j + I_k] \Delta q_k - 0.5 \, T_{kk} \Delta q_k \Delta q_k \qquad (4)$$

$$= - (\Delta q_k)^2 [1/\lambda_k + T_{kk}/2] .$$

In writing equation (4), we consider for simplicity's sake only the change in H due to a change in the kth neuron alone. For most problems of interest, $T_{kk} \leq 0$, and thus for an appropriate choice of λ_k, one can actually have the energy *increase*. This result is the essence of the difference between the continuous and the discrete case: whereas $dH/dt \leq 0$ for the former case, in the latter case we have the distinct possibilities:

$$\text{(i)} \quad \Delta H_k \geq 0 \quad \text{if} \quad \lambda_k \leq 0 \quad \text{and} \quad T_{kk} \leq 2/|\lambda_k| ; \qquad (5a)$$

$$\text{(ii)} \quad \Delta H_k \geq 0 \quad \text{if} \quad \lambda_k \geq 2/|T_{kk}| \quad \text{and} \quad T_{kk} \leq 0 ; \qquad (5b)$$

otherwise $\Delta H_k \leq 0$. Thus a viable mechanism exists for increasing the "energy", and therefore escaping from local minima.

In general, the objective function may contain terms which are not quadratic forms, such as $\sum_j q_j \ln(q_j)$ (which allows one to maximize the configurational entropy of the solution, and often arises in deconvolution problems.) Although such terms do not fit directly into the neural network formalism, we have devised two methods for incorporating them into the computational structure:

(1) Taylor-expand the term to be added, and only retain terms up to a certain order in q. This truncation may not capture some of the essential analytical properties of the original constraint; for example, in the case at hand, the expansion is not necessarily positive, unlike the (q ln(q)) term.

(2) Add terms such as entropy *directly* to the update equation, rather than to the objective function. Thus, add the derivative of the entropy term normalized in such a way that the added term will vanish if the trial solution does indeed have maximum entropy. This method is to be preferred because it generally does capture the essential analytical properties of the constraint.

The second prescription leads to the modified update equation

$$\Delta q_i = \lambda [\sum_j T_{ij} q_j + I_i + \phi_1 c (1 + \ln(c q_i))] , \qquad (6)$$

where ϕ_1 is a Lagrange multiplier that describes the relative weighting and the normalization constant c is chosen to be

$$c = N \exp\{-1\} / \sum_j q_j .$$

This choice insures that for a constant solution ($q_i = \{1/N\} \sum_j q_j$ for all i, i.e., a solution with maximum entropy), the added term will vanish.

EXAMPLE 1: FREDHOLM EQUATION OF THE FIRST KIND WITH CONVEX COST FUNCTION

The problem is defined as follows[3]: We want to invert the Fredholm equation of the first kind

$$g(y) = \int_0^1 x \exp(-xy) q(x) dx , \qquad (7)$$

given the data $g(y)$ at a fixed number of points y_i. In discretized form, the kernel functions are $k(x,y_i) = x \exp\{-xy_i\}$, where we use the values $\{y_i^{-1}\} = \{0.1, 0.3, 0.5, 0.7, 0.9\}$. We shall specify a particular form for $q(x)$, namely $q(x) = 1 + 4 (x - 1/2))^2$, carry out the direct transform in order to synthesize the data at the $\{y_i\}$, and then invert in order to recover the known test function.

The problem at hand is especially ill-posed because the kernels for each i greatly overlap; hence a direct inversion of equation (7) is not possible. Instead, additional constraints such as maximizing entropy, smoothness, or other a priori knowledge about the solution must be used to insure a physically plausible answer. We chose the objective function

$$H(\mathbf{q}) = \sum_i [(g_i - g^d_i)/g^d_i]^2 + \phi_1 \sum_j q_j \ln(cq_j) + \phi_2 [\sum_j q_j - F_{tot}]^2 . \qquad (8)$$

The first term measures the "goodness-of-fit" to the data; the second constraint selects out a plausible solution with maximum configurational entropy; and the third constraint arises if (typically) $\mathbf{q}$ obeys a conservation law, e.g. $F_{tot} \equiv \int q(x) dx$. For example, if $q(x)$ is a measure of the monoenergetic photon flux at energy x, then the last term requires the flux in the solution to match the observations. The ϕ's are Lagrange multipliers that describe the relative weighting of the terms. Thus the update equation for this example reads

$$\Delta q_i = \lambda [\sum_j T_{ij} q_j + I_i + \phi_1 c \{1 + \ln(cq_i)\}] , \qquad (9)$$

where

$$T_{ij} = - \sum_l k_{li} k_{lj}/(g^d_l)^2 - \phi_2 , \quad I_i = \sum_l k_{li}/g^d_l + \phi_2 F_{tot} ,$$

and the entropy normalization, c, is as defined above. We compared the performance of simulated annealing, neural network processing, and a standard Polak-Ribiere conjugate gradient method for solving equation (13) for *noise-free* observations. Excellent agreement was obtained in all cases, but the computation time varied substantially; these results are summarized in Table 1.

TABLE 1
RELATIVE COMPUTATION TIME

METHOD	INTEGRAL EQUATION	FUNCTION MINIMUM	GLOBAL MIN. FOUND (%)
SIMULATED ANNEALING	150	2005	75
CONJUGATE GRADIENT	1.2	1	56
NEURAL NETWORK	1	1.8	76
		4.2	100

EXAMPLE 2: PROBLEMS WITH NON-CONVEX OBJECTIVE FUNCTIONS

The objective function in equation (8) is convex, and so possesses only one minimum -- the global minimum. In contrast, the real power of the neural network lies in its ability to escape from local minima and to find the global minimum. In order to fully illustrate this capability, we search for the global minimum of the following non-convex function:

$$f(x,y) = [4 - 2.1x^2 + x^4/3]x^2 + xy + [4y^2 - 4]y^2, \qquad (10)$$

with $\{-3 \leq x \leq 3, -2 \leq y \leq 2\}$. It is easily verified that this function has six minima, of which two are global minima.

Since we want to locate (one of) the global minima of this function, we can use equation (10) directly as the objective function. Only two parameters are then to be determined: The values of x and y at the global minimum. Thus in the framework of the neural network, we only have two "neurons" to consider; and a function with n independent variables would require n "neurons".

In order to solve equation (10) using the neural network algorithm, we partition the function $f(x,y)$ into quadratic terms and terms higher order in x and y. The quadratic terms fit directly into the formalism as outlined above, whereas the higher order terms can be treated in a similar fashion to the entropy term in the previous example. We thus generate the following update equation:

$$\Delta q_1 = \lambda [\sum_j T_{ij}q_j - 2(q_1)^5 + 8.4(q_1)^3], \qquad (11a)$$

$$\Delta q_2 = \lambda [\sum_j T_{ij}q_j - 16(q_2)^3], \qquad (11b)$$

where $T_{11} = -8$, $T_{12} = T_{21} = -1$, $T_{22} = 8$, and $q_1 = x$, $q_2 = y$.

Table 1 also compares simulated annealing, neural network processing, and the conjugate gradient method (which always stops after encountering any minimum) in finding the global minimum of equation (10). The results are striking: Although in this case, the conjugate gradient method was fastest in descending to the global minimum *once* it entered the basin of attraction for the global minimum, it located the global minimum only $\approx$ 56% of the time (for 1000 trials). The effective rate for annealing is instead about two thousand times slower (based on 70 trials) than the conjugate gradient procedure, but has a success rate of $\approx$ 75% of locating the global minimum. Finally, the neural network algorithm instead ran roughly half as quickly than the conjugate gradient search (as long as the latter once again found itself within the attracting basin), but located the global minimum $\approx$ 76% of the time (for 1000 trials); if instead we are willing to tolerate speeds roughly four times slower than the conjugate gradient method, then for *this* problem we can find the global minimum 100% of the time (again based on 1000 trials).

EXAMPLE 3: VARIATIONAL PROBLEMS

Neural networks are also useful for solving variational problems. As a specific example, consider nonlinear pattern formation near the onset of Rayleigh-Benard convection[4]. In this example, we are trying to find the stable convection patterns described by the evolution equation[5]

$$\partial\Psi/\partial\tau = [\ \epsilon - (\nabla^2 + 1)^2\ ]\Psi - \Psi^3 \qquad (12)$$

with boundary condition $\Psi = \partial_n \Psi = 0$ on all lateral walls; ∂_n is the normal derivative, τ is in units of the vertical diffusion time, ϵ is a small parameter related to the Rayleigh number, and Δ^2 is the two-dimensional Laplacian. This fourth-order (in space) partial differential equation can be derived from a Lyapunov functional

$$F(\Psi) = 0.5 \int dx\ dy\ \{\ -\epsilon\Psi^2 + \Psi^4/2 + [(\nabla^2 + 1)\Psi]^2\ \}, \qquad (13)$$

which we seek to minimize. This problem is straightforward to solve using the neural network since the update equation for the Ψ's is given by equation (12). Figure 1 shows the result of minimizing equation (13) via the neural network for a 25x25 mesh cell with aspect ratio equal to 4 in x and y, and $\epsilon = 0.5$. The solutions are in agreement with direct solution of (12)[4].

CONCLUSIONS

Our work shows that "neural network"-like processing is a viable technique for solving continuous optimization problems.

246

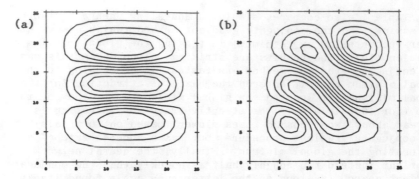

Fig. 1. Results of solving a variational problem (equation 13) for the nonlinear pattern formation near the onset of convection. The contours depict the amplitude of the velocity field. Figure (a) represents a non-optimal local minimum and Figure (b) the global minimum (cf. Fig. 8 of ref. [4]).

Specifically, we have shown that:

    1. Neural networks can be implemented that allow for the "energy" to *increase* (e.g., to climb out of minima).

    2. The neural network heads "straight in" to the solution (due to the strict contraction).

    3. Non-quadratic terms can be easily incorporated into the update equation.

    4. A wide range of problems can be solved using the "neural network" algorithm, including inversion of integral equations and variational problems.

Nevertheless, much work still must be done to determine how efficient the algorithm is. Foremost on the list of unknown qualities is how the efficiency of the algorithm scales with the number and distribution of minima, a problem which we are presently focussing on.

REFERENCES

1. J. J. Hopfield and D. Tank, submitted (1985).
2. J. J. Hopfield, Proc. of Natl. Acad. Sci. (USA) 79, 2554 (1982).
3. S. Twomey, in Introduction to the Mathematics of Inversion in Remote Sensing and Indirect Measurements (Elsevier, N.Y.).
4. H. S. Greenside and W. M. Coughran, Jr., Phys. Rev. A 30, 398 (1984).
5. J. Swift and P. C. Hohenberg, Phys. Rev. A 15, 319 (1977).

PARALLEL STRUCTURES IN HUMAN AND COMPUTER MEMORY

Pentti Kanerva
Research Institute for Advanced Computer Science
MS 230-5, NASA Ames, Moffett Field, CA 94035

ABSTRACT

If we think of our experiences as being recorded continuously on
film, then human memory can be compared to a film library that is
indexed by the contents of the film strips stored in it. Moreover,
approximate retrieval cues suffice to retrieve information stored
in this library: We recognize a familiar person in a fuzzy
photograph or a familiar tune played on a strange instrument.

This paper is about how to construct a computer memory that
would allow a computer to recognize patterns and to recall sequences
the way humans do. Such a memory is remarkably similar in structure
to a conventional computer memory and also to the neural circuits
in the cortex of the cerebellum of the human brain.

The paper concludes that the frame problem of artificial
intelligence could be solved by the use of such a memory if we were
able to encode information about the world properly.

INTRODUCTION

The title of my paper suggests two things: that a memory is
a parallel processor and that human and computer memories are
structurally similar. I will discuss both of these topics.

The memory model discussed in this paper was developed by me,[1]
and I will simply refer to its properties as needed. It is related
closely to the cerebellar models of Marr[2] and Albus[3] and more
distantly to the associative-memory models of Anderson,[4] Hopfield,[5]
Kohonen,[6] Willshaw,[7] the Boltzmann Machine,[8] and others.[9] The
detailed comparison to computer memory is unique to my work.

It is convenient to begin with something about which we can
say that we truly understand it. We can say that about a computer
memory because we specify it in minute detail and build it from very
simple components. The point of my talking about computer memories,
however, is that their organization is remarkably similar to my
model of human memory. A comparison of the two can be useful in
several ways. First, it can help us understand the proposed memory
model. Second, the differences in the two can give us insights into
intelligence, both natural and artificial: Why are some things so
easy for us and so hard for computers? This, in turn, can guide

_____

This research was supported by a gift from the System Development
Foundation to the Center for the Study of Language and Information,
Stanford University, and by NASA Cooperative Agreement No. NCC 2-408.

research in artificial intelligence. Finally, it can suggest ways
to build computer memories for artificial intelligence.

THE ORGANIZATION OF COMPUTER MEMORY

The random-access memory of a computer is an array of storage
locations or registers. A location is identified by its position
in the array—a sequence number—which is called the address of the
location. A location stores information: a fixed-length vector of
bits, a binary word. The word stored in a location is called the
contents of the location. For example, today's small computer may
have a memory with 100,000 locations, each with a capacity of 8 bits,
and a large computer may have several million 64-bit locations. The
main parts of a computer memory appear on the right in Figure 1.

Storing information (a 32-bit word) in memory is called writing
and retrieving it is called reading. They involve two special
registers: the address register, to hold the address of some memory
location, and the datum register, to hold a word that is being
transferred into or out of the memory. In writing, the contents
of the addressed location (exactly one location) are changed (the
datum register is copied into the addressed location, replacing
the location's old contents), and in reading, the contents of the
addressed location are retrieved (they are copied into the datum
register).

To find the addressed location in the memory array, the memory
has a network of circuits called address decoders. In principle,
each storage location has its own address decoder, which will
recognize the location's address and no others. When an address
is placed in the address register, it becomes available to all the
address decoders, but only one will recognize it as its own and make
that location available for a subsequent transfer of a word.

In addition, a computer memory has circuits to control the timing
of data transfers. They are included in Figure 1 for completeness.

PARALLELISM IN COMPUTERS

We usually think of the computer as operating in serial. The
processor executes a program one instruction at a time, and words
(data) are read from or written into memory one at a time. The
nervous system, by contrast, appears to be highly parallel in its
operation. This serial versus parallel operation is commonly taken
to be a major difference between computers and brains. Perhaps too
much has been made of that difference.

First, some parallelism is apparent even in the schematic
picture of a computer in Figure 1: The bits of an address, and of
a word being written or read, are transferred from one part of the
computer to another in a single swoop—in parallel—instead of bit
after bit. Such parallelism is present also in my model for human
memory, only the model memory operates with 1,000-bit words compared
to the computers' words of, say, 32 bits. Parallelism of this
kind—of data paths—appears common also in the brain.

Less apparent in Figure 1 but exceedingly important in reality

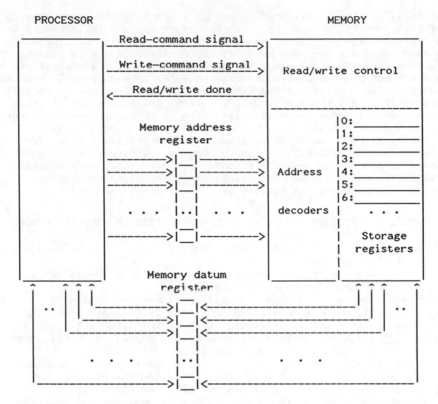

FIGURE 1. A computer organized as a processor and a memory.

is the parallelism in the working of the address decoders. Recall
that the function of the address decoders is to find the addressed
location in the memory array. We may think of it as one primitive
operation, but, in fact, a very large selection network is active
in parallel and, as a result, memory access is fast (the time to
find a location grows with the logarithm of memory size). Address
decoding has this same parallelism in my memory model. The
counterpart of address decoding in human memory is pattern
recognition, which likewise appears to involve much parallel
processing similar to address decoding.

By contrast, in a truly serial computer—a machine with tape for
a memory, such as a Turing Machine—the memory locations are examined
one after another, and it takes a long time to find the addressed
location (the time grows linearly with memory size, i.e., the length
of [the active part of] the memory tape). The absence of parallelism
in such machines makes the accessing of memory very slow.

When computer professionals speak of parallel processing, they
tend to overlook address decoding as an important instance of it and
think only of many programmable processors (see Fig. 1) being active

at once, working on different parts of a problem. In an extreme form
of such parallel processing, the processors operate in lock-step on
different pieces of large, homogeneous data. Matrix calculation is
a good candidate for such parallel processing. In a more general
form of parallel processing, the processors are coupled loosely to
one another and can work quite independently of each other. Address
decoding is a variant of the first, more restricted form of parallel
processing: All the individual, primitive processors--the logic
gates--are performing the same selection function at once, only they
are doing it relative to different parts of the memory array.

A final note about parallelism: It is a practical issue for
computers and, presumably, also for brains. It provides a way to
speed up computation--even millionfold--and to increase a system's
reliability, but it does not affect the class of things that a
computer, in theory, is capable of.

SPARSE, DISTRIBUTED MEMORY AS A GENERALIZATION OF COMPUTER MEMORY

In this section, I will transform a fairly large computer memory into
a model of human memory.

A storage location. Instead of the computer's 32-bit words,
the model memory will have 1,000-bit words; instead of 20-bit
addresses--enough to address one million locations--the model memory
will have 1,000-bit addresses; and instead of each bit slot of a
storage location containing either 0 or 1, the 1,000 bit slots of
a storage location in the model memory each contain a count, a small
integer. Initially, all counts are zero.

Two locations are said to be d bits apart if their addresses
differ by d bits. Such a distance is called the Hamming distance;
it is the number of bit positions at which two binary words differ.
The number of n-bit words that are d bits away from an arbitrary
n-bit word is given by the binomial coefficient 'n choose d'. Hence,
the distribution of distances from an arbitrary address to all
possible addresses is the binomial distribution with parameters
$n = 1,000$ and $p = 1/2$, which is approximated by the normal
distribution with mean 500 and variance 250. For example, in the
model memory, the maximum distance between two locations is 1,000
bits, the mean distance is 500 bits, and .998 of the locations are
at least 451 bits and at most 549 bits away from any given location.

Sparse memory. A computer memory with 20-bit addresses usually
has 2^{20} locations (about one million), so it is natural to assume
that a memory with 1,000-bit addresses should have $2^{1,000}$ locations.
But that is an enormously large number; there are not that many
elementary particles in the known universe. Fortunately, the memory
can be made to work with a relatively small number of locations.
We will assume that the model memory has 1,000,000 actual locations
and that their addresses are a random sample of the $2^{1,000}$ possible
addresses. Thus, the memory is very sparse: The median distance
from a storage location to its nearest neighbor is 424 bits.

Distributed memory. When a computer memory is addressed for writing or reading, exactly one location is selected, and that location either receives or emits a word. But when a sparse memory is addressed, practically never is there an actual location with that exact address. A way to access a sparse memory is to select many locations at once. Given an address for writing or reading, select the locations that are within 450 bits of that address. In this way, nearly 1,000 memory locations closest to the read or write address will be selected (.001 of the address space lies within 451 bits of any given address).

Writing. A 1,000-bit word is stored in the model memory by storing it in all the locations that are within 450 bits of the write address. A word is stored in a location by incrementing and decrementing the location's 1,000 counters. To store 1, one is added to the appropriate counter; to store 0, one is subtracted from it; except when a counter would overflow or underflow, in which case it remains unchanged. Thus, writing a word causes nearly 1,000 times 1,000 counters to be incremented or decremented. These nearly million operations can be carried out in parallel.

Reading. A word is retrieved from the model memory by pooling the contents of the locations that are within 450 bits of the read address and then finding a word that represents the pooled data. Each bit of this 1,000-bit word is determined by the majority rule: If, in the words written into the pooled locations, the bit was more often 0 than 1, the bit read will be 0; otherwise it will be 1. In practice, we add the counters across the pooled locations—in parallel—and compare the resulting 1,000 sums to zero. Since the read word is a statistical reconstruction, it is not necessarily identical to any of the words that has been written in memory.

Reading from the memory is illustrated in Figure 2. If the memory part of Figure 1 were drawn in greater detail, it would look very much like Figure 2.

Capacity. After a word has been written in a location in computer memory, it can be read by specifying the location's address. Is this also true for a sparse, distributed memory? If a word W has been written with A as the write address, can it be read back by using A as the read address? It can, with a very high probability, if the total number of words written in memory is not too large—100,000 for our model memory—and if no other word has been written with an address very similar to A (two addresses are similar to each other if the Hamming distance between them is small). So even though the model memory has 1,000,000 locations, its capacity is somewhat less than 100,000 words.

CONVERGENCE TO THE BEST-MATCHING WORD AND SEQUENCE

The most significant property of the model memory is that it is sensitive to similarity. In other words, approximate retrieval cues can be used to retrieve exact information. This makes the memory a candidate for a model of human memory.

For the model to work in this way, it is necessary that a word read from memory can be used to address the memory. I call this the unifying principle of the theory. It is also necessary that

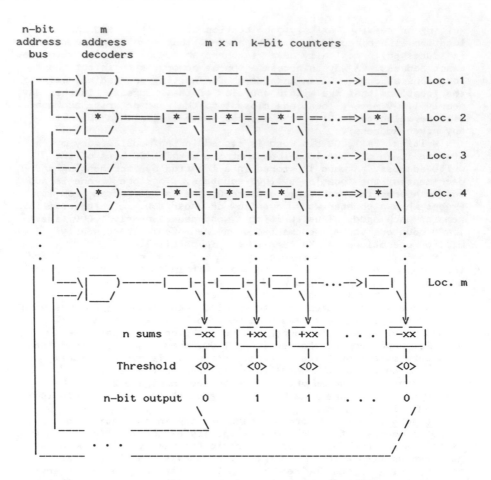

```
n-bit           m
address      address                    m x n  k-bit counters
  bus         decoders
```

FIGURE 2. Sparse, distributed memory and its operation during a
read cycle. The memory has m locations, each made of an address
decoder and n k-bit counters (e.g., m = 1,000,000, n = 1,000, and
k = 6). The contents of the locations closest to the read address
(locations 2, 4, ...; marked with an asterisk *) are added together
(vector addition), and the n sums are compared to a threshold of 0.
The bottom of the figure shows that the output of one read cycle
can become the address of the next. In a write cycle, an n-bit word
is stored in the memory by adding it to all the locations selected by
(and closest to) the write address: A 1 bit is added by incrementing
and a 0 bit by decrementing the appropriate counter. The signal
lines for writing are not shown. They have the same layout as do the
lines for reading, with the data flowing in the opposite direction.

the memory has not been filled to capacity. In the examples below, I will assume that 10,000 words have been written in memory. Since each word is written in nearly 1,000 locations, nearly 1,000 times 10,000 copies have been written in the million locations, or about 10 words per location.

Assume that the word X has been written with X itself as the address. That means that all locations within 450 bits of X (nearly 1,000 locations) will store one copy of X each—their counters are incremented and decremented according to the bits of X. We have already implied that reading with X as the address will retrieve X with very high probability. The reason is that we get back the nearly 1,000 copies of X, and they reinforce each other in the grand sum, plus nearly 1,000 times 10 copies of other words written in those same locations, and they mostly cancel out each other.

The interesting case is when some word X′ that is sufficiently similar to X (within 209 bits of X) is used as the reading address, because the read word X′′ will be more similar to X than X′ is. Reading then with X′′ as the address will retrieve the word X′′′ that is even more similar to X. Fewer than ten iterations of this kind will retrieve X. The statistical argument here is similar to the one above: When the memory is read with an address X′ that is sufficiently similar to some previous write address X, writing and reading will access many common locations, each one of which holds a copy of X, and these multiple copies reinforce each other in reading. When the distance is sufficiently large (over 209 bits), the locations by which writing and reading overlap are too few to let the word X stand out against the background noise from the other words in the pooled data. Iterated reading will then result in a sequence of random 1,000-bit words with little resemblance to any of the words written in memory.

A sequence of words X, Y, Z, ... can be stored in memory as a chain of pointers by using X as the address to write Y, Y as the address to write Z, and so forth. If we then read with X as the address, we will retrieve Y, then read with Y as the address, we will retrieve Z, and so forth. In other words, we can read back the sequence by starting with its first member, just as we would with pointer chains stored in conventional computer memory.

Similarity works here as well. Assuming that the sequence has been stored as a pointer chain and that we read with address X′ that is sufficiently similar to X, we will retrieve a word X′′ that is more similar to Y than X′ is to X. With X′′ as the address we will read a word X′′′ that is even more similar to Z, and a few more iterations will suffice to read exact words of the stored sequence. The statistical argument for this is the same as it was above.

NEUROPHYSIOLOGICAL PARALLELS

Most interesting about the memory model is that realizing it in hardware (see Fig. 2) requires components and circuits that resemble common neural components and circuits in the brain.

An address decoder that responds to a set of addresses that are within a certain distance from a specified address can be realized by a linear threshold function. Linear threshold functions, in turn, have been used commonly to model neurons, and, in fact, a neuron appears to be an ideal address decoder for a storage locations.

A storage location is made of 1,000 counters, the values of which are incremented and decremented to store 1s and 0s. Accordingly, modifiable synapses along the axon of an 'address-decoder neuron' could constitute a storage location.

In reading from memory, corresponding bit locations of many storage locations are pooled to form a single bit of output. An ideal neural structure for that would be a large, flat dendrite plane of a 'read-out neuron', perpendicular to the axons of the address-decoder neurons. It would correspond to a bit plane in computer memory.

In writing into memory, to cause a bit to be written in the very bit locations that are pooled for a single bit of output upon reading, there would have to be a 'write-in neuron' with a branching axon tree that matches the dendrite tree of a read-out neuron.

The structure of the cerebellar cortex[10] follows very closely the above plan. The granule cells correspond to the address-decoder neurons. Their axons, called the parallel fibers, are perpendicular to the flat dendrite trees of the Purkinje cells, and the Purkinje cells provide the only output of the cerebellar cortex. So the Purkinje cells correspond to the read-out neurons, and the synapses of the parallel fibers with the Purkinje-cell dendrites correspond to the bit locations (k-bit counters in Fig. 2). Finally, the climbing fibers correspond to the axons of the write-in neurons, as they pair up with the dendrites of the Purkinje cells.

Whether the cerebellum actually works as a memory is not clear to me. However, climbing fibers similar to those in the cerebellum are common in the brain. In general, any place with a climbing fiber matching the dendrite tree of another cell is a good candidate for a pair of write-in and read-out neurons and hence a good starting point in trying to interpret the function of a neural structure.

SPARSE, DISTRIBUTED MEMORY AS A MODEL OF HUMAN MEMORY

I assume that the function of human memory is to store a model of the world for use by an individual in dealing with the world. The usefulness of the model memory for this function will be discussed in the next two sections.

The 1,000-bit words on which the model memory operates can be thought of as patterns of 1,000 abstract features. We have established that it is possible to store such patterns and sequences of them in the model memory and then retrieve them by cueing the memory with those same patterns or with ones similar to them.

To apply the model to human memory, we will identify a pattern with a moment of an individual's experience. A graphic, even if crude, way to think about it is that an individual has a thousand special neurons in the brain—the equivalent of the 1,000-bit memory address register—and the momentary state of those neurons encodes the individual's subjective experience of the moment. I will henceforth call this address register the mind's _focus_ (actually, it is a combined datum and address register; see Fig. 1). The individual experiences things through the focus. To attend to something, that something (its encoding) has to be in the focus.

The individual is coupled to the world through the senses. The senses feed into the focus, and memory storage and retrieval are through the focus. This is illustrated by Figure 3.

A succession of moments—a section of an individual's life—is then represented by a sequence of patterns. It is natural to store the sequence in memory as a pointer chain, because it can then be retrieved later, as has already been described. The usefulness of such storage will become clear in the next section.

APPLICATION TO THE FRAME PROBLEM

As an application of the memory model, consider the _frame problem_ of artificial intelligence, particularly, of robotics. I will explain the frame problem with an example.

A robot lives in a world. To function in it, the robot maintains an internal model of the world—a data base. In the data base are represented objects of the world (e.g., the robot, a cart, a telephone, room 1, room 2), properties of the objects (e.g., all rooms are stationary, the cart is movable, the telephone is blue), and relations between the objects (e.g., the cart is in room 1, the telephone is on the cart, the telephone's receiver is on hook).

In addition, the model of the world must specify the ways in which things interact when the robot acts on the world, say, moves the cart from room 1 to room 2. What, besides the cart (and the robot), will end up in room 2; what entries in the data base,

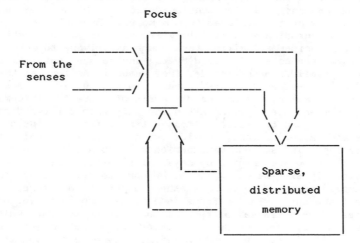

FIGURE 2. The coupling of the memory with the world.

other than the one for the cart (and the robot), must be updated?
Naturally, entries for all the things resting on the cart—i.e., the
telephone—except those tied by a short cord to the wall—again, the
telephone—as they and things resting on them will fall on the floor
of room 1 and will no longer be on the cart (nor will the receiver
be on hook). The story can be made as complicated as one wishes.

Updating the data base as the robot interacts with the world is
known as the frame problem, and it is as yet unsolved in robotics.

How do the higher animals and we humans handle the frame
problem? According to the computational view of mental functions,
the problem is as real to us as it is to the robot. An easy answer
is that we have common sense, which the robot lacks, and we have
gained it through experience. But how does common sense work? How
is experience acquired and how is it used?

The memory model suggests a way to answer these questions.
Assume that the situation in which the individual is at present
resembles ones encountered by it in the past. The consequences of
those past situations are then likely to predict what is about to
happen this time. So the memory should allow the retrieval of those
consequences. According to the memory model, they will be retrieved
automatically if the individual's record of experiences has been
stored as a pointer chain. Furthermore, due to the statistical
nature of memory storage and retrieval, the common parts of those
past consequences will reinforce each other and thus will stand
out in what is retrieved from memory, the other parts being blurred
away. Thus, the likely consequences of the present—a statistical
abstraction based on the individual's past experience—are
automatically brought to the focus.

We are still left with a major problem, namely, encoding.
On that, the memory model suggests the following: First, the
things that come to the focus from our senses are the very things
written in memory—encoding happens outside the focus–memory loop
(abstraction happens in it, as mentioned above, and encoding can
be affected by what is retrieved from memory). Psychological
experiments on encoding support this view, as does our subjective
experience. Consider a familiar task, such as driving to work.
In doing it we are constantly cued by the environment but fill in
most of the detail from the inside, from memory. As long as there
are no surprises—as long as what comes from the outside agrees with
what comes from memory—we are hardly aware of where the information
is coming from and start paying attention only when the two disagree.
Experiences that are controlled almost entirely from the inside, such
as dreams and hallucinations, provide further support to this view
about encoding. The subjective experience in such cases can be very
real and it can be accompanied by physical signs of pleasure or fear,
for example. In extreme cases it may be hard to tell whether the
thing actually happened to us or whether we just "made it up."

Second, the entity in the focus is a high-dimensional vector
of features (a very "large" pattern, a point of an abstract multi-
dimensional space) that encodes everything about that moment, that
is, any specific things that the individual may be attending to as
well as the overall context. In that sense the memory is holistic,

and whatever is retrieved from it is affected strongly by the
context. This agrees with memory experiments with human subjects
that have shown conclusively that recall and recognition are
sensitive to manipulations of context.

Assuming that the function of memory is to store a model of the
world for later reference, we can now see that this model is dynamic:
The present situation (its encoding) brings to focus the consequences
of similar past situations—the organism makes use of its experience.
Referring back to the frame problem, the memory predicts continuously
and automatically what is about to happen.

Notice, however, that this does not solve a basic problem of
robotics, but it only shifts it. Whereas, before, we had chosen
an encoding of the world in a data base but had a problem updating
the data base (i.e., the frame problem), we now have an idea of how
to make the memory work (how to maintain the data base) but are
left with the problem of encoding the data. So has there been any
progress? Possibly. Whereas, before, we knew neither how to encode
information about the world nor how to manipulate this information,
we now have a reasonable candidate for the latter. Before, we had
just assumed that we can encode the information in object and
property lists and in rules of manipulating the lists, but that
approach led to the frame problem. The present research suggests
another approach: Assuming the memory dynamics to be known, how
to encode the data?

SUMMARY AND CONCLUSIONS

I have tried to describe a memory model in sufficient detail to give
an idea of how it works. At this level of description the model has
problems, some of which are avoided by more complete descriptions
of it. Particularly worth pointing out is that the theory is valid
mathematically over a very wide range of dimensions: It works
for patterns with as few as 100 and as many as 100,000 components.
Furthermore, the components, or features, need not be binary.
Therefore, the theory could apply even if a pattern encoding a moment
of human experience were to have 10,000 components instead of 1,000
and the total number of memory locations were in the billions instead
of a million. Such large memories would be practical because of the
massive parallelism in writing and reading, whereas simulating them
on a conventional computer would be utterly impractical.

The real story about human memory will be much more complicated
than any of our models of it so far. However, it helps to understand
simple models of the right kind if we want to develop more
comprehensive models and eventually to understand the real phenomenon
of memory itself. I have used a similar strategy by describing my
memory model in terms of a computer memory, and, presumably, that
helps people who already know how computer memories work.

Below are some statements about human memory that I believe to be
true and that are supported by my model.

1. The mind has a focus. It is associated with the states of
a set of neurons. This set is very small compared to the total
number of neurons in the brain.

2. At any moment, the information in the focus is a minute fraction of the total information stored in memory.

3. The contents of the focus serve as an address to the memory.

4. Retrieval of information from memory is iterative through the focus, making possible the retrieval of information by approximate retrieval cues.

5. We are aware of (conscious of, attend to, perceive) only things that have been brought to the focus. Stable and well—behaved states of the focus correspond to clear mental images, thoughts, and actions.

6. Some associations are inherent——they are based on form (similarity in the model)——and others are learned (frequent juxtapositions in time, experience).

7. The memory (storage) is highly distributed, and images are reconstructed statistically.

8. Writing in and reading from memory are highly parallel.

Besides possibly helping us understand human memory, the present research suggests a new construction principle for computer memory: We could build a sparse, distributed memory. It would be a random—access memory that operates on patterns with a very large number of features——i.e., points of a high—dimensional, abstract space——and it would have dynamic properties resembling those of human memory. To use such a memory in robots, we would have to learn to encode information about the world into high—dimensional feature vectors. I conclude with the suggestion that studying perception and its encoding could be a fruitful line of investigation.

REFERENCES

1. Kanerva, P. Self—propagating Search (Rep. No. CSLI—84—7, Stanford, 1984; MIT Press, in press, 1986 [est.]).
2. Marr, D. J. Physiol. 202, 437 (1969).
3. Albus, J. S. Brains, Behavior, and Robotics (BYTE Books of McGraw—Hill, Peterborough, N.H., 1981).
4. Anderson, J. A. IEEE Trans. Syst., Man, Cybern. 13, 799 (1983).
5. Hopfield, J. J. Proc. Nat. Acad. Sci. (Biophys.) 79, 2554 (1982).
6. Kohonen, T. Self—organization and Associative Memory (Springer—Verlag, N.Y., 2nd ed., 1984).
7. Willshaw, D. In G. E. Hinton and J. A. Anderson (Eds.), Parallel Models of Associative Memory (Lawrence Erlbaum Assoc., Hillsdale, N.J., 1981) p. 83.
8. Hinton, G. E., Sejnowski, T. J., and Ackley, D. H. Boltzmann Machines (Rep. No. CMU—CS—84—119, Carnegie—Mellon, 1984).
9. Rumelhart, D. E., and McClelland, J. L. (Eds.), Parallel Distributed Processing (MIT Press, 1986).
10. Llinás, R. R. Sci. Am. 232, 56 (1975).

BASINS OF ATTRACTION OF NEURAL NETWORK MODELS

James D. Keeler
U. C. San Diego Physics Department and
Institute for Nonlinear Science B-019, La Jolla CA 92093

ABSTRACT

The basins of attraction of memory states of a neural network model are explored by taking random slices through the state space. As the number of stored states increases, the basins of attraction become very complicated. This phenomena substantially reduces the performance of the network as a content addressable memory. It is shown that the performance of the network can be improved by algorithms which smooth the basins of attraction.

INTRODUCTION

Hopfield[1,2] and others[3-5] have proposed using parallel networks of neuron-like elements (neural networks) as content addressable memories (CAMs). These models use the idea of storing memory states as fixed points of the dynamics of the neural network. These fixed points are supposed to be attracting, so that any initial configuration of neurons sufficiently close to the memory state will be attracted onto that memory state. The set of all initial configurations of neurons which lead to a given memory state is called the basin of attraction of that memory state. It is clearly desirable to have a well-behaved basin of attraction around each memory state; if the basin becomes too complicated, the CAM will not be uniformly robust in its error correction. Nevertheless, little attention has been devoted to investigating the basin structure. Developments in nonlinear dynamics indicate that the basins of attraction of a nonlinear system such as this will be very complicated. Indeed, one might even expect the basin boundaries to be fractal.[8,9]

Previous methods of investigating the properties of the CAM could say nothing about the structure of the basins of attraction because the metric used, (the Hamming distance), is inadequate to distinguish among different directions in the state space. Lapedes and Farber[5] have looked at algorithms for sculpting the basins of attraction, but a detailed investigation of the structure of the basins remains to be done.

The problem encountered in investigating the basins of attraction of these systems is that the systems are very high dimensional and there is no simple method of exploring the basins. Previous investigations of basin boundaries have concentrated on one and two dimensional systems. *In the following, I develop an algorithm for looking at sections of basins of attraction by taking two dimensional slices through the state space.* An entire basin is contained within the union of all of these slices. The information gained in a single slice does not contain the information needed to calculate the basin boundary. Nevertheless, these slices yield valuable information about the *local* structure of the basin of attraction near a memory state, and can be a very useful diagnostic tool for investigating the performance of the network. An investigation such as this yields qualitative information about the dynamics of the neural network. An understanding of this dynamics may be useful for designing practical networks.

NEURAL NETWORK MODEL

As a prototype example of a system displaying the essential behavior of interest here, I choose to discuss the discrete version of the Hopfield model.[1] This model has the advantages of being easy to work with, widely known, and fast for numerical simulations. The principles outlined here carry over to the continuous versions of this model as well as other models.

Following Hopfield[2] consider a set of N neurons with the state of the ith neuron, u_i, taking the values ± 1 (i.e. either on or off).[10] The state of each neuron is updated at random times, asynchronously according to the rule:

$$u_i \leftarrow g\left(\sum_{j=1}^{N} T_{ij} u_j\right) \tag{1}$$

where T_{ij} is the connection strength from the jth neuron to the ith, and for the discrete model

$$g(x) \equiv \begin{cases} +1 & \text{if } x > 0 \\ unchanged & \text{if } x = 0 \\ -1 & \text{if } x < 0 \end{cases} \tag{2}$$

Equation (1) defines a dynamical system for the states of the neurons. The goal is to be able to store memory states as stable fixed points of this dynamical system. Each point in this state space can be thought of as an N dimensional vector $\vec{u} = (u_1, u_2, \cdots u_n)$, whose components are ± 1. Suppose we have M randomly chosen memory states which we wish to store in this system of neurons. Denote these memory states as $\vec{S}^\alpha = (S_1^\alpha, S_2^\alpha, \cdots S_N^\alpha)$, $(\alpha = 1, 2, 3, \ldots M)$. One method of storing these states is to start with no connections among the neurons then increment the values of the connection matrix according to the rule:

$$\Delta T_{ij} = S_i^\alpha S_j^\alpha \tag{3}$$

for each $\alpha = 1, 2, 3, \ldots, M$. These stored memory states $\vec{S}^\alpha$ will be attracting fixed points of equation (1) providing certain assumptions are made about the number of stored memory states and their statistical properties.[1]

The space of all initial conditions, $\vec{u}$, is a discrete space with 2^N possible states. The *basin of attraction* of a stored memory state is defined as the locus of all initial conditions in the state space which are attracted to that memory state. The goal of CAM is to have all initial configurations of neurons which are "close" to a given memory state be recognized as that state. In dynamical systems terminology, the goal is to have a well behaved basin of attraction of radius r around each memory state. The conventional way of looking at the basins of attraction of a 2 dimensional dynamical system is to plot all initial conditions in a plane and label each initial condition according to its asymptotic attractor. The boundary of the basin can then be studied as various parameters of the system are changed. For many systems, the basin boundaries display a complicated fractal structure which leads to loss of predictability of the system.[8,9]

To investigate the predictability of the neural network, it would be useful to analyze the basin boundary around each fixed point, but, in attempting this, two problems are encountered: First, the space is discrete rather than continuous, and second, there is no simple way to project the space of initial conditions onto a plane and preserve the local topology. The problem of projection is similar to that encountered in trying to project a sphere onto a plane; there is no way to make this projection and preserve local distances. Since making a sensible projection is intractable, I choose to look at two dimensional slices through the state space instead. Information about the local structure of the basin of attraction can then be gathered from these two dimensional slices.

RANDOM SLICES THROUGH THE STATE SPACE

The topology of this discrete space is rather bizarre. The space consists of 2^N discrete states which lie at the corners of N-cubes in N dimensional space. One way of measuring distance in this space is to use the Hamming distance. The Hamming distance between two states $\vec{a}$ and $\vec{b}$ is defined as the number of neurons at which $\vec{a}$ and $\vec{b}$ differ. Define a geodesic in this space to be a path of shortest distance between these two distinct states. That is, if $\vec{a}$ and $\vec{b}$ are k Hamming units apart, a geodesic is a path which consists of changing the states of each of the k neurons one at a time. Since there is no interior to the cubes, a geodesic is by no means a straight line. The shortest path starts at the corner of one cube (the state $\vec{a}$), hops to an adjacent corner of the cube, then to another corner, and so on until the state $\vec{b}$ is reached. Hopping from one corner to an adjacent corner is equivalent to changing the value of a single neuron in the network and advancing by one Hamming unit. The geodesic is $k!$ degenerate since there are $k!$ ways of getting from $\vec{a}$ to $\vec{b}$ in k steps.

Define two orthogonal directions as follows: Start at a given state $\vec{a}$, then choose another state $\vec{b}$ at random. Choose one direction, (the x-axis), to be some geodesic leading from $\vec{a}$ to $\vec{b}$. Choose the second direction, (the y-axis) to be some other geodesic leading from $\vec{a}$ to $\bar{b}$ where $\bar{b}$ is the compliment of $\vec{b}$. These two directions are orthogonal in the sense that the paths are independent and lead to positions which are as far apart as possible in this space, and these two directions can be used to make a two dimensional slice through the state space.

The way this slice is actually implemented is very simple: Given a state $\vec{a}$, this state will differ by k Hamming units from a randomly chosen $\vec{b}$. Define a pointer to the N neurons so that the neurons are re-labeled by this pointer. Let the first k re-labeled neurons be the k neurons at which $\vec{a}$ and $\vec{b}$ are the same. The last $N-k$ re-labeled neurons are neurons at which they differ. Define the *restricted Hamming distance*, $H_k(x-a)$ between an arbitrarily selected state $\vec{x}$ and the state $\vec{a}$, as the number of neurons at which $\vec{x}$ and $\vec{a}$ differ in the first k of the re-labeled neurons. That is, H_k is the Hamming distance restricted to looking only at the chosen k neurons. Similarly, define $H_{n-k}(x-a)$ as the number of sites at which $\vec{x}$ and $\vec{a}$ differ in the last $N-k$ re-labeled neurons. Hence the point (i,j) in the plane corresponds to a state $\vec{x}$ which differs from the given state $\vec{a}$ by i Hamming units in the first k neurons and j units in the last $N-k$ neurons. If $\vec{a}$ is one of the stored memory states, the slice will represent a section through the basin of attraction around that memory state.

The slice through the space defined above is a rectangle with $k(N-k)$ points in it. The corners of the rectangle are $(k,0) = \vec{b}$, $(0,N-k)=\bar{b}$ and the opposite corner $(k, N-k) = \vec{\tilde{a}}$ This slice preserves local distances: Choose any point in the rectangle and any nearest neighbor is only one Hamming unit away, thus distances are preserved in this slice. Preservation of distances in the plane is the important feature which allows the visual information to be useful; the sections of the basin can be viewed in a two dimensional plane without distortion of distances.

INVESTIGATION OF THE BASINS OF ATTRACTION

To investigate the basins of attraction as a function of the number of states stored in the network, M states were stored and one state was chosen to be the state $\vec{a}$. Each point in the above defined rectangle can be thought of as an initial condition for the dynamical system of equation (1). Remember that each point in the plane represents an N dimensional vector so there are $N-2$ dimensions not shown in these pictures. Each point in the rectangle was iterated according to equation (1) until a fixed point was reached. If the final state was the stored memory state $\vec{a}$, then the initial point was labeled with a symbol, otherwise, it was left blank. The slices were chosen at random by choosing a random pointer. The expected value of k a random initial condition is $k = N/2$, so this value of k was used.

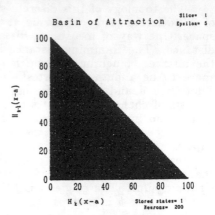

Figure 1. A slice trough the state space as described in the text. 200 neurons were used in the network and only one state was stored. There were 10,000 initial conditions tried in this square. A symbol was plotted if the initial condition was attracted down onto the stored state at (0,0).

For one state stored in the Hopfield network, the slice through the basin of attraction always looks as in figure 1). In this figure, 10,000 initial conditions were taken in the square. All of the points marked with a symbol were attracted to the only stored state, $\vec{a}$. It is clear that $\approx 1/2$ of the initial conditions found their way to this stored state. The other $1/2$ went to $\vec{\tilde{a}}$, because of the symmetric connections in the matrix T.

It would be desirable to have a well-defined ball of Hamming radius r centered at each memory state such that this ball is contained entirely within the basin of attraction. The existence of such a ball insures that the CAM will recall any state within r Hamming units of a stored state. In figure 1) it is clear that such a ball exists and the ball is of radius $r = N/2$.

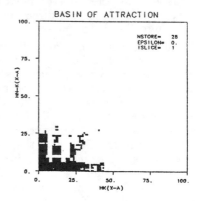

Figure 2. The same slice as in figure 1), with the same stored state as the origin (0,0), but now 28 states have been stored in the network. Notice the large crevices and holes which develop.

This is the maximum radius of attraction expected for the Hopfield model, and corresponds to 50% error in the initial configuration of neurons. The "balls" of constant radius in this slice look like triangles as seen in figure 1). The point (i,j) in the slice is $i+j$ Hamming units away from the stored state at $(0,0)$. Hence, a ball of constant radius in this slice is a line extending from the point $(0,r)$ to $(r,0)$, and this line gives the impression of a triangle.

As more states are stored, the basin changes as shown in figures 2-3). From these figures, it is clear that the basins become very complicated. As more states are stored in the network, the basin section develops large crevices and holes close to the stored memory state as shown in figure 2). These holes degrade the performance of the network by allowing errors to be incurred for small Hamming distances from the stored state. Other random slices were taken which show qualitatively the same behavior and structure as shown in these figures, so it is not misleading to look at only one slice. In figure 3) the radius of the basin has shrunk almost to zero. Note that the degradation of the basin occurs around $M = 0.15N$, and is near the predicted value.[6] This degradation of the basins of attraction is expected to occur in any high-dimensional nonlinear system such as a neural network, and is presumably fractal in the continuous case.

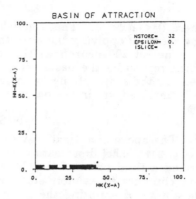

Figure 3. The same slice as in figure 1) and 2), but now the number of states stored in the network has increased to 32 The average radius of the basin has shrunk to almost zero, and the basin is very asymmetrical.

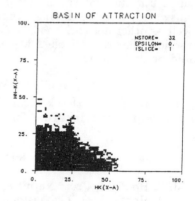

Figure 4. The same slice and number of stored states, (32) as in figure 3), but the unlearning algorithm has been used for a total of 1,000 times with a unlearning strength of $\delta = 0.01$. Notice that the basin has become much more regular and the average radius has increased by a factor of almost 10.

SMOOTHING THE BASINS

Now that the problem has been demonstrated, the immediate question is how to alleviate the problem. The goal is to find an algorithm for modifying the basins of attraction of these memory states to fill the holes and smooth out the rough edges of the boundaries. Space requirements prevent me from presenting elaborate algorithms for modifying the basins, so I present a very simple algorithm here to demonstrate the principle. Other results will be presented in a much longer report. One algorithm that has been used in the past for improving the performance of this CAM is the so-called unlearning algorithm.[2,7] In this algorithm, random initial conditions are given and allowed to relax to a final state. These final states $\bar{u}\,'$ are then used to modify the connection matrix by

$$T(i,j) \leftarrow T(i,j) - \delta u'_i u'_j \qquad (4)$$

where $0 < \delta << 1$. This algorithm is repeated for many initial conditions. The result of applying the algorithm for 1000 initial conditions with $\delta = 0.01$ on a net of 200 neurons is shown in figure 4). Compare this to figure 3), which represents the basin before unlearning. There is a dramatic improvement in the shape of the basins of attraction and the maximum radius of a ball contained within the basins.

DISCUSSION

The above investigation yields information on two important features of the CAM. The first feature is the structure of the basins of attraction around a fixed point of the CAM, and how the basin of attraction becomes very complicated as more states are stored. The second important feature is the ability of sculpting the basins of attraction and alleviating the problem incurred by the complicated structure. Although the results presented here are qualitative, the information obtained by this investigation is important for understanding the underlying dynamics of the CAM, and hence how to improve its performance. This is the first investigation of its kind into the detailed structure of the basins of attraction and many questions remain to be answered. A more detailed analysis should be done for continuous models as well as other network models. Other algorithms for modifying the basins of attraction should be looked at, and some way of quantifying these results should be developed.

ACKNOWLEDGEMENTS

I would like to thank John Hopfield for helpful discussions, and Henry Abarbanel and the Institute for Nonlinear Science at U. C. San Diego for support.

REFERENCES

[1] J. J. Hopfield, Proc. Natn'l Acad. Sci., **79** 1554 (1982)

[2] J. J. Hopfield, D. I. Feinstein, and R. G. Palmer, Nature **304** 158 (1983)

[3] W. A. Little, G. L. Shaw, Math. Biosci. **39** 281 (1978)

[4] T. Kohonen, *Self-Organization and Associative Memory,* (1984), Springer-Verlag, Berlin.

[5] A. Lapedes, R. Farber, " A Self-Optimizing, Nonsymmetrical Neural Net for Content Addressable Memory and Pattern Recognition" Los Alamos Preprint (1985)

[6] D. J. Amit, H. Gutfreund, and H. Sompolinsky, Phys. Rev. Lett. **55** 1530 (1985)

[7] D. Kleinfeld, D. B. Pendergraft, "Unlearning Increases the Storage Capacity of Content Addressable Memories" Preprint

[8] S. W. McDonald, C. Grebogi, E. Ott, J. A. Yorke, Physica **17D** 125 (1985)

[9] E. G. Gwinn, R. M. Westervelt, "Fractal Basin Boundaries and Intermittency in the Driven Damped Pendulum", Preprint.

[10] W. A. McCulloch, and W. Pitts, Bull. Math. Biophys. **5** 115 (1943)

A DRIVE-REINFORCEMENT MODEL OF SINGLE NEURON FUNCTION:
AN ALTERNATIVE TO THE HEBBIAN NEURONAL MODEL

A. Harry Klopf*
Air Force Wright Aeronautical Laboratories
Wright-Patterson Air Force Base, Ohio   45433

ABSTRACT

A neuronal learning mechanism is proposed that accounts for the basic animal learning phenomena that have been observed. Among the classical conditioning phenomena predicted by the neuronal model are delay conditioning, trace conditioning, simultaneous conditioning, conditioned stimulus duration and amplitude effects, unconditioned stimulus amplitude effects, interstimulus interval effects, second and higher order conditioning, conditioned inhibition, habituation and extinction, reacquisition effects, backward conditioning, blocking, overshadowing and serial compound conditioning. The proposed neuronal model and learning mechanism offer a new building block for constructing neural network-like computer architectures for artificial intelligence.

THE NEURONAL MODEL AND LEARNING MECHANISM

It is suggested that the Hebbian[1] neuronal model be modified in the following ways to make it consistent with the animal learning phenomena it is intended to explain: (a) Instead of correlating pre- and postsynaptic levels of activity, changes in pre- and postsynaptic levels of activity should be correlated; (b) Instead of correlating approximately simultaneous pre- and postsynaptic signals, earlier presynaptic signals should be correlated with later postsynaptic signals. More precisely and consistent with (a), earlier changes in presynaptic signals should be correlated with later changes in postsynaptic signals. Thus, sequentiality replaces simultaneity in the model; (c) A change in the efficacy of a synapse should be proportional to the current efficacy of the synapse, accounting for the initial positive acceleration in the classic s-shaped acquisition curves observed in animal learning.

The resulting neuronal model is an extension of the Sutton-Barto[2] model. It may be termed a drive-reinforcement model because it suggests that nervous system activity can be understood in terms of two classes of signals: Drives that are defined to be

*This research was supported by the Life Sciences Directorate of the Air Force Office of Scientific Research under Task 2312 R1. Jim Morgan is acknowledged for the software he wrote for the single neuron simulator employed in this research.

signal levels and <u>reinforcers</u> that are defined to be changes in
signal levels. Mathematically, the learning mechanism of the
drive-reinforcement model may be characterized as follows:

$$\Delta w_i(t) = \sum_{j=1}^{\tau} c_j |w_i(t-j)| [x_i(t-j) - x_i(t-j-1)][y(t) - y(t-1)] \qquad (1)$$

where $w_i(t)$ is the weight or efficacy of synapse i at discrete time,
t; Δw_i is the change in efficacy of synapse i; τ is the longest
interstimulus interval over which conditioning is effective; c_j is
a learning rate constant which is proportional to the efficacy of
conditioning when the interstimulus interval is j; x_i is a measure
of the frequency of action potentials at synapse i; and y is a
measure of the frequency of firing of the neuron.

Two refinements of the above equation further improve the
model's ability to predict animal conditioning phenomena. The first
refinement involves allowing only positive changes in x to
contribute to a change in a synaptic weight, w. Negative changes in
x are set equal to zero for the purpose of calculating changes in
synaptic efficacy. The second refinement involves assigning a new
role to negative changes in x. A negative change in x serves to
reduce an earlier positive change in x if the negative change in x
comes between a positive change in x and a change in y. This way of
utilizing negative changes in x is termed the <u>trace conditioning
mechanism.</u>

A lower bound is set on the absolute values of the synaptic
weights, w_i. The bound is near but not equal to zero because
synaptic weights appear as factors on the right side of equation
(1). It can be seen that the learning mechanism would cease to
yield changes in synaptic efficacy for any synapse whose efficacy
reached zero; i.e., $\Delta w_i(t)$ would henceforth always equal zero. A
nonzero lower bound on the efficacy of synapses models the notion
that a synapse must have some effect on the postsynaptic neuron in
order for the postsynaptic learning mechanism to be triggered.

Computer simulations of the drive-reinforcement neuronal model
are discussed below. In these simulations, each input to the neuron
is made available via both an excitatory and an inhibitory synapse
so that the neuronal learning mechanism has, for each input, both an
excitatory and an inhibitory weight available for modification. In
the classical conditioning experiments that were simulated, the
weights associated with synapses carrying unconditioned stimuli were
assumed to be fixed (nonplastic) and the remaining weights were
assumed to be variable (plastic).

PREDICTIONS OF THE NEURONAL MODEL

It can be demonstrated that the drive-reinforcement neuronal model accounts for the basic animal learning phenomena that have been observed. This includes, within the category of classical or Pavlovian conditioning: delay conditioning, trace conditioning, simultaneous conditioning, conditioned stimulus duration and amplitude effects, unconditioned stimulus amplitude effects, inter-stimulus interval effects, second and higher order conditioning, conditioned inhibition, habituation and extinction, reacquisition effects, backward conditioning, blocking, overshadowing and serial compound conditioning. The neuronal model also provides a basis for understanding instrumental conditioning, thus providing a single-process explanation of classical and instrumental conditioning.

Examples of conditioning phenomena predicted by the drive-reinforcement neuronal model are shown in Figure 1. Space limitations prevent the presentation and discussion of further examples. In the figure, comparisons are made of the predictions of three models: Hebb's[1] learning mechanism, the Sutton-Barto[2] learning mechanism and the drive-reinforcement learning mechanism. For all three models, the following neuronal input-output relationship is assumed:

$$y(t) = \sum_{i=1}^{n} w_i(t)x_i(t) - \theta \tag{2}$$

where $y(t)$ is a measure of the postsynaptic frequency of firing at discrete time, t; n is the number of synapses impinging on the neuron; w_i is the efficacy of synapse i; x_i is a measure of the frequency of action potentials at synapse i; and θ is the neuronal threshold. The synaptic efficacy, w_i, can be positive or negative, corresponding to excitatory or inhibitory synapses, respectively. Also, $y(t)$ is bounded such that $y(t)$ is greater than or equal to zero and less than or equal to the maximal output frequency, $y'(t)$, of the neuron.

The learning mechanism for the drive-reinforcement neuronal model has already been specified in equation (1). The Hebbian learning mechanism may be specified as follows:

$$\Delta w_i(t) = cx_i(t)y(t) \tag{3}$$

where c is a learning rate constant and the other symbols are defined as above. The Sutton-Barto learning mechanism is specified by the following equations:

$$\Delta w_i(t) = c\bar{x}_i(t)[y(t) - \bar{y}(t)] \tag{4}$$

$$\bar{x}_i(t) = \alpha \bar{x}_i(t-1) + x_i(t-1) \tag{5}$$

where $\bar{y}(t) = y(t-1)$ for the theoretical curves shown in the figure,

α is a positive constant and the other symbols are defined as above.

In Figure 1, predicted acquisition (learning) curves are compared for the Hebbian, Sutton-Barto and drive-reinforcement learning mechanisms. A delay conditioning paradigm is utilized; this is a classical (Pavlovian) conditioning paradigm in which the onset of the conditioned stimulus (CS) precedes the onset of the unconditioned stimulus (UCS) and the offset of the CS occurs at or after the onset of the UCS. The CS and UCS were presented once in each trial and the values of the synaptic weights at the end of each trial are shown in the graphs. (Data points are not shown on these graphs because they fall exactly on the computed theoretical curves.)

In Figure 1, a comparison of the predictions of the three models for delay conditioning is made for each of three cases: CS offset at the time of UCS onset in the case of CS_1; CS offset at the time of UCS offset in the case of CS_2 and CS offset after UCS offset in the case of CS_3. Thus, the predicted effects of CS duration are examined for each model. Experimentally, it is known that conditioned excitation (corresponding to positive synaptic weights) is observed in all three cases with the acquisition curve positively accelerating initially and negatively accelerating subsequently. [3,4] It can be seen in Figure 2(c) that the drive-reinforcement model's predictions are consistent with the experimental evidence in all three cases. The Hebbian model predicts conditioned excitation for two of the three cases (CS_2 and CS_3) and no conditioning for the third case (CS_1), as can be seen in Figure 2(a). Furthermore, the predicted acquisition curves for the Hebbian model are essentially linear. The Sutton-Barto model predicts conditioned excitation for one case (CS_1) and conditioned inhibition for the other two cases (CS_2 and CS_3), with acquisition curves that are negatively accelerated. The theoretical acquisition curves shown in Figure 1 were obtained by means of computer simulations of each neuronal model. The CS's and UCS's are represented in the neuronal models as presynaptic signal levels, x_i. The CS's and UCS's have a baseline level of zero. Y is equal to $y(t)$ for the last trial shown on the graph. Parameter values utilized in the computer simulations of the three neuronal models are as follows: $0 \leq y(t) \leq 1.0$, $\theta = 0.0$, $n = 2$, UCS (fixed) weight = 1.0, CS amplitudes = 0.2, UCS amplitudes = 0.5, CS onsets at t = 10, CS_1 offset at 13, CS_2 offset at t = 14, CS_3 offset at t = 15, UCS onsets at t = 13, UCS offsets at t = 14; in (a), c = 0.5, initial plastic weight values = 0.0; in (b) c = 0.5, α = 0.9, initial plastic weight values = 0.0; in (c) initial values of excitatory plastic weights and lower bound = 0.1, initial values of inhibitory weights and lower bound = 0.0, $\tau = 5$, $c_1 = 5.0$, $c_2 = 3.0$, $c_3 = 1.5$, $c_4 = 0.75$, $c_5 = 0.25$. In (c), the relative values for c are consistent with the assumption that a time step represents (nominally) one-half second. For the trace conditioning mechanism, negative changes in x reduce positive changes in x if the trace interval is greater than zero.]

CONCLUSIONS

The accuracy of the drive-reinforcement model's predictions, as shown in Figure 1(c), is typical of its predictions for the other categories of classical conditioning phenomena noted earlier. This provides reason to believe that the drive-reinforcement learning mechanism specified by equation (1) may model the learning mechanism employed by single living neurons. If so, the drive-reinforcement neuronal model and learning mechanism offer a new building block for constructing neural network-like computer architectures for artificial intelligence.

REFERENCES

1. D.O. Hebb, The Organization of Behavior (Wiley, N.Y., 1949).

2. R.S. Sutton, A.G. Barto, Toward a modern theory of adaptive networks: expectation and prediction, Psychological Review, 88, 135 (1981).

3. K.W. Spence, Behavior Theory and Conditioning (Yale University Press, New Haven, 1956).

4. L.J. Kamin, Temporal and intensity characteristics of the conditioned stimulus. In W.F. Prokasy (Ed.), Classical Conditioning: A Symposium (Appleton-Century-Crofts, N.Y., 1965).

270

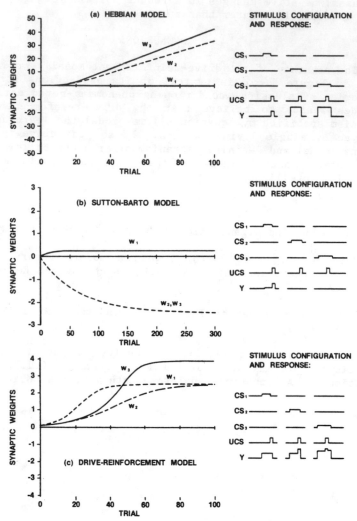

FIGURE 1. A comparison of predicted acquisition (learning) curves for classical (Pavlovian) conditioning for the (a) Hebbian, (b) Sutton-Barto, and (c) drive-reinforcement models of single neuron function. See text for additional explanation and discussion.

REPRESENTATION OF SENSORY INFORMATION IN SELF-ORGANIZING FEATURE MAPS

T. Kohonen and K. Mäkisara
Helsinki University of Technology, SF-02150 Espoo, Finland

ABSTRACT

This paper shows how the internal representations necessary for memorization can be automatically formed in neural networks or artificial systems.

INTRODUCTION

The main functions of the brain are the control of animal functions like blood circulation and respiration, and the control of behaviour. To perform the latter task the behaviour of the environment must be predicted and the brain must be able to form various kinds of internal model at different levels of abstraction. This is the concept of memory in its most general sense.

From above it may be obvious that the two main functions of memory are: 1. Representation of sensory information, 2. Storage of the associations. Most of the earlier and contemporary modelling work has concentrated on the second aspect. It is self-evident, however, that the first one, the "front-end" problem of memory must be solved first. To this end it is important to be able to form compressed mappings from the outer world onto the brain so that the dimensionality of the signal space is reduced without loss of the structures included in it.

A few attempts (cf., e.g. Ref. 1) have been made to transform feature values of sensory signals[2] into spatial patterns on the cortex. In 1981, one of the authors[2] succeeded in "isolating" those functional laws which seem to be essential and sufficient for the formation of such internal models. It turned out that the structure of an internal model can be a two-dimensional "map" or coordinate system which represents topological relationships between various sensory occurences. In effect, such a map "flattens" a high-dimensional signal space in a nonlinear way into two dimensions, keeping only the most important feature dimensions in the representation.

THE NEURAL MODEL

The most important types of neural signal are the trains of action pulses which are propagated down the axons. We hold the view that a single pulse has a negligible information value, whereas the average frequency of neural pulses defines the signal. This frequency is a positive real number denoted by ξ_{ij} or η_i in the following. An approximation of the "transfer function" of a single cell may then be expressed by the functional

$$\eta_i = \sigma\left(\sum_{j=1}^{n} \xi_{ij} \mu_{ij} - \Theta_i\right) \tag{1}$$

where $\sigma(.)$ is a "sigmoid" function, having saturation limits at zero and some positive value $\Theta_{i\,max}$; Θ_i is a threshold or bias, and the μ_{ij} can be positive or negative, corresponding to the excitatory and inhibitory synapse, respectively. In many cases it is sufficient to consider only the linear part of (1) in which case the response of a neuron is the inner product of the incoming signals ξ_{ij} and the connection weights μ_{ij}.

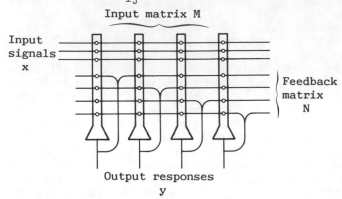

Input matrix M

Input signals
x

Feedback matrix N

Output responses
y

Fig. 1. The basic structure of neural circuits used in brain models.

A neural cell, however, never operates alone. It is usually embedded in a network with abundant feedback connections. The ordinary memory models mainly concentrate on laterally connected neural nets. We have, however, come to the conclusion that even the simplest neural nets should be described with a model (see Fig. 1) in which the neurons receive two kinds of input: the connections from sensory organs or other areas, and the feedback connections within the same area. It seems that the synaptic system of the incoming connections is responsible for feature analysis of the incoming signals and the feedback synaptic system is responsible for the associative memory, respectively. The same network can handle both of these tasks.

The operation of feedback systems is in general very complex, but in principle the neural network can be described by the following three differential equations:

$$\frac{dy}{dt} = f(x,y,M,N) \quad , \tag{2a}$$

$$\frac{dM}{dt} = g(x,y,M) \quad , \tag{2b}$$

$$\frac{dN}{dt} = h(y,N) \quad , \tag{2c}$$

in which the inputs to the cells are denoted by the vector x and the outputs of the cells by the vector y. M and N are connection matrices relating to the different synaptic connections between the cells.

The first differential equation (2a) relates the incoming signals, the feedback signals and the connection matrices. This equation mainly describes relaxation of electrical activities in the network within a time of the order of 100 ms. The parametric equation for the incoming signals (2b) describes a much slower phenomenon (changes occurring within weeks) based on changes in proteins and macroscopic structures. This equation is responsible for formation of the feature maps.

The third equation (2c) is the equation of the feedback connections which are responsible for the associative memory. Here we have the difficulty that the memory traces should be formed rapidly whereas they should be permanent. The accurate treatment of this equation is not possible here and we should also include the control circuits for attention which modulate the plasticity of the neural connections.

RELAXATION OF THE ACTIVITY

We will first consider the relaxation equation (2a) in a memoryless case assuming that the interaction function between the cells is like the "Mexican hat" function seen in Fig. 2 a). At short distances up to about half a millimeter the feedback is positive, and from there maybe up to one millimeter negative. The role of this feedback is to emphasize the signal activity in the net. At still longer distance there is a weak positive feedback which is presumed to relate to associative memory. The function shown here is first assumed constant in time.

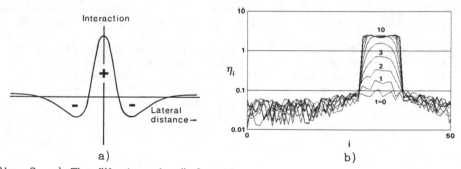

a)   b)

Fig. 2. a) The "Mexican hat" function.
b) Formation of "bubbles" of activity over a one-dimensional network.

In Fig. 2 b) there is a simple example of the effect caused by this function. The input to the network has random spatial distribution with a slight maximum in its average (see curve t=0). Curve t=10 gives the activity after relaxation and the activity has been concentrated around the initial maximum. We call this area of activity the "bubble".

The bubbles can be formed in the same way in two- or multidimensional nets. They have also been measured, e.g., on the cortices of several small mammalia[3].

SELF-ORGANIZATION OF THE INPUT CONNECTIONS

The self-organizing process to be described below creates an ordered feature extraction mapping in the incoming synaptic connections (matrix M). Assume first, for simplicity, that the input is connected in parallel to all cells of a two-dimensional net. In the following discussion, the input weights are assumed modifiable.

Assume that bubble formation for slowly varying inputs has happened first. Now we can write the adaptation equations for the weights in many ways, of which one is

$$d\mu_{ij}/dt = \alpha \eta_i \xi_j - \beta(\eta_i) \mu_{ij}. \tag{3}$$

The first term corresponds to the Hebbian law[4] which states that the changes of the connection are proportional to the product of the incoming signal and the activity of the cell. The second term is a nonlinear, activity-dependent forgetting term and its effect is to stabilize the memory traces[5]. The parameter α is a positive constant.

The process can be simplified considerably by making use of bubble formation and assuming that the constant term in the Taylor expansion of β is zero, which is equivalent with assuming that the forgetting is active:

$$d\mu_{ij}/dt = \text{const.} \; (\xi_j - \text{const.} \; \mu_{ij}) \tag{4a}$$

inside the bubble,

$$d\mu_{ij}/dt = 0 \qquad \text{outside the "bubble".} \tag{4b}$$

Since the scaling of signals in relation to the connection strengths is arbitrary, we may put the second constant (in the parentheses) equal to unity.

It can be shown[5] that for a rather general choice of the form $\beta(.)$ the norms of the vectors tend to constant, identical values. Because of this, matching on the basis of maximum inner products yields the same location for the "bubble" as if the matching were based on the minimum of the vectorial differences between x and the m_i. Combining this result and (4) leads to the simplest self-organizing algorithm: Starting with random initial connections, the following two steps for each learning pattern presented to the network are computed:

Bubble formation:
If $\|x - m_c\| = \min_i \{\|x - m_i\|\}$, then N_c is defined as the set of cells corresponding to the bubble with fixed radius, centered at cell c.

Adaptation:
$$dm_i/dt = \alpha (x - m_i) \qquad \text{for } i \in N_c ,$$

$$dm_i/dt = 0 \qquad\qquad \text{for } i \notin N_c ,$$

where α is the "adaptation" gain ($0 < \alpha < 1$).

As shown by numerous experiments, for best organizing results the radius of the "bubble" and the parameter α should be made time-variable. Experimentally established values for $\alpha = \alpha(t)$ and $N_c = N_c(t)$ can be found from Ref. 5. There may exist biological counterparts of these effects, associated with time-variable neural plasticity and control of lateral inhibition, which we do not discuss here, however.

We have demonstrated by a number of experiments[2,5,6] that the above process is able to form two-dimensional "maps" of the input signals such that the most important feature dimensions that are present in the statistical distribution of x will be displayed as coordinates in the map. Since this would perhaps remain obscure otherwise, the following result is used for illustration.

To find out the usefulness of a neural model, it must be tested with natural signals which have a certain stationary density function of the input denoted $p(x)$. The "maps" have been tested by using natural speech as input[6]. The input signals for the map are obtained by simulating the effect of the inner ear by computing short-time spectra of the acoustic waveform every 10 milliseconds from samples over 25 milliseconds. The responses from the "map" to different acoustic spectra were labeled according to the location of the centroid of the "bubble" and the corresponding phoneme present in speech. The map was shown to learn the responses to the different phonemes in an orderly fashion (Fig. 3). The cells were organized in a hexagonal lattice corresponding to the letters.

a a a ah h $æ$ $æ$ $ø$ $ø$ e e e
o a a h r $æ$ l $ø$ y y j i
o o a h r r r $ŋ$ n y j i
o o m a r m n m n j i i
l o u h v vm n n h hj j j
l u v v p d d t r h hi j
$.$ $.$ u v tk k p p p r k s
$.$ $.$ v k pt t p t p h s s

Fig. 3. Two-dimensional map of Finnish phonemes. The double labels mean mapping of different phonemes onto the same location.

We have used such phoneme maps for the recognition of continuous speech, and a practical microprocessor equipment has already been constructed[7].

One would, of course, like to have a clear analytical explanation and mathematical proofs for this self-ordering process. In principle, this is a Markov process, although a rather complicated one. It is not possible to show here the proof even in the simplest case[5]. Intuitively one may become convinced that eqs. (3) represent a smoothing process and the m_i tend towards a monotonic two-dimensional sequence. The distribution of the asymptotic m_i values, however, can be shown to be a delicate function of the x values; in fact, the set of the m_i can be shown to approximate the density function $p(x)$, and the map represents a hierarchical clustering graph of the x values. In other words, the various cells of the network correspond to various feature detectors which will be distributed into the input space optimally with respect to the statistics of the input signals.

RELATIONSHIP OF SENSORY MAPS TO THE DISTRIBUTED MODELS OF ASSOCIATIVE MEMORY

It is now interesting to see how the three systems described by the differential equations (2) cooperate in the network. The input synaptic system determines a set of features from the input signals. These features are assumed constant in the following discussion and this results in a set of fixed input weights for each cell. In the feedback there exist, however, two different factors: the time-invariant "Mexican-hat" type of feedback enhancing the activity in the net, and the long-range time-variable feedback responsible for the associative memory. The latter can be modeled e.g. by the cross-correlation matrix of the input-output pairs stored in the memory[5,8,9,10,11]

276

$$N = \sum_{k=T_0}^{T} y_k y_k^T, \tag{5}$$

where the y_k are the output patterns which we associate with themselves. This matrix can be used in the differential equation for the
outputs (2a) in many ways but it can be simply added to the
activity-enhancing feedback to show the effect of the memory. This
can be shown with the following simple example. We have again the
linear array of 50 cells but now we have added to the "Mexican-hat"
feedback function a feedback function which corresponds to
memorization of three patterns (y_1, y_2, y_3). Each pattern consists of
two "bubbles" (at locations 7 and 28, 15 and 35, and 22 and 42,
respectively; see Fig. 4 a) created with the same relaxation as
before. When a key pattern containing only one slight maximum (at
cell 15 in Fig. 4 b) is presented to the network, first a "bubble" is
formed at that location. A little later (after t=5) another "bubble"
rises and this corresponds to the "bubble" (at cell 35) associated
with the key in the memory.

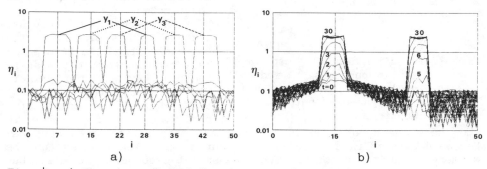

a)   b)

Fig. 4. a) The three "bubble" pairs stored in the memory.
        b) Recollection of a two peak pattern using a one peak key.

REFERENCES

1. C. v.d. Malsburg, Kybernetik, 14, 85-100 (1973).
2. T. Kohonen, Proc. 2nd Scandinavian Conf. on Image Analysis, pp.
    214-220 (1981).
3. D. W. De Mott, Medical Research Engineering, 23-29 (1966).
4. D. Hebb, Organization of Behavior, (Wiley, N.Y., 1949).
5. T. Kohonen, Self-Organization and Associative Memory (Springer,
    1984).
6. T. Kohonen, K. Mäkisara, T. Saramäki, Proc. 7th ICPR, pp. 182-185
    (1984).
7. K. Torkkola, H. Riittinen, Proc. ICASSP 86, pp. 333-336 (1986).
8. T. Kohonen, Helsinki Univ. of Technology Report TKK-F-A130
    (1970).
9. T. Kohonen, IEEE Trans. Comp., C-21, 353-359 (1972).
10. K. Nakano, IEEE Trans. Systems, Man, and Cybernetics, SMC-2,
    380-388 (1972).
11. T. Kohonen, Associative Memory - A System-Theoretical Approach
    (Springer, 1977).

DIFFERENTIAL HEBBIAN LEARNING

Bart Kosko

VERAC, Inc., 9605 Scranton Road, San Diego, CA   92121-1771

ABSTRACT

The differential Hebbian law $\dot{e}_{ij} = \dot{C}_i \dot{C}_j$ is examined as an alternative to the traditional Hebbian law $\dot{e}_{ij} = C_i C_j$ for updating edge connection strengths in neural networks. The motivation is that concurrent change, rather than just concurrent activation, more accurately captures the "concomitant variation" that is central to inductively inferred functional relationships. The resulting networks are characterized by a kinetic, rather than potential, energy. Yet we prove that both system energies are given by the same entropy-like functional of connection matrices, Trace($\dot{E}$ E). We prove that the differential Hebbian is equivalent to stochastic-process correlation (a cross-covariance kernel). We exactly solve the differential Hebbian law, interpret the sequence of edges as a stochastic process, and report that the edge process is a submartingale: the edges are expected to increase with time. The submartingale edges decompose into a martingale or unchanging process and an increasing or novelty process. Hence conditioned averages of edge residuals are encoded in learning though the network only "experiences" the unconditioned edge residuals.

INTRODUCTION

Synaptic connections are causal connections. Their modification is an act of inductive inference. Edge connection strengths are inferred from node (neuron, processing element, etc.) behavior. This suggests that the modification criteria should, at minimum, reflect the logico-causal criteria of scientific method used for attributing a functional relation-ship among variable quantities. And what are these criteria but that the quantities should move or change in the same or opposite directions and that the "cause" temporally precedes the "effect?" Eighteenth century empiricist philosopher David Hume[1] observed that we habitually make causal ascriptions when we observe sustained "constant conjunctions of events." In his System of Logic, nineteenth century empiricist philosopher John Stuart Mill[2] refined Hume's observation. Mill observed that the causality we attempt to inductively infer is simply the "concomitant variation" of the variable quantities, and this formulation, often restated in the jargon of statistical inference, remains the operative notion of causality today.

THE DIFFERENTIAL HEBBIAN LAW

The task is to specify the dynamical equation $\dot{e}_{ij} = f(C, E)$ of the edge e_{ij} that connects node (causal variate) C_i to C_j in a network, where $C^T = (C_1, \ldots, C_n)$ is the node state vector and $E = [e_{ij}]$ is the matrix of edge connections. We assume the transfer equation $\dot{C}_i = g(C, E)$ is given for each node (and is dominated by an inner product of input edges and

nodes):
$$\dot{c}_i = \sum_k c_k e_{ki} + D_i , \qquad (1)$$

where D_i contains all other terms. This functional form, a sort of neural OR gate, predominates in neural modeling. We assume that the righthand side of (1) contains terms _internal_ and _external_ to the network. The simplest internal terms are the inner product minus the current activation C_i. The external terms are an observation or sensor term O_i and an advice or teacher or expert-response term R_i, both of arbitrary structure. Hence $D_i = O_i + R_i - C_i$.

The standard selection of f in $\dot{e}_{ij} = f(C, E)$ is attributed to the "correlation learning" hypothesis of Hebb[3], which is simply that concurrent activation of nodes increases the "synaptic efficacy" or strength of the connection between them. In a nutshell,

$$\dot{e}_{ij} = C_i C_j . \qquad (2)$$

Typically the current strength e_{ij} is subtracted from the righthand side to represent "forgetting" or "memory decay" (or to slow the otherwise exponential growth?). This linear appendage does not affect the current analysis and for notational simplicity we omit it.

Equation (2) is widespread in the neural net literature. It occurs in the famous Grossberg[4-6] equations and the related equations of Hopfield[7] and in similar form in the adaptive equations of most neural modelers. But apart from referencing Hebb's (nonmathematical) conjecture, it seems the operative argument for using (2) is simply that everyone uses it. For surely the problems with (2) warrant investigating alternatives. To begin, (2) promotes spurious causal associations. If any two processors or nodes are active in a network, no matter how big the network, how far apart the nodes, or how independent their patterns of activation, the Hebbian law (2) grows a causal connection between them. Concomitant activation replaces concomitant variation. Worse, the spurious causal attributions tend to grow exponentially fast (as can be seen from the exponential form of the exact solution of (2) when the forget term $-e_{ij}$ is appended). In practice this necessitates "hardclipping" of interconnects both during training and classification sessions. Finally, transfer functions must first be integrated before (analytically) including them in (2). This integration is never easy.

A natural alternative to (2) is the _differential Hebbian law_:

$$\dot{e}_{ij} = \dot{C}_i \dot{C}_j . \qquad (3)$$

The differential Hebbian measures concomitant variation. It imputes causality according to (lagged) conjunctions of event changes. As a _result_, it truly behaves in correlation fashion. For although the functions C_i are nonnegative, their derivatives are not. Hence the connection e_{ij} strengthens iff both nodes agree in sign, hence iff both nodes move in the same direction. (Note this implies concurrent activation.) Negative causality accumulates if they move in opposite directions. Moreover, transfer functions such as (1) can be directly plugged into (3), allowing many properties to be determined analytically. Below we exploit this fact to solve the system (1) and (3) for $e_{ij}(t)$.

The idea of using rates of change in learning laws is spreading. Two especially noteworthy cases are the drive-reinforcement model of Klopf[8-9] and the backward-error-propagation model of Rumelhart[10], Hinton, and Williams. Klopf reports that a wide array of Pavlov-like learning behaviors is accurately predicted (retrodicted) by a change-based law. Rumelhart uses a time derivative of input activation (essentially $\dot{C}_i$) in his "generalized delta law" (subsuming the classical perceptron convergence theorem) to solve the exclusive-or problem, the parity problem, and a variety of others.

KINETIC ENERGY CONNECTIONS

The energy of a network is the sum of the eigenvalues of the product connection matrix $E\dot{E}$. Here $\dot{E}$ is the n-by-n symmetric matrix of connection changes $[\dot{e}_{ij}]$. For Hopfield networks, Abu-Mostafa and St. Jaques[11] have shown that the number of energy minima (memory sites) is no more than n. We conjecture that these and comparable equilibria correspond to the eigenvalues of $E\dot{E}$.

To prove the eigenvalue theorem, define the potential energy (P.E.) and kinetic energy (K.E.) of the network [C, C, E, E] as follows:

$$\text{P.E.} = c^T E c = \sum_i \sum_j c_i c_j e_{ij}, \tag{4}$$

$$\text{K.E.} = \dot{c}^T E \dot{c} = \sum_i \sum_j \dot{c}_i \dot{c}_j e_{ij}. \tag{5}$$

The trick is that if the Hebbian law (2) is in force, then $\dot{e}_{ij}$ replaces $c_i c_j$ in (4). If the differential Hebbian law (3) is in force, then $\dot{e}_{ij}$ replaces $\dot{c}_i \dot{c}_j$ in (5). Hence then P.E. = K.E. !

THEOREM.
$$\sum_{i=1}^{n} \lambda_i = \begin{cases} \text{P.E.} & \text{if } \dot{E} = c\,c^T \\ \text{K.E.} & \text{if } \dot{E} = \dot{c}\,\dot{c}^T \end{cases}, \tag{6}$$

where $\lambda_1, \dots, \lambda_n$ are the eigenvalues of $E\dot{E}$ ($\dot{E}E$).

PROOF. By basic linear algebra, the sum of eigenvalues of $E\dot{E}$ equals the sum of diagonal elements, the trace Trace($E\dot{E}$). Then

$$\text{Trace}(E\dot{E}) = \sum_i (E\dot{E})_{ii} = \sum_i \sum_j e_{ij} \dot{e}_{ji} = \sum_i \sum_j e_{ij} \dot{e}_{ij}$$

equals P.E. or K.E. according as $\dot{E}$ equals $c\,c^T$ or $\dot{c}\,\dot{c}^T$. Q.E.D.

Two comments are in order. First, when a no weight-change analysis is desired--as it often is in Grossberg and Hopfield networks--simply invol the Hebbian or differential Hebbian law to replace $\dot{e}_{ij}$ with $C_i C_j$ or $\dot{C}_i \dot{C}_j$. Second, Trace($E\dot{E}$) can be viewed as an entropy-like functional. For, in learning networks, the edges or neural pathways adapt slowly to the activation or signals flowing through them. I.e., $\dot{e}_{ij}$ tends to be smaller than e_{ij} at any time t. Let us model this property with the hypothesis that $\dot{e}_{ij} \cong \ln e_{ij}$. Now recall that Von Neumann[12] defined the entropy of a (quantum-mechanical) system with Trace(P lnP), where P is a positive semidefinite matrix with Trace(P) = 1, thus generalizing the log-of-probability entropy of Boltzmann and others (including Shannon). Watanabe[13] has kept this entropy measure current by showing how its minimization corresponds to pattern recognition in many cases. On our learning hypothesis, Trace(EE) $\cong$ Trace(P lnP) for suitable P. However, since Trace(E) = Trace(Ė) = 0, E cannot be normalized to 1 and thus an exact identity between the two trace functionals cannot be expected to hold.

CORRELATION AND AN EXACT SOLUTION

Let us interpret the node vector C as a random vector. Let each node C_i be a stochastic process: $C_i : T \times \Omega \rightarrow [0, \infty)$, where T is an index set of time values and (Ω, A, P) is the probability space. Suppose we know the present value of $C_i(t)$ $(C_i(t, \omega))$. Then given no other information, and being true scientific empiricists, what is our best prediction of $C_i(t + 1)$? Surely it is just the present value $C_i(t)$. Let us call this the quasi-martingale assumption:

$$E_P(C_i(t + 1)) = C_i(t) , \tag{7}$$

where E_P is the expectation with respect to probability measure P. If we now use the discrete differential Hebbian, we arrive at

$$e_{ij}(t + 1) = e_{ij}(t) + \Delta C_i(t)\, \Delta C_j(t + 1) \tag{8}$$

$$= \sum_{s=0}^{t} \Delta C_i(s)\, \Delta C_j(s + 1)$$

$$= \sum_{s=0}^{t} [C_i(s) - E(C_i(s))][C_j(s+1) - E(C_j(s+1))],$$

which has the form of a nonnormalized cross-covariance kernel--the key term in the definition of statistical correlation. This identity establishes a direct connection with correlation and our operative definition of causality, the differential Hebbian (3).

Since the differential Hebbian uses derivatives of transfer functions, we can plug in dynamical equations such as (1) and solve directly for e_{ij}. We demonstrate that the network [C, C, E, E] given by (1) and (3) can be easily solved. The trick is that e_{ij} can be pulled out of the inner product in (1). (3) then takes the form

$$\dot{e}_{ij} + a\, e_{ij} = b$$ --a first-order inhomogeneous ordinary differential equation with variable coefficients, one of the most tractable differential equations. Details of this expansion can be found in Kosko[14-15]. Here we state the exact solution:

$$e_{ij}(t) = e^{\int_0^t p(s)\,ds} \left(K + \int_0^t q(s)\, e^{-\int_0^s p(u)\,du}\, ds \right), \quad (9)$$

where the functions p and q contain the grouped terms that occur in the manipulation when C_i and C_j are expanded with (1), which for brevity we omit. The essential point is that p only contains the observation term O_i and contains it linearly and q contains both O_i and O_j and the product $O_i O_j$. The constant K is given by $K = e_{ij}(0)$.

EDGES AS SUBMARTINGALES

The edge e_{ij} is expected to increase in strength in time as information accumulates under the differential Hebbian hypothesis (3). This answers the question where edges tend on average, or, put another way, how connected networks become on average. To formalize such speculations, we interpret e_{ij} as stochastic process--$e_{ij}: T \times \Omega \rightarrow (-\infty, \infty)$-- on the probability space (Ω, A, P). We superscript with t to index the random variables $\{e_{ij}^t\}$.

External input enters the network through the observation terms O_i in (1). A hearty scientific empiricism dictates that, fundamentally, we know nothing of the future flux of experience. Experimentation is the resultant coping device. What does this mean for future values of the random variable O_i^s when we know the present value O_i^s, s < t ? More precisely, suppose we have an increasing sequence of sets of information in the form of a "filtration" of sigma-algebras $A_s \subset A_t \subset A$, s ≤ t. Then what is the conditional expectation $E(O_i^t \mid A_s)$? Surely the most we can assert is the present observed value. This constitutes a network <u>martingale assumption</u>:

$$E(O_i^t \mid A_s) = O_i^s \quad \text{if } s \le t . \quad (10)$$

Since the product $O_i O_j$ occurs in the q term of solution (9), we must further assume that the two observation processes are conditionally independent martingale processes: $E(O_i^t O_j^t \mid A_s) = O_i^s O_j^s$ for s ≤ t. (Technically we also assume the processes are suitably integrable and filtration measurable.)

So is the edge process a martingale if the observation processes are conditionally independent martingales? It turns out that the edge process is a <u>sub</u>martingale process. (The proof, in Kosko[14], uses Jensen's inequality and the convexity of the exponentials in the modified (9).) The expected future value is at least as big as the present value when conditioned on all the information available up to the present time. Hence $E(e_{ij}^t \mid A_s) \ge e_{ij}^s$ for s ≤ t. The proof is facilitated by interpreting the integrals in solution (9) as conditional expectations (since stochastic integrals are always martingales). For instance, the first integral then takes the form $E(p_t \mid A_t)$. This interpretation keeps account of the acquired information and outputs functions not constants.

282

Hence neural or causal networks can be expected to become more connected on average as time passes. The causal consequences of this may seem anti-entropic, perhaps even a violation of the second law of thermodynamics. But in fact total connectivity is the maximum entropy case (where entropy is intuitively interpreted). Order is established by a contrast between edge connections and edge disconnections, as in a military hierarchy or a bureaucratic dictatorship rather than in a voting democracy. Better, in terms of dynamic communication connections, note how a monitored debate tends unidirectionally to a free-for-all discussion.

Another consequence is that edge processes can be decomposed into a sum of a martingale and an increasing process:

$$e_{ij}^t = M_{ij}^t + N_{ij}^t, \tag{11}$$

where M is the martingale process and N is either positive or zero. M represents what stays the same in the connection process. Hence N represents what is new or novel, and, in the spirit of Kohonen[16], we call it the novelty process. A consequence of the Doob-Meyer decomposition (11) is that N, and hence its residual, can be written as follows:

$$N_{ij}^t = \sum_{k=1}^{t} E(e_{ij}^k - e_{ij}^{k-1} \mid A_{k-1}) , \tag{12}$$

$$N_{ij}^t - N_{ij}^{t-1} = E(e_{ij}^t - e_{ij}^{t-1} \mid A_{t-1}) . \tag{13}$$

Of the many things that can be said of this novelty process, perhaps the most significant concerns what it, and hence the edge process, encodes. (13) makes the contribution to differential Hebbian learning. What is significant is that a conditioned average is encoded even though not experienced. The expected edge residual is conditioned on exactly what it should be--all the information accumulated up to the present.

REFERENCES

1. D. Hume, An Inquiry Concerning Human Understanding (1748).
2. J. S. Mill, A System of Logic (1843).
3. D. O. Hebb, The Organization of Behavior (1949).
4. S. Grossberg, Psvch. Rev., 58, 1 (1980).
5. S. Grossberg, Studies of Mind and Brain (Reidel, Boston, 1982).
6. M. A. Cohen and S. Grossberg, IEEE Trans. SMC, 13, 815 (1983).
7. J. J. Hopfield, Proc. Nat. Acad. Sci, 79, 2554 (1982).
8. A. H. Klopf, Proc. 2nd Conf on Comp. with Neural Net. (1986).
9. R. S. Sutton and A. G. Barto, Psych. Rev., 88, 135 (1981).
10. D. E. Rumelhart, G. E. Hinton, R. J. Williams, in Parallel Distributed Processing, Rumelhart and McClelland Eds. (MIT Press, 1986).
11. Y. S. Abu-Mostafa and J. St. Jacques, IEEE Trans. IT, 31, 461 (1985).
12. J. Von Neumann, Mathematical Foundations of Quantum Mechanics (Springer, Berlin, 1932).
13. S. Watanabe, Pattern Recognition: Human and Mechanical (Wiley, NY, 1985).
14. B. Kosko, "Adaptive Inference," subitted for publication (1986).
15. B. Kosko and J. S. Limm, Proc. SPIE, 579, 104 (1985).
16. T. Kohonen, Self-Organization and Associative Memory (Springer-Verlag, Berlin, 1984).

PROGRAMMING A MASSIVELY PARALLEL, COMPUTATION
UNIVERSAL SYSTEM: STATIC BEHAVIOR

Alan Lapedes and Robert Farber

Theoretical Division
Los Alamos National Laboratory
Los Alamos, New Mexico 87545
March 27, 1986

Abstract

Massively parallel systems are presently the focus of intense interest for a variety of reasons. A key problem is how to control, or "program" these systems. In previous work by the authors, the "optimum finding" properties of Hopfield neural nets were applied to the nets themselves to create a "neural compiler." This was done in such a way that the problem of programming the attractors of one neural net (called the Slave net) was expressed as an optimization problem that was in turn solved by a second neural net (the Master net). The procedure is effective and efficient. In this series of papers we extend that approach to programming nets that contain interneurons (sometimes called "hidden neurons"), and thus we deal with nets capable of universal computation. Our work is closely related to recent work of Rummelhart et al. (also Parker, and LeChun), which may be viewed as a special case of this formalism and therefore of "computing with attractors." In later papers in this series, we present the theory for programming time dependent behavior, and consider practical implementations. One may expect numerous applications in view of the computation universality of these networks.

I.   Introduction

Massively parallel systems are the focus of intense interest by numerous researchers spanning many disciplines. They have tremendous practical implications for new kinds of computing[1,2] and may give some insight into how massively parallel biological brains operate.[3,4] They are also theoretically interesting when analyzed as an example of a complex, nonlinear dynamical system.[5] Programming, i.e. controlling, these systems is a difficult task. Recent work of Hopfield[1,2] on a particular system, neural networks, has stimulated intense interest on the part of many researchers. He has shown that nets with symmetrical interactions posess a Lyapunov function, and that such a function is very useful in controlling the network behavior. In particular, he has shown that continuous state neural networks (as opposed to discrete, two state networks) may be used to perform a credible job of solving difficult optimization problems such as the classic Traveling Salesman Problem (see also M. Takeda, G. Goodman[6]). In these types of problems the solution to the optimization problem is a fixed point of the neural dynamics and the problem constraints and the data (e.g. the distances in the TSP problem) appears in the symmetrical neural connections. The solution to the problem, the fixed point of the net, is of course not known a priori, and thus one runs the nonlinear differential equations for the nets until it relaxes to a fixed point in order to find an optimum.

A second class of difficult computation problems, e.g. Content Addressable Memory, associative memory, pattern recognition (and many other forms of high level symbol processing) can also be attacked with neural networks. Here, one knows the desired fixed points a priori (they correspond, for example, to the patterns to be recognized), however, one does not know in general what neural connections will yield the known fixed points. A partial solution to determine the neural connections was given by Hopfield[2], which is based on a symmetric synaptic connection matrix formed by taking outer products of the desired fixed point vectors. Disadvantages of this approach include a very limited degree of control over the basins of attraction, the symmetry of the synaptic connection matrix (which is almost certainly unphysiological), inability to deal with time dynamic behavior, and the fact that interneurons necessary for universal computation are excluded. In spite of these limitations, Hopfield's pioneering work remains an important first step in the theory of computation for neural networks.

The present authors[7] have approached the problem of programming fixed points into a neural net by directly attacking the conditions that the synaptic connection matrix must satisfy if the desired fixed points are to be present. This was accomplished by posing the general problem of determining the connections of a Slave net as an optimization problem, and then using the "optimum finding" abilities of neural nets to define a second net, a Master net, that performed the optimization. This therefore links the two classes of problems currently under investigation in neural network theory. Advantages of this approach include greater control over the basins of attraction, removal of the symmetry restriction in the Slave net, and as we will show, the ability to control interneurons or "hidden neurons" that are necessary for universal computation. The method is effective and efficient[8] and may also be viewed as a learning algorithm for neural networks. We have also extended the method to allow programming of time dependent behavior (paper II of this series).

The extension of our original formalism that is required to deal with hidden neurons is closely related to recent work of Rummelhart et al.,[9] Parker[10], and Le Chun.[11] This work, performed independently of the Master/Slave approach is shown to be a special case of the more general formalism presented below. This comment is not intended in any way to detract from the very fine results of these authors, and we hope that by providing a general setting that the value of such approaches can be better appreciated.

The importance of hidden neurons has, of course, not gone unnoticed by other researchers. It was forcefully pointed out by Minsky et al.[12] that hidden neurons are necessary to solve certain classes of important problems. In fact, it was the difficulty in being able to program these important neurons that led to the decline of interest in the 1970's in the Perceptron formalism of neural networks. Hinton and Sejnowski et al.[13,14] attacked the hidden neuron problem in the early 1980s by developing the Boltzman machine formalism. In this work, simulated annealing[15] for a stochastic two-state neuron model was coupled with the construction of a clever learning algorithm to provide the first solutions for programming massively parallel networks to perform difficult computations. Hinton and Sejnowski and coworkers have provided many beautiful analyses of problems

involving hidden neurons and used their procedure to find many solutions for small-scale model problems. However, there exist two
severe disadvantages to their approach. These are the extreme computational slowness and the inability to directly incorporate time-
varying neural behavior (e.g. direct programming of network limit
cycles etc.).

Recently, Sejnowski has used the Rummelhart procedure (in favor
to the Boltzman machine procedure) to solve a moderately large-scale
problem - that of converting typed text to speech, including correct
context dependent pronunciations of phonemes.[16] As we will see in
the general formalism below, computational slowness is no longer an
issue, and that the work of Rummelhart, Parker, Le Chun, Sejnowski
and the present authors are closely related to the neural networks
popularized by Hopfield, and are examples of computing with attractors in computation universal systems. In view of the widespread
interest by engineers in constructing Hopfield-type nets in hardware
(VLSI or optical)[17], it would seem that actual construction of massively parallel, computation universal devices may closely follow the
development of theory for controlling these systems.

In this paper (I) we confine our attention to developing the
theory for control of static behavior and point out the relations to
work of Rummelhart, Parker, Le Chun. Their work inspired the present
development. However, our results are completely general in that we
allow all possible connections of hidden, input and output neurons,
both forward and backward, including fully recurrent hidden connections. "Clamping" of inputs is not necessary - back connections can
allow inputs to remain "on" as part of the net's fixed-point configuration. Adjustment of basin boundaries may be accomplished in the
manner of Ref. 7. The formalism clearly shows that the method is a
procedure for controlling attractors in Hopfield type nets. In paper
II we present the theory for controlling time dependent behavior. In
subsequent papers we discuss specific applications.

II. Summary of Master/Slave Formalism

The Master/Slave formalism was developed as a general[7] attack on
determining what synaptic connection matrix, T_{ij}, yields desired
fixed points of the neural net equations

$$U_i + \mu \dot{U}_i = \sum_j T_{ij} g(U_j) + I_i \tag{1}$$

Here, U_i represents the membrane potential of the ith neuron, μ is
the cell summation time constant (or the RC time constant of an
electric circuit implementation), T_{ij} is the synaptic connection
matrix, $g(U_j)$ is the sigmoidal firing rate curve of the jth neuron
(or transfer function of an op-amp), and I_i is an external current
(or equivalently an internal threshold). For concreteness we take

$$g(x) = \tfrac{1}{2}(1 + \tanh(\beta x)) \tag{1a}$$

where β determines the slope of the sigmoid curve. The exact form
that the sigmoid assumes is not crucial to the results. These equations were introduced by Hopfield[1,2] in the context of "problem
solving with neural nets" (e.g. the Traveling Salesman Problem), but

related equations have also been written down by many researchers[18,19] as an acceptably crude model of neurophysiology. Hopfield showed that if $T_{ij} = T_{ji}$, then a Lyapunov function, E, exists:

$$E = -\tfrac{1}{2} \sum_{ij} T_{ij} V_i V_j - \sum_i I_i V_i + \sum_i \int^{V_i} g^{-1}(x) \, dx \qquad (2)$$

where $V_i = g(U_i)$. Equation (2) implies that

$$\frac{dE}{dt} = \sum_i g'(U_i) \cdot \dot{U}_i^2 \qquad (3)$$

and since g'(x) is positive semidefinite, E will decrease. E is bounded however, so that it will eventually stop decreasing and $\dot{U}_i$ will be zero, at which point a fixed point is reached. In the high gain limit (large β), the integral in Eq. (2) is negligible and the minima of E that is obtained will be close to a minima of E_1:

$$E_1 = -\tfrac{1}{2} \sum_{ij} T_{ij} g(U_i) g(U_j) - \sum_i I_i g(U_i) \qquad (4)$$

Hopfield chose

$$E_1 = -\tfrac{1}{2} \sum_{s=1}^{m} (V^{(s)} \cdot V)^2 \qquad (5)$$

where $V_i^{(s)} = g(U_i^{(s)})$, $s = \{1,2\ldots m\}$ are the desired choice of m memory states or fixed points. Although this choice works, subject to limitations discussed in the Introduction, it is ad hoc and it is not a general solution to the problem of inserting desired fixed points into the neural dynamics. The T_{ij} determined by Eq. (5) is a sum of the outer products of the fixed-point vectors and is, of course, symmetric.

In Ref. (7) we considered the general problem of finding T_{ij}'s that produce fixed points at desired states $V_i^{(s)}$, $s = \{1,2\ldots m\}$ with no a priori restriction on the T_{ij}. To do so we reformulated Eq. (1) by introducing a new variable, $V_i(t)$ such that

$$U_i = \sum_j T_{ij} V_j + I_i . \qquad (6)$$

It is easily checked that V_i satisfies

$$V_i + \mu \dot{V}_i = g(\sum_j T_{ij} V_j + I_i) \qquad (7)$$

by multiplying Eq. (7) by T_{ij}, summing over i, and comparing to Eq. (1). It may be shown that $\dot{V}_i$ is essentially a short time average of the firing rate $g(U_i)$, and that at a fixed point of Eq. (7) ($\dot{V}_i = 0$) the value of V_i is the firing rate, $g(U_i)$, of the ith neuron.

The requirement that a particular set of m fixed points $V_i^{(s)}$, {s = 1,2...m} of Eq. 7 exists may be expressed as

$$E_1 = \sum_{s,i} [V_i^{(s)} - g(\sum_j T_{ij} V_j^{(s)} + I_i)]^2 \tag{8}$$

with $E_1 = 0$. From Eq. 7 we see that this requires that $\dot{V}_i = 0$ at each $V_i^{(s)}$. The $V_i^{(s)}$ are the known fixed points that one desired to insert into the net and the T_{ij}'s are unknown but subject to Eq. (8). Note that no restrictions such as symmetry have been placed on the T_{ij}'s. In Ref. 7 we rewrote Eq. (8) slightly as

$$E_1 = \sum_{s,i} [g^{-1}(V_i^{(s)}) - \sum_j T_{ij} V_j^{(s)} - I_i]^2 \tag{9}$$

and noticed that finding the T_{ij}'s satisfying $E_1 = 0$ requires finding the minimum of the quadratic form in T_{ij} that results from expanding the square in Eq. (9).

It is desirable to keep T_{ij} bounded and a natural parameterization that accomplishes this is

$$T_{ij} = 2g(U_{ij}) - 1 \tag{10}$$

where $g(U_{ij})$ is another sigmoidal function (between 0 and 1) of the new variable U_{ij}. T_{ij} ranges over -1 to $+1$. I_i may be parameterized in a similar fashion but for simplicity let us take I_i to be zero for the moment. We consider non-zero I_i in the next section, but wish to keep the following summary as simple as possible and will not consider non-zero I_i until later. There will be no problems with non-zero I_i - it just adds some more terms.

Inserting (10) in (9) and expanding the square yields

$$E_1 = -\sum_{ijk\ell} T_{ijk\ell}\, g(U_{ij})\, g(U_{k\ell}) - \sum_{ij} I_{ij}\, g(U_{ij}) + \text{constant} \tag{11}$$

where

$$T_{ijk\ell} = -4 \sum_s \delta_{ki} V_j^{(s)} V_\ell^{(s)} \quad , \quad T_{ijk\ell} = T_{k\ell ij} \tag{12a}$$

$$I_{ij} = \sum_s 4 V_j^{(s)} [g^{-1}(V_i^{(s)}) + \sum_\ell V_\ell^{(s)}] \quad . \tag{12b}$$

Comparison of Eq. (11) with Eq. (4) shows that Eq. (11) is related to the Lyapunov function for <u>another</u> network, the Master net, with membrane potential U_{ij} and synaptic matrix and external currents given by Eq. 12. Adding a suitable integral places Eq. (11) exactly in the same form as Eq. (2) with the Master net neurons indexed by two indices, i and j, instead of just one index. Therefore, solving the two index form of Eq. (1)

$$U_{ij} + \mu \, U_{ij} = \sum_{k\ell} T_{ijk\ell} \, g(U_{k\ell}) + I_{ij} \tag{13}$$

will result in a fixed point of the Master net, $g(\bar{U}_{ij})$ that is a minimum of Eq. (11) and equivalently Eq. (9) (for large β). At the Master fixed point the Master neurons are firing at a sustained rate $g(\bar{U}_{ij})$ and the Slave net T_{ij} will be given by Eq. (10), i.e.,

$$T_{ij} = 2g(\bar{U}_{ij}) - 1 \quad . \tag{14}$$

The Master net firing rates, therefore, modulate the synaptic connections in the Slave.

Simulations[7] show that a good approximation to the global minimum of Eq. (11) is generally found, and that the resulting Slave net T_{ij}'s generally produce the desired fixed points. For further discussion, and extensions to add control over basins of attraction, see Ref. 7. More discussion and an analysis of the efficiency of the above procedure may be found in Ref. 8.

III. Extension of Master/Slave Procedure: Hidden Neurons

The above procedure was originally developed for auto-associative memory.[20] That is, fixed points were inserted in a Slave net such that configurations in the basins of attraction evolved to the desired fixed points. Hetero-associative memory[20], i.e. true associative memory, seeks to find T_{ij}'s such that <u>particular</u> configurations A_1, $A_2 \ldots A_m$ will evolve to other specific patterns B_1, $B_2 \ldots B_m$. Thus, pattern $A_1 \rightarrow B_1$, $A_2 \rightarrow B_2$ etc. A trivial psychological interpretation is that recalling fact A_1 ("a dog") evokes fact B_1 ("dogs bark"). Although this is amusing (and potentially very powerful in, say, a relational data base or massively parallel expert system) the psychological interpretation is the least important aspect of hetero-associative memory.

In our opinion the prime significance of hetero-associative memory, when accomplished with interneurons or hidden neurons, is the ability to perform universal computation. For example, the following bit pattern associations $00 \rightarrow 0$, $01 \rightarrow 1$, $10 \rightarrow 1$, $11 \rightarrow 0$ may only be realized in a neural network with the aid of hidden neurons[12,13,14] and are the logical expression of XOR ("exclusive or"). Negation and other logical functions may also be expressed as simple associations and can only be accomplished with hidden neurons. Therefore networks with hidden neurons are capable of space-bounded universal computation. To realize this capability one must be able to control the attractors in such networks.

In the following we show that a minor extension of the Master/Slave procedure (section II and Ref. 7) allows one to present a collection of inputs and outputs to a Master/Slave neural net, where the outputs may be complicated logical functions of the inputs, so that the net will adapt its attractors to "deduce" an algorithm that reproduces the input/output pairs. The net has limited information capacity and is clearly not just recording the input/output pairs, but is instead developing a collective algorithm to associate input and output. The net has certain capabilities of deduction and generalization (this seems to be related to overloading the memory capacity) and may even be operated in a "backward mode" where it is told the output and can back propagate to "correct" a corrupted input. We allow all possible connections of hidden, input, and output neurons, both forwards and backwards, including fully recurrent connections among the hidden neurons. "Clamping" of inputs is not necessary. Adjustment of basin boundaries, if desired, may be accomplished in a similar manner to Ref. 7. A natural extension of this method, to deal with time dependent behavior is presented in the next paper in this series.

Development of the Master/Slave procedure for hidden neurons begins with Eq. (7) and (8). In the previous section, and in Ref (7), we introduced the g^{-1} function so that the energy expression in Eq. (8) could be written as a quadratic form (Eq. (9)) with known coefficients related to the desired fixed points of the Slave net. This was done for pedagogical purposes to clarify the connection to a Hopfield Lyapunov function for a Master net, which was originally presented as a quadratic form. However, it is not mathematically necessary to do this, and to clarify the relation to Rummelhart et al., we consider Eq.(8) instead of Eq.(9).

Let us now parameterize T_{ij} as before (Eq. 10) and introduce a similar parameterization for I_i that bounds the allowed current values,

$$T_{ij} = 2g(U_{ij})-1 = \tanh(\beta U_{ij}) \tag{15a}$$

$$I_i = 2g(S_i)-1 = \tanh(\beta S_i) \tag{15b}$$

Actually, the β appearing in 15a,b does not have any relation to the β of the Slave net, so for clarity let us retain the notation of Eq. (1a) for the Slave net

$$g(x) = \tfrac{1}{2}(1 + \tanh(\beta x)) \text{ -- Slave net} \tag{16a}$$

and introduce a better notation $g_M(x)$, for the Master net

$$g_M(x) = \tanh(\beta_M x) \text{ -- (Master net)} \tag{16b}$$

with

$$T_{ij} = g_M(U_{ij}) \; \varepsilon[-1,1] \tag{17a}$$

$$I_i = g_M(S_i) \; \varepsilon[-1,1] \; . \tag{17b}$$

Let us now add two integrals to Eq. (8) to produce a Lyapunov function for the Master net that is <u>not</u> a quadratic form:

$$E = E_1 + \sum_{ij} \int^{g_M(U_{ij})} g_M^{-1}(x)dx + \sum_{i} \int^{g_M(S_i)} g_M^{-1}(x)dx \tag{18a}$$

with

$$E_1 = \sum_{s,i} [V_i^{(s)} - g(\sum T_{ij} V_j^{(s)} + I_i)]^2 \tag{18b}$$

and T_{ij}, I_i parameterized as Eq. 17a,b. If

$$U_{ij} + \mu \dot{U}_{ij} = - \frac{\partial E_1}{\partial T_{ij}} \tag{19a}$$

$$S_i + \mu \dot{S}_i = - \frac{\partial E_1}{\partial I_i} \tag{19b}$$

then it is easily verified that

$$\frac{dE}{dt} \leq 0 \tag{20}$$

and an identical argument to the previous section shows that E will decrease to near a minimum of E_1, at which point the Master neurons (with membrane potentials U_{ij} and S_i) are firing at a constant rate determined by the fixed points of 19a,b. The Slave net synapses and currents are then determined by (17a,b) evaluated at the fixed point of (19a,b).

To summarize, the logical argument so far is virtually identical to the previous section and Ref. 7. The only change is that we chose to use the nonquadratic form of Eq. 18 and have also included contributions from currents, I_i. It may be worth emphasizing that the above demonstrates that restrictions to quadratic Lyapunov functions are totally unnecessary, and that Hopfield's "energy minimization" arguments apply to a much wider class of functions than merely quadratic. There seems to be a misconception present in the literature that Lyapunov functions must be restricted to be quadratic, which in turn restricts the class of optimization problems that can be attacked. This restriction is unnecessary.

Up until now we have always assumed that all the $V_i^{(s)}$ fixed-point components were known because they were specified by the user and that there were no hidden neurons. Let us now consider a neural net where a subset of the neurons are arbitrarily labeled as Input, another subset as Output, and the remaining neurons as Hidden (see Fig. 1). Only pairs of associations of Input/Output are now specified by the user, however, we will now show that at a fixed point of the Slave, the states of the hidden neurons are known as well. Knowing the state of all the neurons at the fixed point allows us to determine the synaptic matrix, T_{ij} and currents I_i by the procedure of Ref. (7) (see previous section).

To summarize, the idea is to present the known input/output pairs to the Master net, which then determines Slave net T_{ij}'s and I_i's related to the firing rates of the Master at its own fixed point. A later presentation of an input to the Slave (with the hiddens and outputs set to states midway between "on" and "off") will place the Slave in the basin of attraction of a Slave fixed point

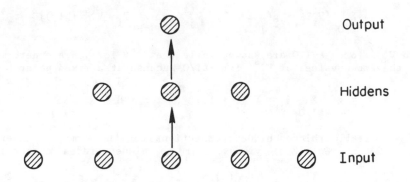

Figure 1

that then completes (i.e. "associates") the appropriate output. Because neural nets will only "complete correctly" if there is sufficient initial information, we would expect that if the inputs do not constitute a majority of the neurons in the net then completion may be incorrect. For these special situations one may disallow back connections to the inputs. However, for most cases the output neurons typically signal "yes/no" answers about complicated input patterns, and therefore completion of the much fewer output neurons will occur. Utilization of the completion capability of neural networks (which is a consequence of the attractor structure) adds new possibilities to neural net information processing.

We will first consider almost all possible connections of the Slave net, both forwards and backwards, including recurrent connections among the Inputs and Outputs. For the moment the only case we will not consider is recurrent connections among the hiddens. Recurrent hidden connections require a slight additional argument that we provide later on so as not to unnecessarily confuse the logical argument. This will then complete all possible cases.

The global minimum of Eq. 18b and 18a expresses the condition that Eq. (7) has fixed points at desired locations, $V_i^{(s)}$. Because the fixed-point values are to be specified only for the Input/Output neurons, we restrict the sum over "i" in E_1 (Eq. 18b) as follows

$$E_1 = \sum_s \sum_{i \varepsilon I/0} [V_i^{(s)} - g(\sum_j T_{ij} V_j^{(s)} + I_i)]^2 \qquad (21a)$$

or expanding the $\sum_j$

$$E_1 = \sum_s \sum_{i \varepsilon I/O} [V_i^{(s)} - g(\sum_{j \varepsilon I/O} T_{ij} V_j^{(s)} + \sum_{j \varepsilon H} T_{ij} V_j^{(s)} + I_i)]^2 \quad . \quad (21b)$$

The $V_j^{(s)}$ for $j \varepsilon I/O$ are known, while the $V_j^{(s)}$ for $j \varepsilon H$ are <u>determined</u> by the known values of $V_j^{(s)}$ for $j \varepsilon I/O$ because at a fixed point

$$V_j^{(s)} = g(\sum_{k \varepsilon I/O} T_{jk} V_k^{(s)} + I_j) \quad , \quad \text{for } j \varepsilon H \quad\quad (22)$$

if we disallow hidden-hidden connections for the moment. Inserting Eq. (22) in Eq. (21b) yields (for no hidden-hidden connections)

$$E_1 = \sum_s \sum_{i \varepsilon I/O} [V_i^{(s)} - g(\sum_{j \varepsilon I/O} T_{ij} V_j^{(s)} + \sum_{j \varepsilon H} T_{ij} g(\sum_{k \varepsilon I/O} T_{jk} V_k^{(s)} + I_j) + I_i)]^2 \quad .$$

$$(23)$$

If we now allow the hidden-hidden connections to be only feed-forward connections then we may split the sum over $j \varepsilon H$ in Eq. (21b) into a sum over $j \varepsilon H_1$ (the first hidden layer), and a sum over $j \varepsilon H_2$ (the second hidden layer), see Fig. 2.

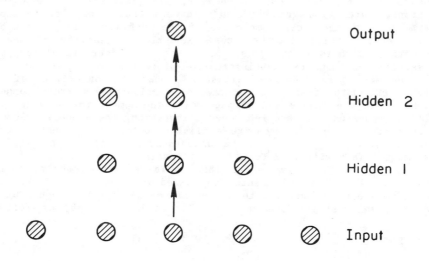

Figure 2

In this case Eq. (21b) becomes

$$E_1 = \sum_s \sum_{i \varepsilon I/O} [V_i^{(s)} - g(\sum_{j \varepsilon I/O} T_{ij} V_j^{(s)} + \sum_{j \varepsilon H_1} T_{ij} V_j^{(s)} + \sum_{j \varepsilon H_2} T_{ij} V_j^{(s)} + I_i)]^2$$

(24)

where V_j for $j \varepsilon H_1$ is as before (Eq. 22) and V_j for $j \varepsilon H_2$ becomes:

$$V_j^{(s)} = g(\sum_{k \varepsilon I/O} T_{jk} V_k^{(s)} + \sum_{k \varepsilon H_1} T_{jk} V_k^{(s)} + I_j)$$

$$= g(\sum_{k \varepsilon I/O} T_{jk} V_k^{(s)} + \sum_{k \varepsilon H_1} T_{jk} g(\sum_{\ell \varepsilon I/O} T_{k\ell} V_\ell^{(s)} + I_k) + I_j) \quad . (25)$$

This clearly generalizes for any number of hidden layers H_1, H_2...H_n.
Eq. (19a,19b) will determine the appropriate synaptic connections and currents given the expressions E_1 (above). It should
already be clear that this procedure incorporates the Rummelhart et
al. algorithm as a special case if one uses a linear function for
$g_M(x)$ (Eq. 16b) and allows only feed forward connections. The sub-
stitutions performed in Eq. 22 → 25 replace the "back propagation" of
Rummelhart while the Eq. (19a,b) are identical to Rummelhart's when
one "smooths the gradient" by overrelaxation (see Sejnowski Ref. 16).
In actual practice, Rummelhart changes the synaptic weights (using an
Euler discretization of Eq. 19) after a presentation of each pattern.
He notes, however, that he is really attempting to perform gradient
descent, which requires weight changes to be made only after the
complete set of patterns is "presented" (as we do above). His simulations verify his claim that changing the weights after presentation
of each pattern does not destroy the gradient descent process. A
more detailed analysis of the relation of our procedure to that of
Rummelhart is given in the next section.
We now consider the case where the hidden-hidden connections are
allowed to be recurrent. This will complete the generalization to
all possible types of connections. The case of feed forward H-H
connections was considered above by appropriately generalizing Eq.
(22). We now return to Eq. (22) and generalize it for the case of
recurrent H-H connections. If we have some recurrent H-H connections
and some purely feed forward (or feed backward) H-H connections then
we would need to add additional terms as in Eq. 24, 25. The separate
cases are additive (one just adds in more terms for the separate
cases), so we see little expository value in writing down a huge
complicated looking expression that sums up all the separate cases.
We therefore return to Eq. (22) and add in only recurrent H-H connections.
In this situation Eq. (22) becomes

$$V_j^{(s)} = g(\sum_{k \varepsilon I/O} T_{jk} V_k^{(s)} + \sum_{k \varepsilon H} T_{jk} V_k^{(s)} + I_j)$$

(26)

for $j\varepsilon H_{recurrent}$. Because there are now feed forward and feed back-ward H-H connections, we are not able to substitute an expression in the known I/O values for the second summation in Eq. (26) as we did in Eq. (25). We therefore leave the second summation as an extra unknown variable, $X_j^{(s)}$, where:

$$X_j^{(s)} = \sum_{k\varepsilon H} T_{jk} V_k^{(s)} \quad .\tag{27}$$

Thus

$$V_j^{(s)} = g(\sum_{k\varepsilon I/O} T_{jk} V_k^{(s)} + X_j^{(s)} + I_j) \quad .\tag{28}$$

We now proceed exactly as in the above and form the analogue to Eq. (23) by substituting Eq. 28 in Eq. (21b).

$$E_1 = \sum_s \sum_{i\varepsilon I/O} [V_i^{(s)} - g(\sum_{j\varepsilon I/O} T_{ij} V_j^{(s)} + \sum_{j\varepsilon H} T_{ij} g(\sum_{k\varepsilon I/O} T_{jk} V_k^{(s)} + X_j^{(s)} + I_j) + I_i)]^2 \quad .$$

$$\tag{29}$$

The unknown $X_j^{(s)}$ appears on virtually the same footing as the un-known, I_j, so that we may parameterize it in a similar fashion to Eq. 17b, add the analogous integral to Eq. 18a and discover all the unknowns by running Eq. 19 supplemented with a third equation for the third variable $X_j^{(s)}$. Of course $X_j^{(s)}$ is not in the range $(-1,1)$, but is instead in the range $(-H,H)$ (see Eq. 27), so that the parameteri-zation similar to 17b uses $H \cdot g_M$ and not just g_M where H = (number of recurrent hiddens).

At the end of the above procedure we know the values of T_{ik} for $(j\varepsilon H)$ and $(k\varepsilon I/O)$, and also for $(j\varepsilon I/O)$ and $(k\varepsilon I/O)$, and for $j\varepsilon(I/O)$ and $(k\varepsilon H)$. We also know all I_j and the $X_j^{(s)}$. All that remains is the determination of the values of T_{ik} for $(j\varepsilon H)$ and $(k\varepsilon H)$. We have all the information we need to do this. In view of the already known variables and Eq. 28, we also know the values of $V_j^{(s)}$ for all j, including $j\varepsilon H$. Equation (27) now determines the unknown values of T_{jk} because we may now form a new E_1,

$$E_1 = [X_j^{(s)} - \sum_{k\varepsilon H} T_{jk} V_k^{(s)}]^2 \quad ,$$

and proceed in the usual fashion to determine T_{ik}. Note the inter-esting consistency condition that arises (over, under, or exact determination of T_{jk}) when one includes recurrent hidden connections.

It is now clear that a straightforward extension of the Master/Slave approach to computing with attractors allows one to program networks capable of universal computation. The only slight complica-tion occurs when one allows recurrent hidden connections. It would be of great interest to have a "theory of computation" for neural networks, which would give some insight into the conditions on types of problems that require recurrent hidden connections (as well as other hidden connections).

IV. Relation to Rummelhart et al.

Rummelhart et al. consider a neural network with a "synchronous update rule" such that the output, O, of a neuron at time (t + 1) is given a function, f, of the other neurons at time t:

$$O_i(t + 1) = f(\sum_j T_{ij} O_j + I_i) \ . \tag{31}$$

They consider primarily feed forward networks, with input feeding to hidden neurons, that in turn feed to output neurons. Recurrent networks are somewhat unnaturally handled by relating them to feed forward networks. They consider an "energy function"

$$E^{(s)} = \tfrac{1}{2} \sum_i (t_i^{(s)} - O_i^{(s)})^2 \tag{32}$$

where $t_i^{(s)}$ is a desired, or target, configuration for the output units (when given an input configuration labeled by s), and $O_i^{(s)}$ is the actual output that occurs. A "learning rule" for T_{ij} was developed by performing gradient descent on $E^{(s)}$ i.e.

$$\Delta T_{ij} = - \frac{\partial E^{(s)}}{\partial T_{ij}} \eta \tag{33}$$

where

$$\frac{\partial E^{(s)}}{\partial T_{ij}} = \delta_i^{(s)} O_j^{(s)} \tag{34}$$

and η is a "learning constant." The $\delta_i^{(s)}$ is given by

$$\delta_i^{(s)} = (t_i^{(s)} - O_i^{(s)}) f'(\sum_j T_{ij} O_j + I_i) \tag{35a}$$

for i ε Outputs and

$$\delta_i^{(s)} = f'(\sum_j T_{ij} O_j^{(s)} + I_i) \sum_k \delta_k^{(s)} T_{ki} \tag{35b}$$

for i ε Hiddens. Equations (35a,b) define a recursive procedure for calculating ΔT_{ij} in their feed forward networks. Because of numerical problems, one can smooth the gradient by overrelaxation (c.f. Sejnowski, Ref. 16), so that Eq. (33) becomes:

$$T_{ij}(t + 1) = \alpha T_{ij}(t) + (1 - \alpha) (- \frac{\partial E^{(s)}}{\partial T_{ij}} \eta) \ . \tag{36}$$

On the other hand, we consider a Slave network with subsets of neurons divided into Input, Hidden and Output classes with no restrictions on the connections. The evolution equation for our Slave network is

$$V_i + \mu \dot{V}_i = g(\sum_j T_{ij} V_j + I_i) \quad , \tag{37}$$

which may be written in an Euler discretized form as

$$V_i(t + \Delta t) = (1 - \frac{\Delta t}{\mu}) V_i(t) + \frac{\Delta t}{\mu} g(\sum_j T_{ij} V_j + I_i) \quad . \tag{38}$$

The units of time are arbitrary, and if we choose to use units of Δt then (38) becomes

$$V_i(t + 1) = (1 - \frac{1}{\mu}) V_i(t) + \frac{1}{\mu} g(\sum_j T_{ij} V_j + I_i) \quad . \tag{39}$$

The relation to Eq. (31) is clear given a slight change in notation. Our Lyapunov function Eq. (18a, b) is virtually identical to Eq. (32) (the integral makes a negligible contribution, but helps remove false minima[1,2]) and our "learning Equation" Eq. (19a, b) is

$$U_{ij} + \mu \dot{U}_{ij} = - \frac{\partial E_1}{\partial T_{ij}} \tag{40a}$$

$$S_i + \mu \dot{S}_i = - \frac{\partial E_1}{\partial I_i} \quad . \tag{40b}$$

Note that the Euler discretized form of Eq. (40) corresponds to Eq. (36) with $\alpha = 1 - \frac{\Delta t}{\mu}$ and with g_M restricted to be a linear function. In this case a few lines of algebra using Eq. 35 show that they are identical. The constant, η corresponds to an overall constant multiplicative factor on E_1. In fact, as we pointed out in Ref. 7, the η's may actually be used to sculpt the basins of attraction by generalizing it to $\eta_i^{(s)}$ (c.f. the constants, $C_i^{(s)}$, in Ref. 7).

It is therefore clear that the two methods (programming attractors versus learning rules) are very closely related. Our evolution equation for the Slave net, Eq. 39) is a Hopfield style neural net equation and differs slightly from the evolution equation of Rummelhart (Eq. (31)) by the addition of a "forgetting" term. The fixed-point conditions for the two evolution equations are, however, identical. Rummelhart's learning equations, Eq. (33-35), are virtually identical to our Eq. (40), if the $g_M(s)$ is specialized to a linear function. The back propagation and recursive manipulations of Rummelhart are replaced by our determination of the fixed-point values of the hidden units in terms of the Input and Ouput. We allow full back propagation to the Inputs (and also recurrent Input connections) to keep the Inputs switched on to their correct values. This can replace the "clamping of inputs" performed by Rummelhart. It allows a new kind of information processing where the output may be set to, say, a "yes" value and information may flow backwards through the net to correct a corrupted input.

Because we have specified the learning problem in terms of evolution to attractors, we require that the network correctly "complete" partial information. As noted above, this is potentially quite powerful, but can be fairly delicate. The network will generally be unable to correctly "complete" partial information unless it has sufficient information to start with. In these situations, we may restrict certain connections to be feed forward, and/or clamp inputs, and perform the task in a manner similar to Rummelhart et al. Another feature of our implementation is the nonlinear form for $g_M(x)$. Restricting $g_M(x)$ to be linear (c.f. Rummelhart[9]) is possible, and changes our algorithm to a straight gradient descent procedure. A nonlinear $g_M(x)$, however, tends to smooth the energy landscape[1] and allows the possibility of annealing in β, This should be helpful in more complicated situations than those considered by Rummelhart, where false minima become a problem.

V. Summary

The Master/Slave approach[7] to controlling attractors in neural networks was extended to the case of networks with hidden neurons. Such networks are extremely important in view of their ability to perform space-bounded universal computations in a massively parallel manner. A collective method of programming such networks was developed as an extension of our original approach. A special case of this method is identical to recent work of Rummelhart[9], Parker[10] and LeChun[11]. Their results were analyzed in the broader context of "computing with attractors" of Hopfield-style neural nets. This results in a unification of these two approaches and clarifies many aspects of the algorithm, in distinction to emphasizing implementation of the algorithm.

All possible types of neural interactions were allowed, including recurrent Hidden connections, and also recurrent Input connections. For certain situations one need not clamp the Slave net Inputs thereby allowing a new kind of information processing. The "learning procedure" is replaced by the Master net evolution equation, which is identical to Rummelhart's learning equation in the case where the Master net is restricted to have a linear firing rate curve. Nonlinear (sigmoidal) Master net firing rate curves are also allowed and this helps to smooth the energy landscape resulting in fewer false minima (c.f. Hopfield[1], in another context). The basins of attraction may also be shaped by hand in the manner of Ref. 7. Computational speed is determined by the time needed to reach a fixed point of the Master Net. This is generally of order μ, which is fractions of a microsecond in hardware.

This analysis therefore results in a formalism for programming attractors in massively parallel Hopfield style neural nets that are capable of performing universal computation. It reduces to previous work of Rummelhart[9], Parker[10], and LeChun[11] in special cases, and they have demonstrated its power in small-scale model problems (see

Sejnowski[16] for applications to a larger-scale problem). Until devices can be built (either optical or VLSI) that allow for synaptic plasticity, one must restrict attention to problems where the synaptic connections can be precomputed and then hard-wired. Sejnowski's Net Talk[16] is such an example. Even in this restricted problem domain one may expect numerous applications in view of the computation universality of these networks.

References
1. J. J. Hopfield, Bio. Cyb. $\underline{52}$, 141 (1985).
2. J. J. Hopfield, PNAS 79, 2554 (1982), PNAS $\underline{81}$, 3088 (1984).
3. G. Hinton, J. Anderson, ed. "Parallel Models of Associative Memory," Erlbaum Assoc., Hillsdale, NJ (1981).
4. D. Rummelhart, J. McClelland, ed. "Parallel Distributed Processing: Exploration in the Microstructure of Cognition," MIT Press (1986).
5. S. Wolfram, Rev. Mod. Phys. $\underline{55}$, p. 601 (1983).
6. M. Takeda, G. Goodman "Neural Networks for Computation," Stanford Univ., E.E. Dept. preprint (1986).
7. A. Lapedes, R. Farber Proceedings: Los Alamos International Conference on Learning, Games, Evolution (May 1985) to be published, Physica D (1986).
8. J. Denker, accepted for publication, Physica D; E. Mjolsness (unpublished).
9. D. Rummelhart, J. McClelland in Ref. 4.
10. D. Parker "Learning Logic" (TR-47) MIT Center for Computational Research in Economics and Management Sciences preprint (1986).
11. Y. LeChun, Proceedings of Cognitiva $\underline{85}$, p. 599 (1985).
12. M. Minsky, S. Papert "Perceptrons," MIT Press (1969).
13. G. Hinton, J. Sejnowski, Proc. IEEE Soc. Conf. on Computer Vision and Pattern Recognition, Washington, D. C., p. 448 (1983).
14. D. Ackley, G. Hinton, and T. Sejnowski, Cog. Sci. $\underline{9}$, 147 (1985).
15. S. Kirkpatrick, C. Gelatt, M. Vecchi, Science $\underline{220}$ (1983).
16. T. Sejnowski and C. Rosenberg, "Net Talk: A Parallel Network that Learns to Read Aloud," Johns Hopkins EECS preprint (1986).
17. D. Psaltis, N. Farhart, Opt. Lett. $\underline{10}$, 98 (1985).
18. See e.g., J. Feldman, J. Cowan, Biol. Cyb. $\underline{17}$, 29 (1975).
19. T. Sejnowski in Ref. 3.
20. T. Kohonen "Self Organization and Associative Memory," Springer-Verlag Press (1984).

NONLINEAR DYNAMICS OF ARTIFICIAL NEURAL SYSTEMS

T Maxwell
Sachs/Freeman Assoc., 1401 McCormick Dr., Landover, Md. 20785

CL Giles
AFOSR, Bldg. 410, Bolling AFB, DC 20332

YC Lee, HH Chen
Energy Research Bldg., U. of Md., College Park, Md. 20742

ABSTRACT

Now that significant progress has been made in developing algorithms for training hidden units, we suggest that it is time to reevaluate the nonlinear discriminate approach, which once fell into disfavor due to the problem of proliferation of high order terms. We show that there are many powerful techniques for reducing the number of spurious terms, and that the high order approach has many advantages over a cascaded slab approach in certain problem areas. Advantages include increased expressive ability, decreased architectural complexity, and dramatically increased learning rates.

INTRODUCTION

In this paper we will introduce and present some background on high order threshold logic units. We will then discuss methods for adapting these units to specific problem areas, addressing the problem of proliferation of high order terms, and examine applications of these units in four examples; learning the exclusive-or, separating horizontal from vertical lines, the TC problem, and a landmark learning problem.

BACKGROUND

Early in the history of pattern recognition research it was known that nonlinearly separable subsets of pattern space can be dichotomised by nonlinear discriminate functions[1]. Attempts to adaptively generate useful discriminate functions led to the study of threshold logic units (TLUs). The most famous TLU is the perceptron[2], which in its original form was constructed from randomly generated functions of arbitrarily high order. Minsky[3] studied TLUs of all orders, and came to the dual conclusion that high order TLUs were impractical due to the combinatorial explosion of high order terms, and that first order TLUs were too limited to be of much interest. More recent work in this area includes studies of "sigma-pi" units by Williams[4] and optical implementations by Psaltis[5].

THRESHOLD LOGIC UNITS

The first order threshold logic unit (TLU) is the familiar unit used by Hopfield[6] and most other neural modelers,

$$y_i(t) = S[\sum_j w_{ij} x_j(t) + I(t)], \qquad (1)$$

where x_j is a set of N inputs from other units, w_{ij} is a set of adaptive weights, and $I(t)$ is an external input. The output of unit i is y_i and S is a sigmoid-like function, typically piecewise linear. This equation admits a straightforward generalization to high order,

$$y_i(t) = S[\sum_j w_{ij}(t) x_j(t) + \sum_j \sum_k w_{ijk}(t) x_j(t) x_k(t) + ... + I(t)]. \qquad (2)$$

Note that in the first order term the weight matrix w has two subscripts and represents second order correlations. In the second order term the weight matrix has three subscripts and represents third order correlations. First order learning rules can be generalized to high order in an analogous fashion. For example, in the case of supervised learning, the second order weight matrix can be updated by

$$w_{ijk}(t+1) = w_{ijk}(t) + (n_i(t)-y_i(t))x_j(t)x_k(t), \qquad (3)$$

where n(t) is the "training signal", or desired output for input x(t).

ADVANTAGES OF A HIGH ORDER APPROACH

As discussed by Minsky[3], single feed-forward slabs of first order TLUs can implement only linearly separable mappings. Since most problems of interest are not linearly separable, this is a very serious limitation. One alternative is to cascade slabs of first order TLUs. The units embedded in the cascade ("hidden units") can then combine the outputs of previous units and implement nonlinear maps. However, training in cascades is very difficult[7] because there is no simple way to provide the hidden units with a training signal. Multi-slab learning rules require thousands of iterations to converge, and sometimes do not converge at all, due to the "local minimum" problem.

These problems can be overcome by the use of single slabs of high order TLUs. The high order terms are equivalent to prespecified hidden units, so that one high order slab can take the place of many slabs of first order units. Since there are no hidden units to be trained, the extremely fast and reliable single-slab learning rules can be used. A single slab of high order TLUs requires only a single iteration (one presentation of the complete pattern set) of the learning rule (equation 3) to learn an arbitrary map, but requires 2^N terms for this level of generality.

Other advantages of the high order approach discussed elsewhere are the existence of convergence proofs for recursive iteration[8], and the increase in expressive ability, especially in the implementation of complex inference structures[9] (see below).

EXAMPLE 1: LEARNING THE EXCLUSIVE-OR

The exclusive-or (see below) has been the classic "difficult" problem for first order units because it is the simplest non-linearly separable problem. Training a hidden unit to perform this function requires thousands of iterations of the fastest learning rules[7]. An alternative method is to use a fixed hidden unit, or equivalently, train a second order unit,

$$y(x) = sgn[w_1x_1 + w_2x_2 + w_{12}x_1x_2] \qquad (4)$$

using the learning rule

$$w_i' = w_i + (n(x)-y(x))x_i \qquad (5)$$
$$w_{12}' = w_{12} + (n(x)-y(x))x_1x_2,$$

EXCLUSIVE OR

n(x)	x_1	x_2
1	1	1
-1	-1	1
-1	1	-1
1	-1	-1

where n(x) is the correct output for input x = (x_1,x_2), and sgn[x] is the signum function, which is defined to yield +1 when x ≥ 0 and -1 when x < 0. The second order term is equivalent to a hand crafted hidden unit, so that the one second order unit takes the place of a two layer cascade. Since there is no correlation between x_1 or x_2 and n(x), the w_i terms average to zero in the learning process. However, since x_1x_2 is perfectly correlated with n(x), w_{12} is incremented positively at each step of the learning procedure, so that the training process converges in one iteration to the solution $w_1 = w_2 = 0$ and $w_{12} > 0$. This unit will learn an arbitrary binary map in one iteration of the learning rule, equation 5.

PROLIFERATION OF HIGH ORDER TERMS

As noted above, the implementation of an arbitrary map with a single slab of high order units requires 2^N high order terms per unit, which is generally intractable. Fortunately, keeping all these terms is usually unnecessary, since only those terms which capture the essential nonlinearities of the problem are required. Well defined methods exist for tailoring a high order adaptive unit to a specific problem area. The first step is to eliminate all terms higher then some degree k. A kth order unit can solve a kth order problem (see Minsky[3]). Further methods[8-11] include averaging over symmetry groups, and the use of short range and sparse interconnects. For problems of order higher then three or four, however, it is likely that even with these techniques cascading will still be necessary to curb the proliferation of high order terms. Since the well known algorithms for learning in cascades are all easily generalized to high order, we suggest that the inclusion of high order terms might also improve performance in this area.

SOLVING PROBLEMS WITH INVARIANCES

Problems with invariances can be solved very efficiently with high order units that have the same invariances. Invariant TLUs can be generated by the method of averaging over transformation groups.[11,12] This process helps solve the problem of proliferation of high order terms by eliminating a large number of terms which are not appropriate to the problem.

For example, imposing shift invariance on the output of a high order TLU is equivalent to imposing the constraint that the correlation matrix depend only on relative coordinates, not absolute coordinates. This constraint can be imposed by averaging over the translation group,

$$w_{jkl} \rightarrow \sum_t w_{j+t,k+t,l+t} = \sum_n w_{n,n+dkj,n+dlj} \rightarrow w_{dkj,dlj}, \qquad (6)$$

where $t = n-j$, $dkj = k-j$, and $dlj = l-j$.
In this case a symmetry reduction from N^3 to N^2 elements is effected. The resulting dynamics for a third order unit are

$$y(x) = step[\sum_{dkj} \sum_{dlj} w_{dkj,dlj} \sum_j x_j x_{j+dkj} x_{j+dlj}], \quad step[x] = \{ \begin{matrix} 1 : x \geq 0 \\ 0 : x < 0 \end{matrix} \} \quad (7)$$

with the corresponding learning rule (in the case of supervised learning),

$$w_{dkj,dlj}' = w_{dkj,dlj} + (n(x)-y(x)) \sum_j x_j x_{j+dkj} x_{j+dlj}. \qquad (8)$$

This technique can be used to hand-craft units whose output is invariant under the action of an arbitrary transformation group. High order units must be used because imposing shift invariance on a second order correlation matrix leaves too few degrees of freedom to solve any but the most trivial problems. Alternative methods of implementing invariances, involving cascades[7,13] or preprocessing[22] have been explored by other researchers.

COMPETITIVE LEARNING

Geometrical features can be represented as invariants of transformation groups. Competing group-invariant TLUs will discriminate between equivalence classes under the same group without training. Thus feature detectors can be constructed from high order TLUs by averaging over the appropriate groups.

For example, consider the problem of separating horizontal from vertical lines. These equivalence classes under translation will be naturally separated by a pair of competing second order translation-invariant TLUs. Without training, one unit will have maximum sensitivity to horizontal lines and the other will have maximum sensitivity to vertical lines. After training, the network generalizes to sort arbitrary patterns according to their degree of horizontalness or verticalness.

EXAMPLE: TC PROBLEM

This problem involves distinguishing between a T and a C with shift and rotation invariance. The training patterns consist of single letters inscribed in 3x3 squares in arbitrary position and orientation in a larger (10x10) space. This problem can be solved without training using a pair of competitive shift and rotation invariant units as discussed above. Here we will examine an alternate procedure, utilizing supervised learning. In this case the solution requires a single third order shift invariant TLU with short range interconnects (a 3x3 window). Dynamics are governed by equations 7 and 8 with the dkj, dlj sums limited to the window. In general, for an mxm window, this architecture will require less then m^{2k-2} adaptive weights, where k is the order of the TLU. The training procedure converges to a correct set of weights in one to ten iterations. Note that all known methods of solving this problem involving training hidden units in cascades require more units and adaptive weights, and utilize learning rules which usually take at least thousands of iterations to converge. We expect these advantages to become more pronounced as the networks are scaled up in size.

COMPLEX INFERENCE STRUCTURES

Second order correlations are well suited to inference structures of the form $A_i \rightarrow B_i$, mapping a set of patterns A_i to a set of patterns B_i. The more general case, $A_i \xrightarrow{C_i} B_i$, in which the relation between A_i and B_i is modulated by C_i, is easily handled by third order correlations, which naturally capture the structure of this type of relationship. These "shunting" interactions are commonly observed in neurophysiology, when one synapse modifies the action of another synapse or an entire dendritic tree[14],[15]. For example, consider the process of recognizing an object independent of viewpoint. As proposed by Hinton[17], let A_i represent a set of viewpoint dependent retina-based features and B_i represent a set of viewpoint independent object-based features. Then C_i will represent the set of possible orientations of the object. A wide range of relationships of this type can be implemented in neural networks using third order correlations. Another example is discussed in the following section.

LANDMARK LEARNING

This problem is an extension of a landmark learning problem that has been extensively discussed by Barto, Anderson, and Sutton[18]. Consider an organism ("bug") at point * in an environment consisting of two landscapes (Fig. 1).

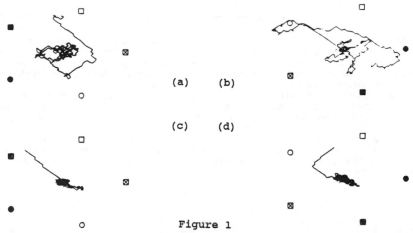

(a) (b)

(c) (d)

Figure 1

Each landscape is characterized by a set of six landmarks L_i. The bug receives a scalar "olfactory" input d_i from each of the landmarks, given by

$$d_i = \text{ramp}[\ R - \text{dist}(L_i, *) \ ] \qquad \qquad \text{ramp}[x] = \{ \begin{matrix} x \ \text{if} \ x \geq 0 \\ 0 \ \text{if} \ x < 0 \end{matrix} \}, \qquad (9)$$

where R is the range of the interaction, and dist(x,y) is the Euclidian distance between points x and y. The bug's dynamics are governed by the equations

$$y_i(t+1) = S[\ \sum_j w_{ij}(t)d_j(t) + \sum_j \sum_k w_{ijk}(t)d_j(t)d_k(t) + I(t) \ ], \qquad (10)$$

where I(t) is a random variable, and y_i generates the movements of the bug:

$$y_1 = \{ \begin{matrix} +1 : \text{move north} \\ -1 : \text{move south} \end{matrix} \} \qquad y_2 = \{ \begin{matrix} +1 : \text{move east} \\ -1 : \text{move west} \end{matrix} \}.$$

The task of the bug is to associate with each input vector the action which maximizes a payoff. In this case the central landmark (tree,L_0) acts as an "attractant", so that an increase in the signal from the tree, $d_0(t)$, represents a positive payoff. Associations are formed via the equations

$$w_{ij}(t+1) = a_1 w_{ij}(t) + (d_0(t+1) - d_0(t))y_i(t)d_j(t) \qquad (11a)$$
$$w_{ijk}(t+1) = a_2 w_{ijk}(t) + (d_0(t+1) - d_0(t))y_i(t)d_j(t), \qquad (11b)$$

where a_i is a forgetting parameter which serves to keep the weights bounded.

Equation 11a generates first order representations composed of second order correlations between the positions of single landmarks and optimal actions. Equation 11b generates second order representations composed of third order correlations between the positions of pairs of landmarks and optimal actions. These correlations are then used to generate movement toward the tree from any point in the environment via equation 10. Forming a representation of a single landscape is an order one problem, since it requires second order correlations between single landmarks and optimal actions. Learning representations of more then one landscape is a higher order problem, since the associations will be context dependent.

Since the weights are initialized to zero, the bug begins by randomly wandering about the environment, driven by the I(t) term in equation 10. At each step, however, associations are built via equation 11 and gradually the random walk gives way to direct movement toward the tree from any point in the environment (Fig. 1a). When the landscape is changed, the bug will initially try to generalize what it has learned in the first landscape to the second, resulting in incorrect actions (Fig 1b). If the bug now remains in the new landscape for a sufficiently long period of time, the first order representation formed in the first landscape will be completely forgotten and a new first order representation will be formed. However, if the landscape is changed periodically (with period less then the bug's memory span) the correlations which are not invariant under the change in landscapes will be washed out, and the bug will form a representation that is valid in both landscapes (Figs 1c,1d). Thus a first order representation will be formed of the invariant aspects of the environment, and an invariant second order representation will be formed of those aspects which change with the shift in landscapes. Once this higher order representation is formed, the bug will be using pairs of landmarks rather than single landmarks to navigate in the changing environment. By including the appropriate higher order terms, one can easily generalize this system to enable the bug to look at triplets, quadruplets, etc., of landmarks, useful in cases in which landmark pairs are not sufficient.

CONCLUSIONS

Early in the evolution of neural network theory it was decided that the use of nonlinear discriminants was limited due to the combinatorial explosion of high order terms. In the effort to develop practical networks that could learn arbitrary nonlinear maps, adaptive learning research has concentrated on layered networks of first order TLUs. A great deal of work has been done in this direction[7,19-21], and most algorithms proposed tend to be very slow. In light of these results, we feel that it is appropriate to take another look at the high order approach. As shown in this paper, in certain problem areas the use of single slabs of high order units has pronounced advantages over the use of cascades of first order units, including decreased architectural complexity and dramatically increased learning rates, especially if some "hand-crafting" is allowed. Also, many promising techniques exist for curtailing the explosion of high order terms, some of which are discussed briefly in this paper. We feel that the incorporation of high order terms in existing cascade architectures will increase their expressive power without the decrease in learning rates that additional slabs generate. Current research is constantly suggesting new architectures and applications for nonlinear units[23].

We would like to thank Andy Barto for his helpful comments and advice on the presentation of this material.

REFERENCES

1. NJ Nillson, Learning Machines, McGraw-Hill, NY, 1965.
2. F Rosenblatt, Principles of Neurodynamics, New York, Spartan, 1962.
3. ML Minsky, S Papert, Perceptrons, MIT Press, Cambridge, MA. 1969.
4. RJ Williams, Institute for Cognitive Science, ICS Report No. 8303, UCSD, La Jolla, Ca., 1983.
5. D Psaltis, Neural Networks for Computing Conf., Snowbird, Utah, 1986.
6. JJ Hopfield, Proc. Natl. Acad. Sci. USA, 79, 2554-2558, 1982.
7. DE Rumelhart and JL McClelland, Parallel Distributed Processing, (Bradford Books/Mit Press, Cambridge, MA.), to appear, 1986.
8. YC Lee, G Doolen, HH Chen, GZ Sun, T Maxwell, HY Lee, CL Giles, Proc. Evol. Games, & Learn., Los Alamos, May 1985.
9. T Maxwell, CL Giles, YC Lee, HH Chen, U. of Md. Tech. Rept. UMLPF 86-035, April 1986.
10. HH Chen, YC Lee, T Maxwell, GZ Sun, HY Lee, CL Giles, Neural Networks for Computing Conf., Snowbird, Utah, April 1986.
11. T Maxwell, CL Giles, YC Lee, HH Chen, U. of Md. Tech. Rept. UMLPF 86-036, May 1986.
12. W Pitts, WS McCulloch, Bull. Math. Biophy., 9, 127, 1947.
13. K Fukushima, S Miyake, Pattern Recog., 15, #6, 455-469, 1982.
14. C Koch, Neural Networks for Computing Conf., Snowbird, Utah, 1986.
15. The Brain, Scientific American Books, 1979.
16. WA Little, Neural Networks for Computing Conf., Snowbird, Utah, 1986.
17. GE Hinton, Proc. IJCAI, vol. 2, 683, 1981.
18. A Barto, CW Anderson, RS Sutton, Bio. Cyb., 43, 175-185, 1982.
19. A Barto, P Anandan, IEEE Trans Sys, Man, Cyber, 13, 360-375, 1974.
20. C Anderson, PhD thesis, Comp. & Inf. Sci, U. Mass. Amherst, 1986.
21. T Sejnowski, Johns Hopkins Univ. Tech. Rept. JHU/EECS-86/01, 1986.
22. D Psaltis, Proc. Opt. Comp. Symp., Optoelectronics & Laser appl. in Sci. & Eng., SPIE, 625, Los Angeles, Jan. 1986.
23. Proc. Neural Networks for Computing Conf., Snowbird, Utah, 1986.

A Neural Network Model for the Mechanism of Pattern Information Processing

Sei Miyake and Kunihiko Fukushima
NHK Science and Technical Research Laboratories
1-10-11, Kinuta, Setagaya, Tokyo 157, JAPAN

ABSTRACT

We propose a new multilayered neural network model which has the ability of rapid self-organization. This model is a modified version of the cognitron. It has modifiable inhibitory feedback connections, as well as conventional modifiable excitatory feedforward connections, between the cells of adjoining layers. We also discuss the role of context information for pattern recognition, and propose a symbol-processing model which can send the context signal to the pattern-recognizing network.

INTRODUCTION

Previously, one of the authors proposed a self-organizing multilayered neural network model, the cognitron[1].

In the cognitron, in which information flows only forwards, every cell in the deepest layer usually becomes selectively responsive to one of the learning patterns. However, if the cognitron is trained with stimulus patterns closely resembling one another, the cells in even the deepest layer often come to respond to two or more patterns.

In order to endow the cognitron with the ability of rapid self-organization and to improve pattern-selectivity, we have modified its structure and developed a new multilayered neural network model[2].

BASIC STRUCTURE

As shown in Fig.1, the new network has modifiable inhibitory feedback connections, as well as the conventional modifiable excitatory feedforward connections, between the cells of adjoining layers.

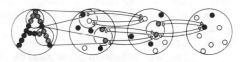

● : large output   O : no response
◉ : output is suppressed by the feedback inhibition

Fig.1 Structure of the network

One of the problems is whether these reciprocal connections exist in the brain. The correctness of this assumption is suggested by anatomical and physiological studies. It has been found that reciprocal connections exist in the visual area of the cortex in the monkey[3].

FUNCTION OF FEEDBACK INHIBITION

According to the experiment by Bechtereva, a certain subcortical neurons may change respons, depending on whether the incoming stimulus information is familiar or not[4].

In this experiment, a subject listened to a stimulus word, and reproduced the word verbally by request, a short time after its presentation. During this procedure, the response of subcortical cells was observed, and it was found that certain subcortical cells specifically changed their responses depending on the familiarity of the stimulus word to the subject.

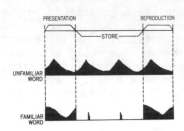

Fig.2 Bechtereva's experiment

As shown in Fig.2, when an unfamiliar word, for instance, foreign word or word-like signal, was presented, the cell continued to respond until the word had been completely reproduced by the subject. On the other hand, when a familiar word is given, the response of the cell was suppressed immediately after the presentation.

This fact, together with the mechanism of feature-extraction from visual patterns, leads us to assume that neuronal activity is emphasized by novel features, and that the response to previously memorized features is suppressed by some feedback inhibition mechanism.

To consider the effect of feedback inhibition, we make a rough estimate of the behavior of the new network.

When a stimulus feature which is already familiar to the network elicits a large output from a feature-extracting cell in a certain layer, the feature-extracting cell immediately sends inhibitory signal back to the cells in the preceding layer, and suppresses their outputs which are relevant to that feature.

When an unfamiliar stimulus is given to the network, however, no inhibitory signal is fed back from the feature-extracting cells, because no feature-extracting cells have been trained to respond to the novel feature yet. Hence, the cells in the preceding layer, if once excited, will continue to yield large sustained output without receiving any feedback inhibition.

Modifiable synapses in the new network are reinforced

in a way similar to those in the cognitron, and synaptic connections from cells yielding a large sustained output are reinforced. Since familiar features do not elicit a sustained response from the cells of the network, only circuits which detect novel features develop. The network therefore quickly acquires favorable pattern-selectivity by the mere repetitive presentation of a set of learning patterns.

COMPUTER SIMULATION

During self-organization, four stimulus patterns "X", "Y", "T", "Z" shown in Fig.3 were repeatedly presented to the input layer U_0 in a cyclic manner.

Fig.3 Learning patterns

Fig.4 shows the responses of the cells in each layer. As may be seen from the response of input layor at $\tau=2$ for each presentation, the parts of the patterns which correspond to previously memorized features are suppressed by receiving feedback inhibition. On the other hand, cells corresponding to novel features keep responding, because there is no feedback inhibition. Hence, circuits for extracting novel features develop quickly.

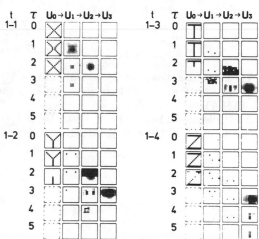

Fig.4 Response of the network

Figure 5 shows the patterns to which each cell in the deepest layer came to respond, after the completion of self-organization. In the conventional model without feedback connections, there are still 36 cells which respond to more than one pattern, while in the new network there are only 5 such cells.

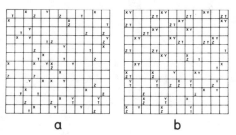

a. New model
b. Conventional model
Fig.5 Pattern-selectivity

SYMBOL–PROCESSING NETWORK

The capability of pattern recognition of the network would be much more improved by the aid of "context". As an initial step to the solution of this problem, we also propose a symbol–processing network which is connected at the back of the pattern–recognizing network described above.

Figure 6 schematically shows the hibrid model. After completion of word–learning, even if an imperfect word is given to the pattern–recognizing network, the missing letter in the word can be interpolated by the feedback signal from the symbol–processing network.

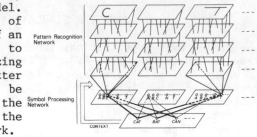

Fig.6 Symbol–processing model

REFERENCES

1. K. Fukushima, Biol. Cybernetics 20, 121 (1975).
2. S. Miyake, K. Fukushima, Biol. Cybernetics 50, 377 (1984).
3. M. Wong–Riley, Brain Res. 147, 159 (1978).
4. N. P. Bechtereva, Brain Res. Monogr. 4, 311 (1979).

A Preliminary Analysis of Recursively Generated Networks

Eric Mjolsness
Department of Computer Science
Yale University
New Haven, CT 06520

David H. Sharp
Theoretical Division
Los Alamos National Laboratory
Los Alamos, N.M. 87545

Abstract
This paper outlines an approach to the automated generation of structured networks of threshold elements or "neurons" for computational purposes. The structured networks are obtained by recursive operations on primitive elements, constrained to produce networks whose recursive description is relatively simple. A preliminary analysis of the performance of structured networks on a specific continuous coding problem is presented, and compared to the performance of unstructured networks on the same problem. This comparison shows that recursively structured networks can exhibit efficient learning and high performance, while permitting generalization to larger instances of the same problem.

I. Introduction

Along with the recent interest in networks of threshold elements ("neural" networks) as a way of expressing and implementing parallel computations, there is an understandable tendency to approach the design of such networks with some of the same methods that have proven useful in analysing individual computing networks: analogies with simple dynamical systems, collective phenomena, and algebraic objective functions (e.g. Hinton and Sejnowski[1], Lapedes and Farber[2]). There are also well-established methods for designing circuits, algorithms, software and computer systems which are based on principles of hierarchical design. We aim to combine these apparently disparate approaches, mapping hierarchical design to scaling ideas (similar to the renormalization group formalism described in Wilson and Kogut[3]) for dynamical systems, in the context of an especially simple computation.

The use of hierarchical design leads to structured networks, that is ones describable with only a few parameters, generated by recursive operations on primitive elements. Given a structured and an unstructured network which perform the same computation, we prefer the structured one since it is easier to discover (being an element of a much smaller search space) and it is more likely to suggest a generalization to larger network and problem sizes. If such generalization can be automated, it would provide an interesting example of dynamical learning.

In this paper we generate structured networks which solve a variation of an encoding problem considered by Hinton and Sejnowski[1]. This problem is formulated in section II, and section III contains a description of the search strategies used to obtain several types of structured networks. In section IV, the performance of structured and unstructured networks is compared. The comparison involves both testing the networks' performance on the encoding problem and, to evaluate the prospects for dynamical learning in this task, examining the networks' generalizability to larger problem sizes. The results support the idea that it is possible to find a dynamical method for adjusting the free parameters of a structured threshold network which is capable of solving (or learning to solve) the test problem and which is sufficiently simple that it could plausibly be tried out on other problems as well.

II. A Continuous Coding Problem

We consider the following problem. A unary input (a picture-like representation of one number) is to be converted into some code which has extensive continuity properties: a small change in the input number corresponds to a small Hamming distance between the two output codes (as in a Gray code), a small Hamming distance between two code words corresponds to a small difference between the corresponding numbers (graceful degradation under code word corruption – a kind of fault tolerance), and in general the unsigned difference between two numbers is to be a monotonic function of the Hamming distance between the two codes. If $\{s_i = 0 \text{ or } 1\}$ (with only one $s_i = 1$ at a time) is a set of N threshold elements representing the unary input, and $\{s_\alpha = \pm 1\}$ is a set of $A = O(\log N)$ threshold elements representing the code word output, then

$$s_\alpha = \text{sgn}\left(\sum_i F_{\alpha i} s_i\right) \qquad = \sum_i F_{\alpha i} s_i, \text{ if } F_{\alpha i} = \pm 1$$

represents both the threshold network's operation and the code. This is because, if $s_i = 1$ and $s_j = 0$ for $j \neq i$, the output code word is $s_\alpha = F_{\alpha i}$. (We assume that $F_{\alpha i} = \pm 1$.) Thus the columns F_{*i} of F are the possible code words corresponding to the possible inputs i, $1 \leq i \leq N$. Likewise, each output element s_α may be thought of as a "feature detector". The rows $F_{\alpha *}$ of F are the features which may be detected; in other words, they are the sets of input values i which would excite the given output element. The desired continuity and monotonicity properties are enforced by minimizing the function

$$E = \frac{1}{N^2} \sum_{ij} \left[\left| \frac{i-j}{N} \right|^p - \frac{1}{A} d_{\text{Hamming}}(F_{*i}, F_{*j}) \right]^2 \tag{1}$$

for $p > 0$. Since the Hamming distance between two code words is

$$d_{\text{Hamming}}(F_{*i}, F_{*j}) = \frac{1}{2} \sum_{\alpha=1}^{A} (1 - F_{\alpha i} F_{\alpha j})$$

we have

$$E(F) = \frac{1}{A} \sum_{\alpha=1}^{A} \frac{1}{N^2} \sum_{i,j=1}^{N} F_{\alpha i} F_{\alpha j} \left(\left| \frac{i-j}{N} \right|^p - \frac{1}{2} \right)$$

$$+ \frac{1}{4A^2} \sum_{\alpha,\beta=1}^{A} \left(\frac{1}{N} \sum_{i=1}^{N} F_{\alpha i} F_{\beta i} \right)^2 + \text{constant} \tag{2}$$

$$= \sum_\alpha E_1^{(p)}(F_{\alpha *}) + \sum_{\alpha \beta} E_2(F_{\alpha *}, F_{\beta *}) + \text{constant}$$

and as indicated E may be divided into a single-feature term and a feature-interaction term. The latter term is minimized for orthogonal features.

Aside from being a useful test case, this encoding problem could be of some interest for non-automatic network design: one is trying to represent a (possibly) meaningful quantity by a correlation (a Hamming distance), and correlations are easily computed with threshold element networks.

To minimize $E(F)$, the most direct approach would be to use a hill-climbing or simulated annealing algorithm. The disadvantages of these methods in general are that they produce irregular networks (hard to understand and to generalize to much bigger networks) and that they are often unacceptably slow. As an alternative, we will examine very compact recursive descriptions of the feature matrix $F_{\alpha i}$, and we will minimize E as a function of these recursive descriptions.

III. Recursively Generated Networks

We will introduce two different kinds of concise recursive descriptions of feature matrices. The second is more elaborate and optimized than the first, but large improvements remain to be made here. In fact, we immediately restrict ourselves to recursive descriptions of the rows of the feature matrix, or "features", $F_{\alpha*}$.

For example, a feature may be recursively built up by duplication (d_+) and duplication with negation (d_-):

$$F_{\alpha*}^{(1)} = (+1)$$
$$F_{\alpha*}^{(2)} = d_+(+1) = (+1, +1)$$
$$F_{\alpha*}^{(3)} = d_- d_+(+1) = (+1, +1, -1, -1)$$
$$d_+ d_- d_+(+) = \ + + - - + + - - \ \text{(shorter notation)}$$

so that a feature with N signs is represented by only $\log N$ signs $(d_\pm)$, with great economy. The set of feature vectors so generated corresponds to the Walsh functions, and we would like to find those $A = O(\log N)$ out of N Walsh functions which minimize $E(F)$. With luck, there will be a pattern which generalizes to larger N and A. For example the binary encoding is represented by a set of features

$$F_{\alpha*} = d_+^a d_- d_+^b, \qquad a + b + 1 = \log_2 N$$

and a binary-reflected Gray code has

$$F_{\alpha*} = d_+^a d_- d_- d_+^b, \qquad a + b + 2 = \log_2 N.$$

The Walsh functions represent features with such economy that it is possible to exhaustively search all the features expressible with d_+ and d_-; there are only N of them. Since the recursive descriptions have been applied only to the features $F_{\alpha*}$, and not to the entire matrix F, an ad hoc search method for feature matrices was used in place of exhaustive search which would take N^A steps – the usual combinatorial explosion which we are able to avoid (only) for single features.

Instead, a beam search of beam width w was used to build up a set of feature matrices (the best w matrices found so far) one feature at a time. The procedure involved trying out each of the N possible concisely-described features as the addition to each of w matrices and keeping, out of the resulting Nw candidate matrices, the w matrices of lowest score $E(F)$. Finally, the beam width w was varied somewhat and the experiment was repeated, to insure that the results were not substantially affected by the beam width.

This procedure forces a serious asymmetry between the α and i indices of F (excused perhaps by the asymmetric index ranges: $A = O(\log N)$) and is thus just a first step towards a dynamics of recursive network descriptions; the α index is not described recursively.

The second kind of recursive description allows, in addition to the $d_\pm$ operations, the concatenation of two features whose relationship is more distant: they share common subfeatures. We consider concatenation with and without negation, producing part-sharing trees as in Figure 1. (This figure shows a feature which occured in the best network produced by the search procedure discussed below, for $p = .5$, $N = 32$, and $A = 10$.) In order to prevent a combinatorial explosion in the number of such trees considered, the sub-features to be concatenated (the "parts" of features) are forced to be shared by the requirement that the tree structure have small width at each level. In detail, the part-sharing trees are put into equivalence classes based on the sequence of numbers of nodes per level $(a_1, b_1, \ldots a_l, b_l)$, and only those equivalence classes of cardinality less than some threshold are kept:

$$\text{cardinality} = \prod_{k=1}^{l} 2^{b_k - 1} S_{a_k}^{(b_k)} < \text{maximum}$$

$$S_a^{(b)} = \text{Stirling number of the second kind}$$
$$= \text{number of partitions of } a \text{ elements into } b \text{ groups}$$

This rule may be efficiently implemented as tree-pruning applied during the recursive construction of the allowed features. Features are combined using a beam search as before, and it must be checked that neither the beam width nor the cardinality threshold affect the results.

At this point the outline of the part-sharing trees experiment should be clear, but there is an important modification which vastly speeds up the search process with little or no degradation in performance as measured by $E(F)$. The modification involves constructing the trees from the top down, as before, but testing the partially built trees and removing from consideration those of relatively poor single-feature score. (A more complicated criterion, based on the frequency of occurrence of a partially built feature inside of low-scoring feature matrices, did not help much.) In what follows, at generations $k = 1, 2, 3, 4,$ and 5 the fraction of the features kept was 1, 1, .5, .3, and .3, respectively; these fractions seem to be about the minimum necessary to retain the optimal feature matrices.

This modification is a plausible one: the bottom portions of a part-sharing tree consist of "shared parts", usually large blocks being built up, and the top regions of such a tree are "templates" for the coarse organization of a feature. We therefore find that it is possible to build good features using relatively few templates whose merit has been shown in previous, smaller encoding networks. In this sense the network "learns" from previous experience with easier problems.

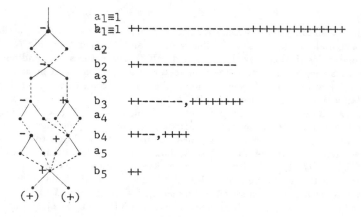

Figure 1. A part-sharing tree from a computed feature matrix for $N = 32, A = 10, p = .5$. Solid lines show concatenation of feature vectors and dotted lines indicate multiple use of one feature vector. Signs attached to the tree nodes indicate concatenation with $(-)$ and without $(+)$ feature negation. The bottom-most signs are always taken to be positive, to reduce the redundancy in the number of part-sharing trees which represent one feature. The middle column records the number of nodes at each level in the structure; case illustrated has $a_2 = 2$ and $b_2 = 1$. The feature vectors corresponding to the nodes at a given level in the tree are shown on the right.

IV. Results

The minimal scores $E(F)$ obtained by hill-climbing, by the semi-recursive network search procedure for Walsh features, and by the more sophisticated semi-recursive search procedure for part-sharing trees are compared in Figure 2. All three methods are comparable for $p <\approx .3$; outside this range hill-climbing is superior to the Walsh functions as well as being much faster as described below. Furthermore, the part-sharing trees score remains very close to that of hill-climbing over most of the range of p (for $0 < p <\approx .7$). The most surprising result, however, is the extreme orderliness of the Walsh function semi-recursive networks. Figure 3 shows this for $p = .01$ (or any p sufficiently near to 0): features are ranked by the number of same-sign blocks (or the number of sign-changes as i varies), and the top A features are chosen to produce the optimal semi-recursive network. This ranking may also be obtained by examining the single-feature scores, a point of some interest for improving the dynamical system which is to produce F. For $p \neq 0$ the pattern of the optimal semi-recursive network is modified by the inclusion of multiple copies of the top-ranked feature, $d_- d_+^a$. These patterns are so strong that an algorithm to generate good $p <\approx .3$ networks recursively would be computationally far cheaper than the hill-climbing method.

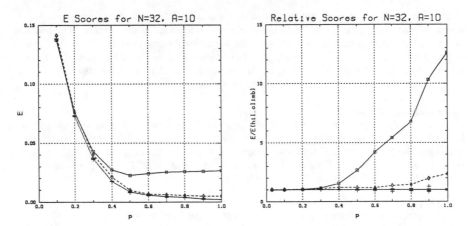

Figure 2. Relative and absolute E scores (see equation (1)) for the continuous encoding problem with $N = 32$ and $A = 2\log_2 N$. The three optimization methods compared are hillclimbing (dotted lines), beam search over features constructed by duplication with and without negation (solid lines), and beam search over part-sharing trees (dashed lines). These methods are discussed further in the text. (a) Absolute E scores for $N = 32$. (b) Scores relative to the average of five hillclimbing runs, for $N = 32$. The corresponding plots for $n = 16$ are almost indistinguishable from these ones.

The running time of the search over part-sharing trees on a conventional serial computer (a Pyramid 90X) is much faster than that of the Walsh function beam search described earlier, and a factor of 10 slower than hill-climbing (measured times were 120 sec, 43.2 sec, and 3.5 sec respectively for $(N, A, p) = (16,8,.1)$; 356 sec, 939 sec, and 24.2 sec respectively for $(N, A, p) = (32,10,.03)$; and 900 sec, unmeasured and 103 sec respectively for $(N, A, p) = (64,12,.3)$). We have several comments on this timing data. The Walsh function beam search is impractically slow, but the resulting networks are so highly patterned that one can generalize them to all larger sized networks and thereby produce large networks far more quickly than a hill-climbing procedure could. However, our real interest is see how this generalization could automatically come out of a more fully recursive description of the network, without human intervention. The slow search over part-sharing trees is 10 times slower than hill-climbing, but is far faster than the slow search over Walsh function networks and produces better networks. The question of the orderliness or pattern of the optimal part-sharing networks remains to be understood and is the key to any automatic

feature vector	description	rank	blocks
++++++++++++++++++++++++++++++++	$d_+d_+d_+d_+d_+$	∞	1
++++++++++++++++----------------	$d_-d_+d_+d_+d_+$	1	2
++++++++----------------++++++++	$d_-d_-d_+d_+d_+$	2	3
++++++++--------++++++++--------	$d_+d_-d_+d_+d_+$	3	4
++++--------++++++++--------++++	$d_+d_-d_-d_+d_+$	4	5
++++--------++++----++++++++----	$d_-d_-d_-d_+d_+$	5	6
++++----++++--------++++----++++	$d_-d_+d_-d_+d_+$	6	7
++++----++++----++++----++++----	$d_+d_+d_-d_+d_+$	7	8
++----++++----++++----++++----++	$d_+d_+d_-d_-d_+$	8	9
++----++++----++--++++----++++--	$d_-d_+d_-d_-d_+$	9	10
++----++--++++----++++--++----++	$d_-d_-d_-d_-d_+$	10	11

Figure 3. Highly patterned optimal features for $p = .01$. Except for the top feature, which is chosen last, the top A features are chosen to form a feature matrix. The pattern of $d_\pm$ operations is best seen by reading down vertical columns of aligned operators; the signs alternate in pairs and successive powers of two. (The feature $d_+d_+d_+d_+d_+$ is placed first to emphasize this pattern, though it is the exception.) The ranking also corresponds to the number of same-sign blocks in the feature; coarse features are chosen first.

generalization procedure.

Aside from the important question of the presence or absence of generalizable patterns in the part-sharing tree feature descriptions, the most pressing problem for investigation is probably the replacement of the beam search by which features $F_{\alpha*}$ are brought together to make a feature matrix F with a fully recursive description of F. We speculate that the two problems may have to be solved jointly.

Conclusion

By recursively building up a neural network to perform the continuous encoding task, one can obtain networks described by so few parameters (so highly structured) that a unique generalization to all larger networks and problem sizes is suggested. When the parameter p in the continuous encoding task is less than $\approx .3$, the resulting network performs well. For most of the range in p, we can obtain good performance by considering less rigid semi-recursive descriptions whose proper generalization to large networks is not yet clear. These networks involve the use of shared standard "parts" and the re-use of previously tested "templates" for large-scale network organization; the templates, at least, are learned from experience with smaller versions of the same encoding task.

References

1. G. Hinton and T. Sejnowski, Proc. 5th Annual Conference of the Cognitive Science Society, 1983.
2. A. Lapedes and R. Farber, Los Alamos preprint LA-UR-85-4037, 1985.
3. K. Wilson and J. Kogut, Phys. Rept. 12C Number 2, 1974.

ERROR CORRECTION AND ASYMMETRY IN A BINARY MEMORY MATRIX

A. Moopenn, S.K. Khanna, John Lambe, and A.P. Thakoor

Jet Propulsion Laboratory
California Institute of Technology
Pasadena, CA 91109

ABSTRACT

Memory content and error correction requirements for moderate size associative memories based on binary interconnection matrices have been investigated. Of particular interest is the use of additional asymmetric connections to correct a limited number of one-bit errors in stored words. Each error correction is carried out by instructing a "wrong" word to go to a nearby correct word. By using a local inhibition scheme, space in the network is "reserved" for introduction of the needed asymmetry. This overall scheme has been implemented in hardware and is found to be error-free in its operation. The error correction is thus introduced with no increase in complexity. These results project an error-free storage with an efficiency of ~25% for binary matrices in a megabit range memory capacity. The mechanism of error correction and the special role of the local inhibition scheme in it are discussed.

INTRODUCTION

Associative memories[1-5] based on Hopfield's neural network model[6] have been a subject of growing interest over the recent years. Hopfield's model network, with its parallel architecture and distributive algorithm, also exhibits some other unique capabilities such as pattern recognition, subject classification, and error correction. Electronic hardware[1-3] and optical implementations[4,5] of networks based on Hopfield's model have been designed and built, which demonstrate several of the associative capabilities of such a network. In this paper, we examine the information storage capacity of an electronic neural network incorporating a binary interconnection matrix. The role of dilute vector coding in the form of local inhibition for optimal information storage and the use of asymmetry for destabilizing false memories to yield an error-free associative memory are discussed.

In a binary connection memory matrix, the information to be stored is coded in the form of N-dimensional binary vectors V^s, (V_i^s = 0 or 1, i = 1 to N). The storage prescription for "memorizing" R such vectors into the memory matrix is given by the clipped sum of outer products of the vectors;

$$T_{ij} = \begin{cases} 1 & \text{if } \sum_{s=1}^{R} V_i{}^s V_j{}^s > 0 \\ 0 & \text{otherwise} \end{cases} \qquad (1)$$

If such a memory is prompted with one of the stored vectors, V^t, the output of the matrix would be given by the matrix-vector product:

$$V_i = \sum_{j=1}^{N} T_{ij} V_j{}^t \sim M V_i{}^t + f_i(V^t) \qquad (2)$$

where M is the vector strength (the number of "1"s) of the input test vector V^t, and $f_i(V^t)$ is the difference between the quantities $\sum T_{ij} V_j{}^t$ and $M V_i{}^t$. The first term in eq. 2 may be considered as the memory output "signal" and the second term as the output "noise." If V_i is now thresholded at M, the correct auto-association would be obtained when $f_i(V^t) < M$ for all $i=1,\dots,N$; so that one has

$$V_i{}^t = \Theta_M(V_i) \qquad \text{where} \qquad \Theta_M(x) = \begin{cases} 1 & \text{if } x \geq M \\ 0 & \text{if } x < M \end{cases}$$

It is clear from the above discussion that unless the set of binary input vectors is judiciously chosen (e.g. an orthogonalized set), the ability of such a binary matrix memory to yield correct auto-associations would be seriously questioned. This problem has been studied by Willshaw et.al.[7,8] and Palm[9]. They have shown that by limiting the vector strength of the binary input vectors to a small fraction of N, the retrieval error rate of binary matrix memories can be reduced significantly with reasonable storage capacity. At the optimum vector strength of $\sim\log_2 N$, the amount of information that can be retrieved from a symmetric binary matrix memory (with a single-element error per vector) approaches $\sim 0.35\ N^2$ bits in the limit of large N. These results confirm the need for dilute vector coding (restricted number of "ones" in each binary word) for optimal information storage in a binary memory matrix.

LOCAL INHIBITION

In the following, we describe one particular form of dilute vector coding which "guarantees" that the noise term, f_i, in eq.(2) is always less than the vector strength, M (we assume that the stored vectors are all of the same strength). This coding scheme is based on the idea of a local inhibition rule which dictates that when one element (component) of a binary vector takes on the value "1", some of the other elements of the vector must take on the value "0". One simple form of local inhibition consists of decomposing a binary vector into a certain number of groups with only one element in each group taking on the value "1". All stored vectors coded in this form would have equal vector strength, M, as determined by the number of groups. The connection matrix T constructed from a set of such vectors therefore consists of M^2 sub-matrices or blocks with the distinctive property that in the diagonal blocks, the off-diagonal elements are always zero. Evidently, this particular property, arising from the use of local inhibition and dilute coding, ensures that the noise term $f_i(V^t)$ would always be less than M.

The local inhibition scheme can easily be implemented in hardware. This is shown schematically in Fig. 1 for a group of four "neurons". Each neuron receives an inhibitory signal I_i, proportional to the number of neurons in the group which are in the high output state. Our hardware simulation study based on local inhibition has indeed shown that the stored vectors always correspond to stable states of the memory. The stability and the associative recall properties of a binary memory matrix incorporating local inhibition in hardware have been described elsewhere[1].

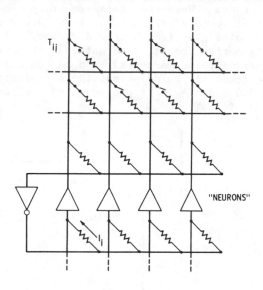

Fig. 1 Electronic implementation of the local inhibition scheme for a group of four "neurons."

FALSE MEMORIES

Although the use of local inhibition assures us that the stored vectors are always stable states of such a binary memory matrix, the memory performance is still limited by the appearance of false memories. From an energy picture, the false memories correspond to spurious satellite minima of the energy function of the memory system. In a general sense, if the memory system is considered as a logic machine, the false memories may have a more important role as a set of additional logical statements which are consistent with the original prescribed statements. However, in an associative memory, the false memories are undesirable and need to be eliminated.

To see how retrieval errors due to the false memories may be minimized, we have examined the dependence of the mean number of false memories, F, on memory loading, dilute vector coding, and the size of the memory matrix. The memory loading, η, is defined here by $\eta = R \times I_w/N^2$, where I_w is the information content of a stored vector ($I_w = M \times \log N/M$ bits), and R is the number of stored vectors. Figure 2 shows the mean relative number of false memories (F/R) for $N \cong 126$ and vector strengths M = 5, 7 and 14. The false memory statistics were obtained from an optimized computer search which identified all false memories generated from a set of random binary vectors stored in the memory matrix. The near exponential increase in the mean fraction of false memories with memory loading is evident. For memory loadings between ~0.10 and ~0.20, the memory matrix with vectors coded with the optimum strength exhibits the least fraction of false memories. The false memory statistics improve with increasing vector length N. Figure 3, shows false states data for N = 126, 256, and 513, with optimium vector strength $M = \log_2 N$ in all three cases. When the false memories comprise a small fraction of the input vectors (F/R < 10%), they are generally no more than two Hamming units away from a stored vector. If it is assumed that all false memories are of this nature, (i.e. single-element errors) the number of false memories could be calculated to a good approximation[10]. From this calculation, the mean fraction of false memories for N = 1000 and M = 10 is projected to be less than 10%, for a memory loading of ~25%.

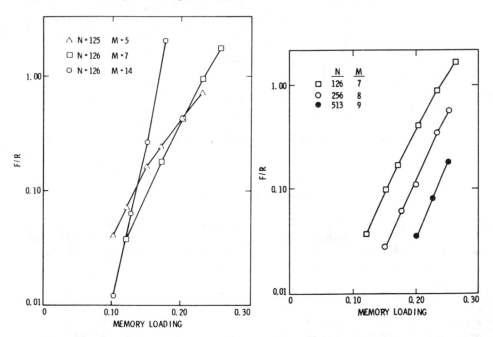

Fig. 2 False memory error rate curves for $N \cong 126$

Fig. 3 False memory error rate curves for N = 126, 256 and 513

ASYMMETRY FOR ERROR CORRECTION

It is possible to suppress false memories by introducing a slight asymmetry to the memory matrix T. The basic idea is to add the necessary connections to the memory matrix so that each false memory is "mapped" to a nearby stored vector. Let V^f denote a false vector and V^s, a nearby vector onto which V^f is to be mapped. The mapping matrix is $\Delta T_{ij} = \Theta_1(V_i{}^s V_j{}^f - T_{ij})$. The required asymmetric connections are then those elements of ΔT_{ij} which are nonzero. With local inhibition enforced, the false memories would no longer be stable states. Moreover, since the false memories are generally single-element errors for small memory loadings, the extra connections needed are among the unused connections located in the diagonal blocks of the matrix T. Fig. 4 shows an example of false state destabilization that has been demonstrated in hardware with a programmable 32 x 32 binary matrix memory. With N=32 and M=4, the binary vectors are represented by 4 digits, each specifying which of the eight elements in a group takes on the value "1". For the given set of random vectors shown, the false memories generated are (4118) and (4158). They are mapped to the nearby vectors (4116) and (5158), respectively. The extra asymmetric connections (identified as "x") needed for the (4118) → (4116) and (4158) → (5158) mappings appear in the first and fourth diagonal blocks, respectively. With these additional connections, the false memories are no longer observed as stable states. When the memory is initialized to either one of the false memories, it always converges to the nearby state onto which the false state is mapped.

Prescribed vectors:
(1455) (4116) (3336) (2234)
(8764) (4252) (7542) (4618)
(5158) (7115) (3784) (6642)

Induced vectors:
(4118) (4158)

Mappings:
  4118 → 4116
  4158 → 5158

Fig. 4 Example of false memory destabilization using asymmetric connections in a 32 x 32 binary matrix.

```
1,...... ..,1.... ...,1... ....1...
.1,..... .1,..... .,1..... ..1,....
..1,.... ,.1..1.. .,1..... ..1,..1
...1,... 11...1.. 1...,1.. .1...,1.1
...x1... 1.......  ....1... ........1
......1, 1...,1.. 1..1,... .1..,1..
......1. .1..1... 1..1,... .1..1...
.......1 ......1. ...,1... ...,1...

..,11.1. 1....... 1...,1.. ....11.1
.1,1.... .1,..... ..1,1... .1.1.,..
1,...... ...1,... ..1,.... ...,1...
...,.,1 ...,1... ...,1... .....1..
...,1.1. ....,1.. 1..1,... .1..,1..
......1 ......1. ...1.1 .1..1...

...1..1. 1...,1.. 1....... ....11.1
.ii.... .ii..... ..1,.... .1.i.1.
1.:ii... 11.:11... ...1,... .1.,1.1
.......1 11.1.,.. ....1... .1..1...
..1.... ......1. .......1 ...1...

..:1.ii. .i.:ii.. ...:ii... .i.:....
,ii...i i,.,.i. ,.1.,1.i ...i,..
1..:..1i 1,.1.... ....1... .....1.,
..ii... 1.1..... 1.1..... ....1.x
..:ii... 1.....1. 1...,1.. .......i
```

Although the emphasis of this paper is on single-element error correction, the method can be applied to two-element error correction as well[10]. The overall extent to which asymmetry can be employed for correcting larger number of false memories, however, needs to be examined further since the increasing number of asymmetrical connections could lead to the destabilization of some of the other valid states. It is anticipated that for a binary matrix memory with N=1000 and F/R < 10%, all false memories could be efficiently corrected with asymmetric connections, to yield error-free operation with a memory loading of up to ~25%.

CONCLUSIONS

The requirements for an error-free associative memory based on a binary connection matrix model have been addressed in this paper. It has been shown that the implementation of a local inhibition scheme, as a form of dilute vector coding for optimal information storage, guarantees prescribed states to be stable states of the memory. In addition, the use of extra asymmetric connections has been demonstrated to provide an effective means for correcting a limited number of single-element errors, and thereby increasing the performance of the memory. These results project an error-free operation with information storage efficiency approaching 25% for binary matrix memories in the mega-bit range memory capacity.

ACKNOWLEDGEMENT

The work reported in this paper was performed by the Jet Propulsion Laboratory, California Institute of Technology, and was sponsored by the Defense Advanced Research Projects Agency, through an agreement with the National Aeronautics and Space Administration.

REFERENCES

1. J. Lambe, A. Moopenn, and A. Thakoor; Proc. AIAA/ACM/NASA/IEEE Computer in Aerospace V Conference, 160 (1985).
2. R.E. Howard, L.D. Jackel, W.J. Skocpol; Microelectronic Engineering, Vol. 3, 3 (1985).
3. L. D. Jackel, R. E. Howard, H. P. Graf, B. Straughn, and J. S. Denker; J. Vac. Sci. Technology, B4, 61 (1986).
4. D. Psaltis and N. Farhat; Optics Letters, Vol. 10, 98 (1985).
5. N. Farhat, D. Psaltis, A. Prata, and E. Peck; Applied Optics 24, 1469 (1985).
6. J.J. Hopfield, Proc. Natl. Acad. Sce., Vol. 79, 2558 (1982).
7. D. J. Willshaw, O.P. Buneman, and H. C. Longuet-Higgins; Nature, Vol. 222, 960 (1969).
8. D.J. Willshaw, "Models of Distributed Associative Memory," Thesis, Univ. of Edinburgh (1971).
9. G. Palm; Biol. Cybernetics, Vol. 36, 19 (1980).
10. A. Moopenn, J. Lambe, and A. P. Thakoor; to be published.

A MACHINE FOR NEURAL COMPUTATION OF ACOUSTICAL PATTERNS WITH APPLICATION TO REAL TIME SPEECH RECOGNITION.

P. Mueller
University of Pennsylvania, Dept. of Biochemistry
and Biophysics, Phila., PA 19104-6059.

J. Lazzaro
Dept. of Computer Science, California Institute of Technology,
Pasadena, CA 91125.

ABSTRACT

400 analog electronic neurons have been assembled and connected for the analysis and recognition of acoustical patterns, including speech. Input to the net comes from a set of 18 band pass filters (Q_{max} 300 dB/octave; 180 to 6000 Hz, log scale). The net is organized into two parts, the first performs in real time the decomposition of the input patterns into their primitives of energy, space (frequency) and time relations. The other part decodes the set of primitives.

216 neurons are dedicated to pattern decomposition. The output of the individual filters is rectified and fed to two sets of 18 neurons in an opponent center - surround organization of synaptic connections ("on center") and ("off center"). These units compute maxima and minima of energy at different frequencies.

The next two sets of neurons compute the temporal boundaries ("on" and "off") and the following two the movement of the energy maxima (formants) up or down the frequency axis. There are in addition "hyperacuity" units which expand the frequency resolution to 36, other units tuned to a particular range of duration of the "on center" units and others tuned exclusively to very low energy sounds.

In order to recognize speech sounds at the phoneme or diphone level, the set of primitives belonging to the phoneme is decoded such that only one neuron or a non-overlapping group of neurons fire when the sound pattern is present at the input. For display and translation into phonetic symbols the output from these neurons is fed into an EPROM decoder and computer which displays in real time a phonetic representation of the speech input.

INTRODUCTION

The analysis of acoustical patterns requires decoding of rapidly changing relations between energy, frequency and time variables. The analog and parallel computational methods used by neurons allow the simultaneous real-time evaluation of these relations and are therefore ideally suited for this task.

We give here a brief description of a neural network assembled for the analysis and recognition of acoustical patterns including speech. Ultimately the project is aimed at the real or compressed time recognition of continuous speech at the phoneme level and is based on our earlier work on neural computation of acoustical patterns (1,2).

NEURON PROPERTIES

The network shown in figure 1 comprises 400 electronic neurons which can be connected in any desired fashion with individually controlled synaptic gain and time constants. The circuit diagram and the steady state and dynamic imput-output relations of the neurons are given in figure 2. Each neuron has a single summing input and a separate positive and negative output. Fanout capability at unity transfer gain is >500:1. Threshold and minimum output at threshold are adjustable. Synaptic gain and time constants are determined by individual resistors and capacitors.

Figure 1

This assembly contains 400 electronic neurons mounted in groups of 24 on panels seen in back and at right. Each neuron contains an LED indicating its output amplitude. Synaptic connections are made on the circuit boards at bottom, which contain provisions for 10^5 RC connections. Additional connections can be made on cross point boards. All synaptic gains and time constants are determined by plug-in resistors and capacitors.

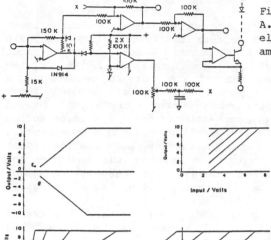

A. Circuit diagram of the electronic neuron. Operational amplifiers are TL084.

B. Steady state input-output relations of the circuit shown in A. Top left: At unity gain, the output has a threshold, θ and a minimum output at threshold, E_o. Above threshold the I-O relation is linear up to a maximum. Both inhibitory and excitatory output are plotted. Top right: The minimum output at threshold, E_o, can be internally controlled without affecting the threshold or the slope of the linear range. In the limit the circuit can act as a boolean switch. Bottom left: Variation of the synaptic gain (input resistance) controls the threshold and the slope of the linear range. Bottom right: Internal adjustment can shift the threshold into the negative region so that the neuron is "on" in the absence of an input.

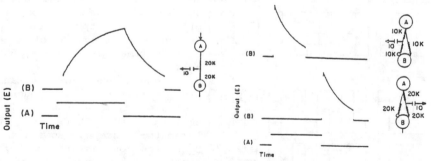

C. For computation of relations involving the time domain excitatory or inhibitory inputs are low pass (RC) coupled. Left: Low pass coupling of the excitatory input causes a delayed and exponentially rising output from neuron B in response to a square output from neuron A. Notice the nonlinearity in output B at the threshold due to E_o. Right: Combinations of direct and low pass coupled excitation and inhibition generate a typical "on" response (top) whereas the inverse arrangement (Bottom) generates an "off" response. In each case the time from the beginning or end of the input is transformed into a potential.

PATTERN DECOMPOSITION

18 band pass filters with a Q_{max} of 300 dB/octave and a log frequency scale of 180 to 6000 Hz form the input to the net. 216 neurons are dedicated to the initial pattern decomposition in terms of energy, space (frequency) and time primitives. The filter output is rectified and fed into 2 rows of 18 neurons in a center-surround organization typical of biological sensory input stages. The "on center" units receive excitatory inputs from a center filter and inhibitory connections from the surrounding filters with gains that decrease logarithmically from the center. Another row of 18 neurons has the inverse organization ("off center"). These two rows compute in essence the positive and negative second derivatives of energy with respect to space (frequency) and recognize maxima (formants) and minima of energy in the frequency space.

The next two sets of 18 units compute the temporal changes of the energy amplitude $\pm dE/dt$ in the corresponding "on center" and "off center" units and correspond to biological "on" and "off" units.

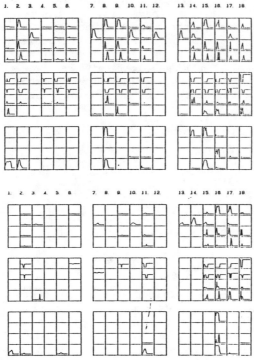

Figure 3

Outputs from the 216 neurons comprising the decomposition section for the phoneme "a" (father) (top) and s (sit), bottom. The outputs of the 216 neurons are fed into multiplying A/D converters and plotted after storage. In each box is plotted the output from one neuron as a function of time (total time per box = 800 ms).

Inputs are from 18 Band pass filters. Frequency increases from neuron 1 (180 Hz), to 18 (6000 Hz) in logarithmic steps. Rows 1 and 2 are the "on center" and "off center" units, rows 3 and 4 the "on" and "off" units, rows 5 and 6 their complement and rows 7 and 8 the "motion" detectors, which respond to changes in frequency (see text).

In general different phonemes or diphones have characteristic patterns of activity.

"On" units receive a direct excitatory input and a low pass filtered RC-coupled inhibitory input from their corresponding "on center" unit. For the "off" units inhibition is direct whereas excitation is RC coupled (see figure 2).

Changes in formant frequency which represent local movement of the energy maxima along the frequency axis in either direction ($\pm$dE/ds) are computed in the next two sets of units. Detection of directional motion is achieved by neural "and gating" of neighboring "on" and "off" units (1) which in this analogue requires the representation of the complement of "on" and "off" units. Thus there are 36 units whose activity is normally high and is suppressed by the activity in the "on" or "off" units. In addition there is a set of 18 "on center" neurons responding exclusively to low energy sounds, another set of "hyperacuity" units which expand the frequency resolution to 36 and a set of "on center" units which respond only to sounds with durations less than 80 ms. Two examples of typical outputs from the decomposition section for the phoneme "a" (father) and "s" (sit) are shown in figure 3.

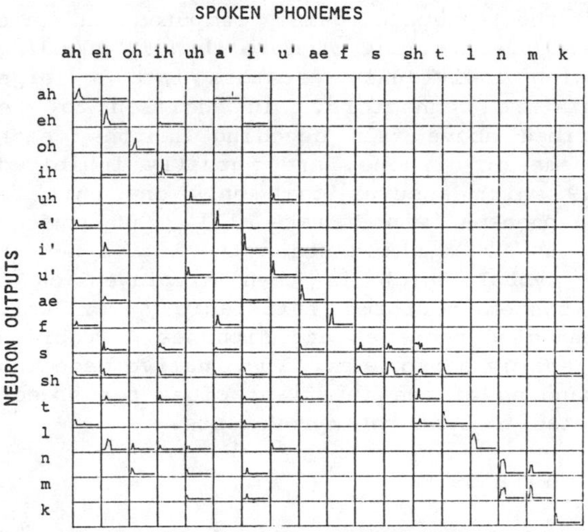

SPOKEN PHONEMES

Figure 4

Computed outputs from the decoding neurons for 18 phonemes. Synaptic connections from the 216 decomposition neurons to individual decoding neurons each tuned to a different phoneme were computed using the decoding scheme described in the text. The output of these neurons for each of the phoneme was evaluated for the entire duration of the phoneme. Mutual inhibitory connections between the decoding neurons were not included in the computation. They would suppress activity in neurons tuned to phonemes other than the current one.

PATTERN DECODING

A particular acoustical pattern such as a phoneme or a diphone generates a specific pattern of activity in the net. These patterns are decoded by individual neurons, each tuned to the presence of a particular phoneme. In order to determine the synaptic inputs and gains to the decoding neurons, 80ms segments of the outputs from the decomposition units are stored after A/D conversion in a digital computer which computes the decoding connections for 40 ms time slices using the following global decoding strategies where each decoding neuron receives 216 inputs.

1). Generally all inputs from the primitive neurons to the decoding neurons are positively weighted by the probability of a primitive being a member of the set active during an average phoneme as well as by their average output amplitude divided by the sum of all outputs.

2). Non-members and week members (units whose output is on average below 10% of maximal) are inhibitory and weighted by their "distance" from the set boundaries.

3). All "off" units that become active at the end of the phoneme or during transition to another phoneme are inhibitory with gains determined by their average amplitude.

4). Those complement units ($\sim$ "on" units) used for motion detection which are blocked during a phoneme are also inhibitory.

The expected performance of the decoding matrix is evaluated by computing the outputs from the decoding neurons for different phonemes. An example of this computation for 18 phonemes and 18 decoding neurons is shown in figure 4.

Using the values of the synaptic gains computed in this fashion the physical connections to a set of more than 50 decoding neurons are made by plugging individual resistors into a large cross point array of miniature connectors. In addition to the decoding connections described above, all decoding neurons, each tuned to a different phoneme or diphone, are mutually inhibited with a gain factor of 0.9 which assures that only one unit is active at any time during a phoneme (winner take all). Output from the decoding neurons is to an EPROM which codes the active neuron into its proper phonetic symbol which is then displayed on a screen. At present the system performs satisfactory for well articulated isolated or connected phonemes and diphones. Problems are still presented by low energy consonants. Alternative decoding schemes and better understanding of the invariant clues for speech perception (3) can be expected to lead to improvements.

REFERENCES

1. P. Mueller, T. Martin and F. Putzrath, General Principles of Operations in Neuron Nets with Application to Acoustical Pattern Recognition. Biological Prototypes and Synthetic Systems, Vol. 1, E.E. Bernard and M.R. Kare, Ed. pp. 192-212, Plenum, New York (1962).
2. P. Mueller, Principles of Temporal Pattern Recognition in Artificial Neuron Nets. Artificial Intelligence, S-142, The Institute of Electrical and Electronics Engineers, Inc., New York (1963).
3. S.E. Blumstein and K.N. Stevens, Perceptual Invariance and Onset Spectra for Stop Consonants in Different Vowel Environments. J. Acoust. Soc. Am. 76(2): 648-662, 1980.

A COMPARISON OF ALGORITHMS FOR NEURON-LIKE CELLS

David B. Parker

925 Oak Lane #4, Menlo Park, CA 94025

ABSTRACT

There are many different ways to connect neuron-like cells into large scale learning networks. These different patterns of connections are called architectures. One problem in designing an architecture is deciding what types of neuron-like cells to base it on. By examining the properties of learning networks that are independent of their architectures, this paper proposes that there is at least one type of cell which can be used in any reasonable architecture to give it nearly optimal performance. Cells of this type implement an algorithm called second order least mean square (2nd order LMS, for short).

INTRODUCTION

A large scale learning network is composed of many neuron-like cells of one or more types. The pattern of connections between the cells is called the architecture of the network. One example of a network architecture is the back propagation scheme of Parker[1], and of Rumelhart, Hinton and Williams[2]. One problem in designing an architecture is deciding what types of neuron-like cells to base it on. The main result of this paper is that there is at least one type of cell which is architecture independent, i.e. which can be used in any reasonable architecture to give it nearly optimal performance.

To establish this result, we need to examine the properties of learning networks which are independent of their architectures. Every reasonable learning network is commonly expected to possess the following three desirable properties:

1. The network should be tolerant of imperfections in its components.
2. Learning should be distributed throughout the network.
3. When learning something new, the network should remember as much of its old training as possible.

These three architecture independent properties lead to three constraints which must be satisfied by architecture independent cells:

1. The cells must be tolerant of non-linearities in the components that implement them. Some architectures rely on the non-linearities of the components.
2. To distribute learning throughout the network, the learning rates of the cells should be roughly the same. For example, we can look at the worst case — if one cell learns very quickly and the other cells don't learn at all, then learning isn't evenly distributed throughout the network. Since the statistical properties of the signals in a learning network may be hard to predict, this constraint means that the learning rates of the cells should be independent of the statistical properties of these signals.
3. To remember as much previous training as possible, the cells should change as little as possible when learning something new. For most types of cells, learning means changing the values of weights contained in the cells. As these weights change, they follow some path through weight-space. This constraint means that the path should be a straight line.

These constraints need not apply to every cell in the network. For example, Parker[1] has proposed an architecture that requires two types of cells. Of these two types, one does the learning while the other propagates training signals. These constraints would only apply to the cells that do the learning.

We can now examine existing learning algorithms to see if they satisfy these constraints. The five algorithms we will consider in this paper are:

Algorithm	Continuous and Discrete Formulas	Preconditions for Convergence in the Mean
LMS	$$\frac{\partial \vec{w}(t)}{\partial t} = 2\mu e(t)\vec{x}(t)$$ $$\vec{w}_{n+1} = \vec{w}_n + 2\mu\Delta t e_{n+1}\vec{x}_{n+1}$$	$0 < \mu\Delta t < \dfrac{1}{\text{tr}(A^*)}$
gradient descent	$$\frac{\partial \vec{w}(t)}{\partial t} = \alpha\left(\vec{a}(t) - A(t)\vec{w}(t)\right)$$ $$\vec{w}_{n+1} = \vec{w}_n + \alpha\Delta t(\vec{a}_{n+1} - A_{n+1}\vec{w}_n)$$	$0 < \alpha\Delta t < \dfrac{2}{\text{tr}(A^*)}$
2nd order LMS	$$\frac{\partial^2 \vec{w}(t)}{\partial t^2} = \alpha\left(\mu e(t)\vec{x}(t) - A(t)\frac{\partial \vec{w}(t)}{\partial t}\right)$$ $$\vec{w}_{n+1} = \vec{w}_n + \alpha\mu(\Delta t)^2 e_{n+1}\vec{x}_{n+1}$$ $$+ (I - \alpha\Delta t A_{n+1})(\vec{w}_n - \vec{w}_{n-1})$$	$0 < \alpha, \mu$ $(\frac{1}{2}\mu\Delta t + 1)\alpha\Delta t < \dfrac{2}{\text{tr}(A^*)}$
exact least squares	$$\vec{w}(t) = A(t)^{-1}\vec{a}(t)$$ $$\vec{w}_{n+1} = A_{n+1}^{-1}\vec{a}_{n+1}$$	none
updating exact least squares	$$\frac{\partial \vec{w}(t)}{\partial t} = A(t)^{-1}\mu e(t)\vec{x}(t)$$ $$\vec{w}_{n+1} = \vec{w}_n + A_{n+1}^{-1}\mu\Delta t e_{n+1}\vec{x}_{n+1}$$	$0 < \mu\Delta t < 2$

TABLE I The algorithms that we will compare. The notation tr(A^*) means the trace of the matrix A^*.

1. Least mean square (LMS, for short). This algorithm is also known as steepest descent, or the Widrow-Hoff algorithm. It is the most commonly used learning algorithm. A variation of this algorithm was first used in the ADALINE's of Widrow and Hoff[3].
2. Gradient descent. The LMS algorithm is derived from this one.
3. Second order least least mean square (2nd order LMS, for short). A variation of this algorithm was first used in the learning-logic function cells of Parker[1].
4. Exact least squares. All of the other learning algorithms are basically variations of this one.
5. Updating exact least squares. This is the updating form of the exact least squares algorithm, and is the algorithm from which 2nd order LMS can be derived.

Even though non-linear variations of these algorithms would be used in practice, we will use the linear versions because they are sufficient for the purposes of this comparison.

We will find, after performing the comparison, that two of the algorithms — 2nd order LMS and updating exact least squares — satisfy the constraints. Of these two, 2nd order LMS is the simplest to implement.

DESCRIPTION OF THE ALGORITHMS

The five types of cells that implement the five learning algorithms have many features in common:

- All receive an arbitrary number, p, of input signals $x_1(t)$, $x_2(t)$, ..., $x_p(t)$, denoted by the vector $\vec{x}(t)$;
- All emit a single output signal $f(t)$;
- All contain p weights $w_1(t)$, $w_2(t)$, ..., $w_p(t)$, denoted by the vector $\vec{w}(t)$;

Algorithm	Complexity	Expected Convergence	Preconditions for Expected Convergence
LMS	$O(p)$	$\mathrm{E}(\vec{w}_{n+1}-\vec{w}^*) = [I-2\mu\Delta t A^*]\mathrm{E}(\vec{w}_n-\vec{w}^*)$	none
gradient descent	$O(p^2)$	$\mathrm{E}(\vec{w}_{n+1}-\vec{w}^*) = [I-\alpha\Delta t A^*]\mathrm{E}(\vec{w}_n-\vec{w}^*)$	$\vec{a}(t), A(t) \to \vec{a}^*, A^*$
2nd order LMS	$O(p)$ or $O(p^2)$	$\mathrm{E}(\vec{w}_{n+1}-\vec{w}^*) = \left[1-\dfrac{\mu\Delta t}{1+\mu\Delta t}\right]\mathrm{E}(\vec{w}_n-\vec{w}^*)$	$\dfrac{4\mu}{\alpha(\mu\Delta t+1)^2} \ll \mathrm{tr}(A^*)$ $(\mu\Delta t+1)\alpha\Delta t < \dfrac{2}{\mathrm{tr}(A^*)}$ $A(t) \to A^*$
exact least squares	$O(p^3)$	$\mathrm{E}(\vec{w}_{n+1}-\vec{w}^*) = 0$	$\vec{a}(t), A(t) \to \vec{a}^*, A^*$
updating exact least squares	$O(p^3)$	$\mathrm{E}(\vec{w}_{n+1}-\vec{w}^*) = [1-\mu\Delta t]\mathrm{E}(\vec{w}_n-\vec{w}^*)$	$A(t) \to A^*$

TABLE II Complexity and convergence properties of the algorithms. The notation $\vec{a}(t), A(t) \to \vec{a}^*, A^*$ means that $\vec{a}(t)$ and $A(t)$ approach a^* and A^*. From equations (4) and (5) in the text, we can deduce that the time constant for $\vec{a}(t), A(t) \to \vec{a}^*, A^*$ is μ.

- All calculate the output signal $f(t)$ as the function $f(t) = w_1(t)x_1(t) + w_2(t)x_2(t) + \cdots + w_p(t)x_p(t) = \vec{w}(t)^T \vec{x}(t) = \vec{x}(t)^T \vec{w}(t)$;
- All receive an error signal $e(t)$ which can be thought of as the difference between a desired output signal $d(t)$ and the actual output signal $f(t)$: $e(t) = d(t) - f(t) = d(t) - \vec{x}(t)^T \vec{w}(t)$ (if a cell receives its error signal externally it is performing supervised learning; if it generates its own error signal it is performing un- or self-supervised learning);
- All adjust their weights $\vec{w}(t)$ to try to make the actual output signal $f(t)$ equal to the desired output signal $d(t)$ (which is equivalent to minimizing the average value of $e(t)^2$);
- All have a parameter μ which is approximately inversely proportional to the amount of past time over which the average value of $e(t)^2$ is minimized.

Some of the cells use the following quantities:

- A convergence factor α which controls the rate at which the cell converges;
- A vector $\vec{a}(t)$ of average products between the desired output signal and the inputs:

$$\vec{a}(t) = \mathrm{avg}(d(t)\vec{x}(t)) = \begin{pmatrix} \mathrm{avg}(d(t)x_1(t)) \\ \vdots \\ \mathrm{avg}(d(t)x_p(t)) \end{pmatrix} ; \tag{1}$$

- A matrix $A(t)$ called the average input correlation matrix:

$$A(t) = \mathrm{avg}(\vec{x}(t)\vec{x}(t)^T) = \begin{pmatrix} \mathrm{avg}(x_1(t)^2) & \cdots & \mathrm{avg}(x_1(t)x_p(t)) \\ \vdots & \ddots & \vdots \\ \mathrm{avg}(x_p(t)x_1(t)) & \cdots & \mathrm{avg}(x_p(t)^2) \end{pmatrix} \tag{2}$$

The vector $\vec{a}(t)$ and matrix $A(t)$ are approximations to the expected values $\vec{a}^* = \mathrm{E}(d(t)\vec{x}(t))$ and $A^* = \mathrm{E}(\vec{x}(t)\vec{x}(t)^T)$. The weights $\vec{w}(t)$ in all five types of cells approach the same expected values $\vec{w}^*$. The method of least squares gives these expected values as

$$\vec{w}^* = \mathrm{E}(\vec{x}(t)\vec{x}(t)^T)^{-1}\mathrm{E}(d(t)\vec{x}(t)) = A^{*-1}\vec{a}^*. \tag{3}$$

330

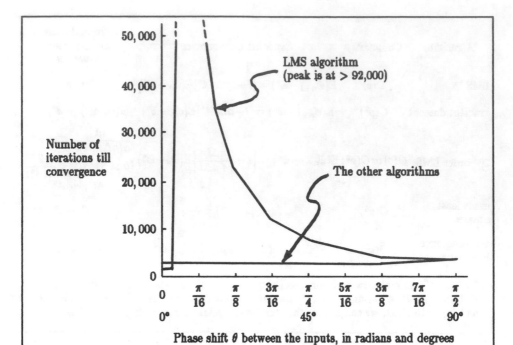

FIGURE 1 An example illustrating the convergence rates of the five types of cells. Each type of cell is trained to perform the desired function $d = -0.2x_1 + 0.4x_2$. The inputs are $x_1 = \sin t$ and $x_2 = \sin(t + \theta)$. Different values of the phase shift θ are used in different training runs to give the results shown here. Varying θ varys the eigenvalues of the average input correlation matrix A^*. The performance of the LMS algorithm illustrates the dependence of its convergence rate on those eigenvalues.

Table I lists the five algorithms and their preconditions for convergence in the mean. The algorithms can calculate $\vec{a}(t)$ and $A(t)$ using the continuous or discrete formulas

$$\frac{\partial \vec{a}(t)}{\partial t} = \mu\left((d(t)\vec{x}(t) - \vec{a}(t))\right) \quad \text{or} \quad \vec{a}_{n+1} = \vec{a}_n + \mu\Delta t(d_{n+1}\vec{x}_{n+1} - \vec{a}_n), \tag{4}$$

$$\frac{\partial A(t)}{\partial t} = \mu\left((\vec{x}(t)\vec{x}(t)^T - A(t))\right) \quad \text{or} \quad A_{n+1} = A_n + \mu\Delta t(\vec{x}_{n+1}\vec{x}_{n+1}^T - A_n). \tag{5}$$

The version of the 2nd order LMS algorithm presented in Table I requires $O(p^2)$ processing elements to implement a p-input cell — or $O(p^2)$ units of time to simulate the cell on a serial computer. Another version of the 2nd order LMS algorithm has a complexity of only $O(p)$, with the same expected convergence properties as the $O(p^2)$ version (at least for the convergence properties relevant to this discussion). The $O(p)$ algorithm can be obtained from the $O(p^2)$ algorithm by replacing the accumulated matrix $A(t)$ by the instantaneous matrix $\vec{x}(t)\vec{x}(t)^T$. The continuous and discrete formulas for the weights then become

$$\frac{\partial^2 \vec{w}(t)}{\partial t^2} = \alpha\left(\mu e(t) - \vec{x}(t)^T \frac{\partial \vec{w}(t)}{\partial t}\right)\vec{x}(t), \tag{6}$$

$$\vec{w}_{n+1} = \vec{w}_n + (\vec{w}_n - \vec{w}_{n+1}) + \alpha\Delta t\left(\mu\Delta t e_{n+1} - \vec{x}_{n+1}^T(\vec{w}_n - \vec{w}_{n-1})\right)\vec{x}_{n+1}. \tag{7}$$

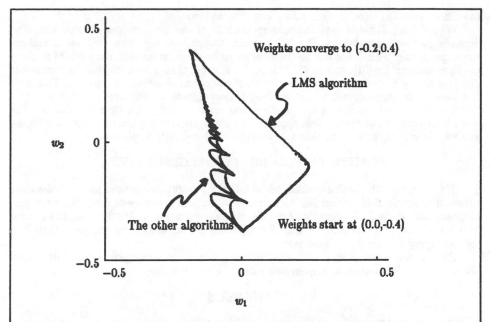

FIGURE 2 An example illustrating the paths taken by the weights in the five types of cells. Each type of cell is given two weights, w_1 and w_2, initialized to $w_1 = 0.0$ and $w_2 = -0.4$. Each cell is then taught the desired function $d = -0.2x_1 + 0.4x_2$, with $x_1 = \sin t$ and $x_2 = \sin(t + \theta)$. The value of θ for this particular graph is $\pi/8$. The path taken by the LMS weights follows the eigenvectors of the average input correlation matrix A^*; the weights in the other cells move, on average, in a straight line from their initial to their final values.

Table II summarizes the complexity and convergence properties of the algorithms[4,5].

CONSTRAINT 1: TOLERANCE OF NON-LINEARITIES

The LMS, 2nd order LMS, and updating exact least squares algorithms are relatively tolerant of non-linearities. These algorithms use the error signal $e(t)$ to adjust their weights. Even if the weights are biased due to non-linearities, the error signal always indicates the correct direction in which to change them.

On the other hand, the gradient descent and exact least squares algorithms use only the accumulated quantities $\bar{a}(t)$ and $A(t)$. If there are non-linearities in accumlating these quantities, then there will be systematic errors in the weights.

CONSTRAINT 2: UNIFORM CONVERGENCE RATE

The LMS algorithm fails to guarantee a uniform convergence rate. This is illustrated by the example in Figure 1. The reason can be found by examing the expected convergence of the LMS algorithm as given in Table II. The expected value of the difference between the actual values of the weights and their ideal values decays exponentially with time, with time constant $2\mu A^*$. Since the matrix A^* appears in the time constant, the convergence rate depends on the eigenvalues of A^*. If the magnitude of one or more of these eigenvalues is much smaller than the others, then the algorithm can take a long time to converge.

We see from Table II that the convergence rate of the gradient descent algorithm also depends on the eigenvalues of A^*. In this case, though, the algorithm still has a uniform convergence rate. This is because the value of α can be made larger than the value of μ. Since the time constant for $\bar{a}(t)$, $A(t) \rightarrow \bar{a}^*$, A^* is μ, if α is larger than μ then the time to accumulate $\bar{a}(t)$ and $A(t)$ will dominate the time for the gradient descent algorithm to converge. This leads to a uniform convergence rate for the gradient descent algorithm, with time constant μ.

The rest of the algorithms have uniform convergence rates, with time constant μ. The exact least squares algorithm would appear to have a faster convergence rate, but, as with the gradient descent algorithm, the time to accumulate $\bar{a}(t)$ and $A(t)$ dominates.

CRITERIA 3: STRAIGHT CONVERGENCE PATH

The weights in the LMS algorithm fail to follow a straight convergence path, for the same reason that the LMS algorithm fails to guarantee a uniform convergence rate. Since the convergence rate of the LMS algorithm depends on the eigenvalues of A^*, the weights will tend to follow the eigenvectors of A^* instead of going straight from their initial values to their final values. Figure 2 illustrates this behavior.

For the same reasons that the other algorithms have uniform convergence rates, their weights will on average follow straight convergence paths.

CONCLUSION

The results of the comparison are summarized as follows.

Algorithm	Tolerant of non-linearites?	Uniform convergence rate?	Straight convergence path?	Complexity
LMS	yes	no	no	$O(p)$
Gradient descent	no	yes	yes	$O(p^2)$
2nd order LMS	yes	yes	yes	$O(p)$ or $O(p^2)$
exact least squares	no	yes	yes	$O(p^3)$
updating exact least squares	yes	yes	yes	$O(p^3)$

The two algorithms which satisfy the three constraints for architecture independence are the 2nd order LMS algorithm and the updating exact least squares algorithm. Of these two, the 2nd order LMS algorithm is less complex to implement. Since the convergence properties of the 2nd order LMS algorithm are nearly identical in practice to those of the exact least squares algorithm, using cells based on the 2nd order LMS algorithm may help an architecture to perform as well as it possibly can.

REFERENCES

1. D. B. Parker, Learning-Logic, TR-47, Center for Computational Research in Economics and Management Science, MIT (April 1985)
2. D. E. Rumelhart, G. E. Hinton and R. J. Williams, Learning Internal Representations by Error Propagation, ICS Report 8506, Institute for Cognitive Science, University of California, San Diego (Sept 1985)
3. B. Widrow and M. E. Hoff, IRE WESCON Conv. Rec., pt. 4., 96-104 (1960)
4. B. Widrow and E. Walach, IEEE Trans. Inf. Theory, 211-221 (March 1984)
5. D. B. Parker, Convergence Properties of a New Class of Adaptive Least Squares Algorithms, submitted to IEEE Trans. Acou. Spch. Sig. Proc.

G-Maximization: an Unsupervised Learning Procedure for Discovering Regularities

Barak A. Pearlmutter
Geoffrey E. Hinton

Department of Computer Science
Carnegie-Mellon University
Pittsburgh, PA 15213

May, 1986

Abstract

Hill climbing is used to maximize an information theoretic measure of the difference between the actual behavior of a unit and the behavior that would be predicted by a statistician who knew the first order statistics of the inputs but believed them to be independent. This causes the unit to detect higher order correlations among its inputs. Initial simulations are presented, and seem encouraging. We describe an extension of the basic idea which makes it resemble competitive learning and which causes members of a population of these units to differentiate, each extracting different structure from the input.

Introduction

There are two distinct classes of theory about how to modify weights in networks of neuron-like units. Supervised learning theories like perceptrons and their more recent generalizations[1] assume that there is a special input to a unit which indicates how it ought to behave or how it ought to modify its behavior. Unsupervised learning theories assume that weight modification is based solely on the inputs to the unit and its actual responses. Typically, these theories have first suggested an intuitively plausible weight modification rule[2, 3, 4, 5] and then investigated the consequences of this rule. This paper presents an alternative approach in which the learning procedure is derived from a principle which specifies how the unit should behave. The principle proposed[6] is that the unit should respond to patterns in its inputs that occur more often than would be expected if the activities of the individual input lines were assumed to be independent. This is equivalent to saying that the unit should respond to higher order statistical regularities in its ensemble of input vectors.

At first sight, it seems that a unit would have to keep a record of its history of input vectors in order to discover higher order regularities. As we shall see, however, it is only necessary to keep track of two variables for each weight and a few more at the level of the whole unit.

Consider a unit with to 8 input lines, each of which alone is active 1/2 of the time, 4 of which are completely unrelated to any of the others, and the other 4 of which are highly correlated, either all being on at the same time or all being off. Our unit would ignore the 4 inputs with no higher order structure and latch onto the regularity present in the other 4. It could do this by developing strong positive weights to the 4 correlated inputs and a very high positive threshhold which could only be overcome if all those 4 input lines were on. Because these lines are correlated, the unit would come on with probability 1/2, but a statistician who assumed the lines were independent would predict that the unit would come on with probability 1/16.

The One Unit Case

Consider a unit which takes the weighted sum of its binary inputs s_i, runs that sum, x, through a logistic function σ to get y, and generates output 1 with probability y and output 0 with probability $1-y$.

$$x=\sum_{i=0}^{n} w_i s_i \quad \text{and} \quad y=\sigma(x)$$

where $\sigma(x)=1/(1+e^{-x})$, s_i is the state of the ith input, and the w_i are the weights. Note that $d\sigma(x)/dx=\sigma'(x)=\sigma(x)(1-\sigma(x))$ resembles a gaussian. Let the unit be exposed to some stationary probability distribution over the 2^n possible input vectors. Given this input distribution P, the unit has expected output of

$$\langle y \rangle = \sum_{\alpha} P(\alpha)\sigma(x^{\alpha}).$$

Imagine someone who thought that the various inputs to the unit were statistically independent. Suppose this person recorded the first order statistics of the input lines, $p_i=\sum_{\alpha} P(\alpha)s_i^{\alpha}$ where s_i^{α} is the state of the ith input line for the αth input vector. Assuming independence, this person would expect the input vectors to follow the distribution P' and would predict the unit to have an expected output $\langle y \rangle'$:

$$P'(\alpha)=\prod_{i} \begin{cases} p_i & \text{when } s_i^{\alpha}=1 \\ 1-p_i & \text{when } s_i^{\alpha}=0 \end{cases} \qquad \langle y \rangle'=\sum_{\alpha} P'(\alpha)\sigma(x^{\alpha}).$$

The unit is detecting an interesting feature of the actual input distribution to the extent that the actual expected output differs from this predicted expected output. More formally, we can measure how many bits of information this person gains about the actual input distribution P when told that the actual expected output of the unit is $\langle y \rangle$.[7]

$$G=\langle y \rangle \log\frac{\langle y \rangle}{\langle y \rangle'}+(1-\langle y \rangle)\log\frac{1-\langle y \rangle}{1-\langle y \rangle'}$$

This measure tells us how good a feature detector the unit is, so in order to develop our unit into a good feature detector we can hill climb in G by modifying the w_i, the only parameters under our control.

$$\frac{\partial G}{\partial w_i}=\frac{\partial G}{\partial \langle y \rangle}\frac{\partial \langle y \rangle}{\partial w_i} + \frac{\partial G}{\partial \langle y \rangle'}\frac{\partial \langle y \rangle'}{\partial w_i}$$

$$=[\log\frac{\langle y \rangle}{\langle y \rangle'}-\log\frac{1-\langle y \rangle}{1-\langle y \rangle'}]\sum_{\alpha} P(\alpha)\sigma'(x^{\alpha})s_i^{\alpha} + [\frac{1-\langle y \rangle}{1-\langle y \rangle'}-\frac{\langle y \rangle}{\langle y \rangle'}]\sum_{\alpha} P'(\alpha)\sigma'(x^{\alpha})s_i^{\alpha}$$

Note that the "prime" notation is used in different senses in P' and σ'. We proceed by accumulating the quantities on the right hand side of the above equation and using these to modify the weights with the simple rule $w_i^{new} = w_i^{old} + \varepsilon \partial G / \partial w_i$.

To accumulate these right hand side quantities, we sample the distributions P and P'. We call the phase during which P is sampled the "structured phase" because the higher order structure is present in the ensemble of input vectors; to sample P' we introduce an "unstructured phase" in which the input lines are statistically independent. Thus, for each unit we accumulate $\sigma(x^\alpha)$ during the structured phase to give us $\langle y \rangle$ and the same quantity during the unstructured phase to give us $\langle y \rangle'$. In addition, during the structured phase we accumulate $\sigma'(x^\alpha)s_i^\alpha$ for each weight to give us $\partial \langle y \rangle / \partial w_i$, and the same quantity during the unstructured phase to give us $\partial \langle y \rangle' / \partial w_i$. Initially, we set the weights to small random values to help break symmetry.

A Simulation

If the input vector is a 10 by 10 array and the distribution is composed of single, randomly oriented, randomly positioned, black-white edges on this "retina," a unit typically develops into on-center off-surround detector (or vice versa) as in figure 1. To understand why this is so, consider the regularity captured in the input. If the input was random uncorrelated noise, like static on a TV screen, this unit would almost always come on, so $\langle y \rangle'$ is almost 1. However, the center of the receptive field is frequently not on the bright side of the edge, so $\langle y \rangle$ is (in this case) about 0.7. The unit is capturing the fact that nearby pixels tend to have the same value. It is interesting to note that, given this input distribution, changing the signs of all the weights would leave G unchanged.

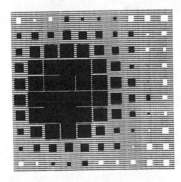 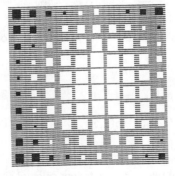

Some typical detectors developed in response to randomly oriented black-white edges on a 2D field.

Figure 1:

Multiple Units

The above treatment deals only with a single unit. Were we to have a number of such units, they could all develop to detect the same feature. We need some force that will cause them to differentiate. One obvious method is to require each pair of units to be pretty much uncorrelated. If r_{ab} is the correlation between units a and b, rather than maximizing G we can maximize a new measure,

$$G^* = (1-k)\sum_a G_a - k\frac{1}{2}\sum_{a \neq b} r_{ab}^2, \qquad \frac{\partial G^*}{\partial w_{ai}} = (1-k)\frac{\partial G_a}{\partial w_{ai}} - k\sum_{b \neq a} r_{ab}\frac{\partial r_{ab}}{\partial w_{ai}}$$

where k is a constant controlling the relative importance of making good feature detectors and making the feature detectors uncorrelated, and r and its derivatives are computed as follows:

$$r_{ab} = q_{ab}^{11} q_{ab}^{00} - q_{ab}^{10} q_{ab}^{01}$$

$$q_{ab}^{11} = \sum_{\alpha} P(\alpha)\sigma(x_a^\alpha)\sigma(x_b^\alpha), \qquad q_{ab}^{10} = \sum_{\alpha} P(\alpha)\sigma(x_a^\alpha)(1-\sigma(x_b^\alpha))$$

$$q_{ab}^{01} = \sum_{\alpha} P(\alpha)(1-\sigma(x_a^\alpha))\sigma(x_b^\alpha), \qquad q_{ab}^{00} = \sum_{\alpha} P(\alpha)(1-\sigma(x_a^\alpha))(1-\sigma(x_b^\alpha))$$

Once again we do gradient descent on a measure by sampling the distribution and accumulating quantities as we sample, notably these q_{ab}^{xx} and their derivatives with respect to each weight,

$$w_{ai}^{new} = w_{ai}^{old} + \varepsilon \frac{\partial G^*}{\partial w_{ai}}$$

$$= w_{ai}^{old} + \varepsilon[(1-k)\frac{\partial G_a}{\partial w_{ai}} - k\sum_{b \neq a} r_{ab}\frac{\partial r_{ab}}{\partial w_{ai}}]$$

$$= w_{ai}^{old} + \varepsilon[k\frac{\partial G_a}{\partial \langle y_a\rangle}\frac{\partial \langle y_a\rangle}{\partial w_{ai}} + k\frac{\partial G_a}{\partial \langle y_a\rangle'}\frac{\partial \langle y_a\rangle'}{\partial w_{ai}}$$

$$-(1-k)\sum_{b \neq a} r_{ab}[q_{ab}^{11}\frac{\partial q_{ab}^{00}}{\partial w_{ai}} + q_{ab}^{00}\frac{\partial q_{ab}^{11}}{\partial w_{ai}} - q_{ab}^{10}\frac{\partial q_{ab}^{01}}{\partial w_{ai}} - q_{ab}^{01}\frac{\partial q_{ab}^{10}}{\partial w_{ai}}]]$$

so we accumulate:

for each pair of units:		for each w_{ai} and other unit b:	
q_{ab}^{11}	$\sigma(x_a^\alpha)\sigma(x_b^\alpha)$	$\partial q_{ab}^{11}/\partial w_{ai}$	$\sigma'(x_a^\alpha)\sigma(x_b^\alpha)s_i^\alpha$
q_{ab}^{10}	$\sigma(x_a^\alpha)(1-\sigma(x_b^\alpha))$	$\partial q_{ab}^{10}/\partial w_{ai}$	$\sigma'(x_a^\alpha)(1-\sigma(x_b^\alpha))s_i^\alpha$
q_{ab}^{01}	$(1-\sigma(x_a^\alpha))\sigma(x_b^\alpha)$	$\partial q_{ab}^{01}/\partial w_{ai}$	$(1-\sigma'(x_a^\alpha))\sigma(x_b^\alpha)s_i^\alpha$
q_{ab}^{00}	$(1-\sigma(x_a^\alpha))(1-\sigma(x_b^\alpha))$	$\partial q_{ab}^{00}/\partial w_{ai}$	$(1-\sigma'(x_a^\alpha))(1-\sigma(x_b^\alpha))s_i^\alpha$

If we have n input bits and m units, our original scheme (without decorrelation) takes 2 units of storage for each unit, to hold $\langle y\rangle$ and $\langle y\rangle'$, and 2 for each weight, to hold $\partial\langle y\rangle/\partial w_i$ and $\partial\langle y\rangle'/\partial w_i$. Assuming we wish to decorrelate each pair of units, our new decorrelation scheme requires an additional 4 values for each of the $m(m-1)/2$ pairs of units, and an additional $4(m-1)$ for each weight. Although simulations show that this decorrelation method is effective, we find it heavyhanded and implausible.

With a simple approximation we can greatly simplify the decorrelation. Given that the units come on rarely, the decorrelation scheme described above can be approximated by mutual inhibition. For instance, if units are on only one time in a thousand then two decorrelated units will come on together only one time in a million, which is negligible.

Mutual inhibition between rarely active units also eliminates higher order correlations (which are not precluded by explicit pairwise decorrelation.) We have not yet simulated this mutual inhibition technique.

It is interesting to note that Boltzmann machines[8] handle higher order correlations in a way that is both principled and space-efficient (but slow.) At thermal equilibrium, a Boltzmann machine communicates information about higher order correlations via local pairwise interactions. This allows it to develop weights which ensure that the higher order correlations between its units are the same in two different phases. Notice that a Boltzmann machine learns by making its spontaneously generated output be as similar as possible to the required structured output, whereas G-Maximization learns by making its response to structured input be as different as possible from its response to unstructured input.

Further Elaborations

If different units are connected to different subsets of the total set of input lines, they will tend to detect different things. This means that decorrelation or mutual inhibition is only needed for nearby units.

If we know what we want a unit to detect, we can supervise it by adding an extra input to the unit and initializing the weight on that input to a high value. We then turn this input bit on when the feature we are interested in is present and off when its not. The unit's other weights will develop to detect this feature, unless such a feature isn't really present in which case the weight to our extra input will decrease until the unit can ignore it and pick up some real feature.

If one desires the feature detectors to be rarely active,[9] one can add another term to the G measure to impose this additional constraint. We let

$$G^{**} = (1 - k_1 - k_2)G - k_1 \frac{1}{2} \sum_{a \neq b} r_{ab}^2 - k_2 \frac{1}{2} \sum_a (\langle y_a \rangle - d)^2$$

where d is the desired activity level. The corresponding modification to the learning procedure is simple, requiring no additional state.

$$\frac{\partial G^{**}}{\partial w_{ai}} = \cdots - k_2 (\langle y_a \rangle - d) \frac{\partial \langle y_a \rangle}{\partial w_{ai}}$$

A unit which is forced to be rarely active will tend to maximize G by responding to very high order regularities.

If we want a unit to be helpful for deciding which of two distributions gave rise to the input vector, we can replace the structured and unstructured phases by these two distributions.

338

Relation to Hebbian Learning

A careful examination of the single unit case reveals that the learning rule resembles Hebbian learning. An intuitive way of looking at the process is as follows. We define a marginal case to be one in which the total input, x, to the unit is on the steep part of the logistic function, where $\sigma(x)$ is higher than usual (if we assume that units are rarely active.) We assume that $\langle y \rangle$ is higher than $\langle y \rangle'$. If an input line, i, is involved in more marginal cases during the structured phase than during the unstructured phase, raising w_i will raise $\langle y \rangle$ more than it raises $\langle y \rangle'$ so it will normally raise G. If we identify the unstructured phase with sleep[10] we expect Hebbian learning during wake and reverse Hebbian learning during sleep.

Acknowledgements

This research was supported by contract N00014-86-K-00167 from the Office of Naval Research. Barak Pearlmutter is a Hertz Fellow. We thank Richard Szeliski for useful discussions.

References

1.   Rumelhart, D. E., Hinton, G. E., & Williams, R. J., "Learning internal representations by error propagation", in *Parallel distributed processing: Explorations in the microstructure of cognition*, D. E. Rumelhart, J. L. McClelland, & the PDP research group, eds., Bradford Books, Cambridge, MA, Vol. I, 1986.

2.   Hebb, D. O., *The Organization of Behavior*, Wiley, New York, 1949.

3.   Marr, D., "A theory of cerebellar cortex", *Journal of Physiology (London)*, Vol. 202, 1969, pp. 437-470.

4.   Von der Malsburg, C., "Self-organizing of orientation sensitive cells in striate cortex", *Kybernetik*, Vol. 14, 1973, pp. 85-100.

5.   Rumelhart, D. E. and Zipser, D., "Competitive Learning", *Cognitive Science*, Vol. 9, 1985, pp. 75-112.

6.   Hinton, G. E., "Implementing semantic networks in parallel hardware", in *Parallel Models of Associative Memory*, G. E. Hinton & J. A. Anderson, eds., Erlbaum, Hillsdale, NJ, 1981.

7.   Kullback, S., *Information Theory and Statistics*, Wiley, New York, 1959.

8.   Ackley, D. H., Hinton, G. E., Sejnowski, T. J., "A learning algorithm for Boltzmann machines", *Cognitive Science*, Vol. 9, 1985, pp. 147-169.

9.   Barlow, H. B., "Single units and sensation: A neuron doctrine for perceptual psychology?", *Perception*, Vol. 1, 1972, pp. 371-394.

10.   Crick, F. & Mitchison, G., "The function of dream sleep", *Nature*, Vol. 304, 1983, pp. 111-114.

TENSOR NETWORK THEORY AND ITS APPLICATION IN COMPUTER MODELING OF THE METAORGANIZATION OF SENSORIMOTOR HIERARCHIES OF GAZE

A.J. Pellionisz

Dept. Physiol. & Biophys. New York Univ. Med. Ctr. NY. 10016

THE CHALLENGE

Neuronal networks are, in fact, used for "computations" in living organisms, producing what we call brain function (eg. sensorimotor coordination and intelligent representation). Neither the networks, nor their functions are fully known as yet, however. Based on what principle does the Central Nervous System (CNS) accomplish these tasks, and whether its mathematical understanding and subsequent or simultaneous technological implementation will lead to utilizable socioeconomical applications, are questions increasingly in the forefront of the interest of neuroscience community at large[1-3], and of its special field of brain theory, which is intimately tied to the artificial intelligence community and computer science and industry[4-7]. Activities range from mathematical analysis[8-11] to rehabilitation medicine[12,14]. The overlap of neuroscience with other disciplines created interdisciplinary subfields; Neurobotics[3,13,14], Neurophysics[16-18], and Neurophilosophy[19]. The new scientific revolution attracts neuroscientists spanning from molecular biologists[20,21] through mathematicians, engineers and physicists to philosophers. The implications warant an increasing awareness of their vital importance by government-agencies worldwide.

A GEOMETRICAL APPROACH TO BRAIN FUNCTION: TENSOR NETWORK THEORY

Motivated by the need of functionally interpreting the structure of existing neuronal networks[22], such as those in the *cerebellum*, this author strives for finding the basic general principle of the organization of "neuronal networks", and gainig a conceptual and formal grasp on what they "compute". The approach exposed here concentrates on sensorimotor neuronal networks (as in the cerebellum) and on the mathematical question of the axioms of their computations. Tensor network theory of the central nervous system may be summarized[1,2,14,15] by stating its axiom that the brain relates to the external world by expressing physical objects (invariants), both in a sensory and motor manner, in systems of coordinates that are *intrinsic* to the organism. Such general, typically non-orthogonal and overcomplete, frames of reference are physically obvious in sensory and motor parts of the CNS. Sensory and motor representation is identified in tensor network theory by *covariant vectors* [23] (with measurement-type orthogonal-projection components) and *contravariant vectors* (with physically executable parallelogram-type components), respectively. Thus, the *metric tensor operation*, which transforms these representations to one another was identified as a basic functional characteristics of sensorimotor networks eg. the cerebellum[29]

Beyond offering a formalism for describing neuronal computations of intrinsic vector-components of physical invariants, this approach conceptually features brain function as comprising functional geometries (via metric tensors, implemented by neuronal networks) in the internal CNS representation-spaces, both in sensorimotor and connected manifolds.

COMPUTER MODEL OF THE METAORGANIZATION OF GAZE SENSORIMOTOR HIERARCHIES

A quantitative example of this approach is a tensor model of gaze. To maintain a stable image in fixation, head & eye must compensate for passive movements or, in tracking, for the

AB.	NAME	YAW	PITCH	ROLL	AB.	NAME	YAW	PITCH	ROLL
(A)	Retinal Ganglion Cells				(D)	Neck Muscle Motoneurons			
MD	Medial direct.	-.155	.988	.000	LC	Longus Capitis	.129	.960	-.251
DR	Dorsal direct.	.988	.155	.000	CL	Cleidomastoid.	.047	.607	-.793
LT	Lateral direct.	.244	-.970	.000	ST	Sternomastoid.	.137	.590	-.796
VT	Ventral direct.	-.999	-.039	.000	LA	Longus Atlantis	-.395	.385	-.834
(B)	Vestibular Canal Neurons				QD	Dors. Sup. Obl.	.094	.318	-.943
					QV	Ventr. Sup. Obl.	-.099	.287	-.953
HO	Horizontal	-1.000	.000	.000	LI	Longiss.Capitis	-.290	.281	-.915
AN	Anterior	.080	.693	.717	OQ	Obliq. Inferior	-.930	-.144	-.339
PO	Posterior	.080	-.693	.717	SP	Splenius	-.456	-.568	-.685
(C)	Eye Muscle Motoneurons				CM	Complexus	-.133	-.571	-.810
LR	Lateral Rectus	-.117	-.966	.232	OC	Occipito-Scap.	-.321	-.693	-.645
MR	Medial Rectus	.001	.968	-.251	BV	Biventer	-.061	-.948	-.312
SR	Super. Rectus	-.851	-.027	-.524	RD	Rectus Medialis	-.263	-.958	.116
IR	Infer. Rectus	.891	.068	.449	RN	Rectus Minor	-.218	-.963	.156
SO	Super. Oblique	.513	-.034	-.858	RM	Rectus Major	-.183	-.957	-.227
IO	Infer. Oblique	-.565	.158	.810					

Table I. *Data, from computerized anatomy, to define coordinate systems intrinsic to tensorial expressions of gaze* . Rotational axes of (A): a mammalian retinal sensory frame[24], (B): frame of vestibular canals[25], (C): a motor frame of eye muscles[26], (D): neck-muscles[27] .

movements of the target. In perfect gaze, the displacement and its compensation are identical; an example when *an identical invariant is expressed in various intrinsic coordinates.*

How various vectorial expressions within and among these frames are transformed by the CNS is the subject of tensor network theory: **A 3-step tensorial scheme** was elaborated to transfer covariant *sensory* vector to contravariants in a *motor* frame[2,28,32]

1) **Sensory metric tensor** (g^{pr}), transforming a covariant reception vector (S_r) to contravariant perception (S^p, lower and upper indices denote co- and contravariants):

$$S^p = g^{pr} \cdot S_f \quad \text{where} \quad g^{pr} = |g_{pr}|^{-1} = |\cos(\Omega_{pr})|^{-1}$$

where $|\cos(\Omega_{pr})|$ is the table of cosines of angles among sensory unit-vectors.

2) **Sensorimotor covariant embedding tensor** (C_{ip}) transforming the sensory vector (S^p) into covariant motor intention vector (m_i). Covariant embedding is a unique operation, regardless a dimensional inconsistency of the sensory and motor space (including over-completeness2), but results in a non-executable expression[23]:

$$m_i = C_{ip} \cdot S^p \quad \text{where} \quad C_{ip} = U_i \cdot W_p \quad \text{where}$$

U_i and W_p are the i-th sensory unit-vector and p-th motor unit-vector.

3) **Motor metric tensor**[1,2,23] that converts intention m_i to executable contravariants; $m^e = g^{ei} \cdot m_i$ (where g^{ei} is computed as g^{pr} was for sensory axes in 1).

In case of overcompleteness, of either or both sensory and motor coordinate systems (as in **A,C,D** in Fig. 1), tensor network theory hypothesizes[23] that the CNS uses the **Moore-Penrose generalized inverse** (MP) of the unique covariant metric[2,15,30]:

$$g^{j,k} = \sum_m \{1/L_m^+ \cdot |E_m \rangle \langle E_m|\}, \quad \text{where}$$

E_m and L_m^+ are the m-th Eigenvector of $g_{j,k}$ and its Eigenvalue (replaced by 1 if it was 0).

This 3-step scheme is used to compute tensors of a sensorimotor reflex[2,28]. For the 4 gaze reflexes of **Fig.1**, *each expressing an invariant both in a sensory and a motor frame* , the above calculation yields tensor-matrices as shown (by patch-diagrams only) in **Fig.2.**

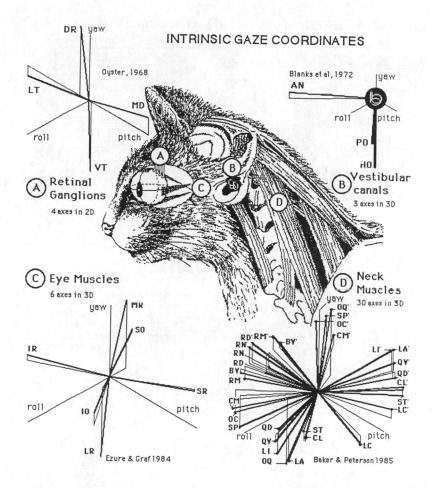

INTRINSIC GAZE COORDINATES

Fig. 1. *Coordinates intrinsic to gaze sensorimotor neuronal networks.* Gaze is expressed by rotations of the head and eye via the neck and eye muscles, so that they compensate for rotations measured by the retinal ganglion cells and by the vestibular semicircular canals. Both the dual (retinal and vestibular) sensory apparatus and the dual (oculomotor and neck-motor) executor systems operate along rotational axes determined by the structure of the organism.In order to express gaze, neuronal networks must measure and produce physical invariants (movements) in these typically non-orthogonal, overcomplete intrinsic frames of reference, by covariant sensory and contravariant motor vectors. Since the frames have been established by quantitative anatomy (cf. Table I.), the problem that we face is to quantitatively interpret how the CNS establishes relationships among various vectorial representations rendered to of a physical invariant such as gaze-displacement. *Tensor tranformations* yield a general interpretation as well as a means of calculation by tensor-matrices, implemented in the CNS by the system of interconnections in *neuronal networks*.

342

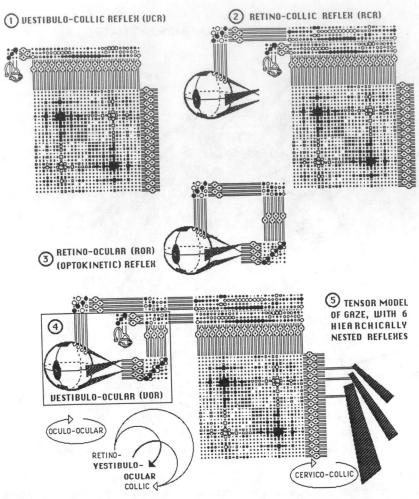

① VESTIBULO-COLLIC REFLEX (VCR)

② RETINO-COLLIC REFLEX (RCR)

③ RETINO-OCULAR (ROR) (OPTOKINETIC) REFLEX

④

VESTIBULO-OCULAR (VOR)

⑤ TENSOR MODEL OF GAZE, WITH 6 HIERARCHICALLY NESTED REFLEXES

OCULO-OCULAR

RETINO-VESTIBULO-OCULAR COLLIC

CERVICO-COLLIC

Fig.2. *Metaorganization, in six developmental steps, of sensorimotor reflexes of gaze.* Three neuronal networks, required for tensor-transformations in *each* sensorimotor reflex, eg. from vestibular- to neck-motor vector in (1), were calculated[2,28,31] by the 3-step tensor scheme. Resulting tensor-matrices are shown by three patch-diagrams in each sensorimotor reflex-arc; VCR(1), RCR(2), ROR(3), VOR(4). These networks are to develop in a definite sequence; in the VCR(1) the motor metric, sensorimotor embedding and sensory metric develop, as described by metaorganization[31]. RCR(2) builds hierarchically on the existing neck-motor metric, and the retinal metric is used also for ROR(3). VOR(4) is built on top of this hierarchy, using the vestibular metric available from VCR. Since the VOR is the only gaze reflex which is *not* a closed-loop sensorimotor system[28], its development must use the already available RCR,VCR,ROR networks. (The oculo-ocular & cervico-collic motor metrics whose development *preceed* those of the 4 gaze reflexes are also indicated in scheme 5).

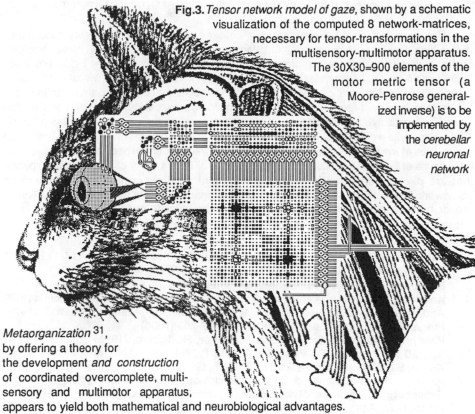

Fig.3. *Tensor network model of gaze,* shown by a schematic visualization of the computed 8 network-matrices, necessary for tensor-transformations in the multisensory-multimotor apparatus. The 30X30=900 elements of the motor metric tensor (a Moore-Penrose generalized inverse) is to be implemented by the *cerebellar neuronal network*

Metaorganization [31], by offering a theory for the development *and construction* of coordinated overcomplete, multi-sensory and multimotor apparatus, appears to yield both mathematical and neurobiological advantages.

As for theory and implementation, advantages result from its employing the MP formula, which a) yields the proper inverse if the space is complete, b) yields a least-squares minimum-energy formula, c) can be generated by the CNS via the process of metaorganization (also yielding the sensory metric and sensorimotor covariant embedding networks), by the utilization of reverberative oscillations[31], d) the theoretical prediction *has been experimentally shown* to conform with the CNS in gaze control[32] and in coordination of human arms[33].

As for Neuroscience, tensor network theory may be useful by functionally interpreting existing neuronal networks. As suggested, the proposed metric-type function can be implemented for sensory modalities by the tectum[30], for motor vectors by the cerebellum[29]. Structural features of the proposed tensor-transformation matrices in sensorimotor reflexes, eg. the three tensor-transformations in the VCR and VOR, appear to match structural properties of the CNS, where eg. vestibular signals are known to be transformed from the semicircular canals to vestibular nuclei, from there to premotor nuclei, and to oculomotor nuclei, before reaching eye muscles[28,32]. While it may require an enormous cooperative effort, quantiative anatomy, experimental network analysis and tensor network theory may gradually reveal not only quantitative operational features of neuronal circuitries, but also basic principles of brain function. This work starts on the proving ground of Brain Theory, sensorimotor coordination, where the physical entities that are the objectives of neuronal computation are most evident. Principles and techniques learnt from these studies, if truly general, could be helpful for understanding intelligent representations in the neocortex.

FUNCTIONAL GEOMETRIES IN CONNECTED CNS REPRESENTATION HYPERSPACES

Metaorganization of gaze networks is an example for creating sensory- and motor metric-type networks. They comprise functional geometries to match the physical geometry of the structure of sensory and motor apparatus. Neuronal networks in the brain, however, incorporate functional geometries that not only passively react to given physical geometries, by compensating for modifications occurring in their relation to the environment as in gaze, but *impose intelligent function* on both the sensorimotor apparatus, and ultimately on the world[15]. Such active modification, to be intelligent, requires the brain to comprise a functional geometry, *a world model, with a geometry that is homeomorphic*. Intelligent functional geometries of the CNS are presently largely unexplored, but are expected to transgress the boundaries of Euclidean or Riemannian manifolds [15]. Thus, a study of sensorimotor spaces that are directly connected to known external structural geometries may be essential homework. Then the generalized principle of metaorganization and the general formalism of tensor network theory may be applied to tackle the ultimate question of how intelligent geometries develop in the CNS, or can be developed to extend them.

REFERENCES

1   **Pellionisz, A.J.** In: *Brain Theory* (Palm G, Aertsen A,eds) Springer, 114-135 (1986)
2   **Pellionisz, A.J.** *J. Theoretical Biology* 110:353-375 (1984)
3   **Loeb, J.W.** *Trends in Neuroscience* 5:203-204 (1983)
4   **Kohonen, T.** *Associative Memory*, Springer Verlag Heidelb.-New York-Berlin (1977)
5   **Anderson, J.R, G.H.Bower** *Human Associative Memory*, Winston, Wash.DC(1973)
6   **Fukushima, K.** *Biological Cybernetics* 36:193-202 (1980)
7   **Palm, G., A. Aertsen**, *Brain Theory* , Springer, Heidelberg, (1986)
8   **Marr, D., T. Poggio** , *Science* 194:283-287 (1976)
9   **Amari, S.**Biological Cybernetics 26:175-185 (1977)
10   **Palm, G.** *Neural Assemblies*, Springer, Heidelberg-New York-Berlin (1982)
11   **Grossberg, S.** *Studies of Mind & Brain*, D.Reidel, Dordrecht (1982)
12   **Mann, R.W.** *Ann Biomedical Engineering* 9:1-43 (1981)
13   **Hogan, N.** *J. Neuroscience* 4(11): 2745-2754 (1984)
14   **Pellionisz, A.J.** *IEEE Conf. on Systems Man & Cybernetics.* pp.411-414 (1985)
15   **Pellionisz, A.J.** *J. Theoretical Neurobiology* 2(3):185-211 (1983)
16   **Cooper, L.N.**In:*Coll. Prop. of Physic. Syst.* Acad. Press New York, pp.252-264(1974)
17   **Hopfield, J.J., D.I. Feinstein, R.G. Palmer**, *Nature* 304:158-159 (1983)
18   **Scott, A.C.** *Neurophysics*, Wiley, Interscience (1977)
19   **Churchland, P.S.** *Neurophilosophy*, MIT Press (1986)
20   **Crick, F.** *Scientific American* 241:219-232 (1979)
21   **Edelman, G.M., V.B. Mountcastle**, *The Mindful Brain*, MIT Press (1978)
22   **Pellionisz, A.J., J. Szentagothai**, *Brain Research* 49:83-99 (1973)
23   **Pellionisz, A.J., R. Llinas**, *Neuroscience* 5:1125-1136 (1980)
24   **Oyster, C.W.** *J. Physiology* (Lond.) 199:613-635 (1968)
25   **Blanks,R.H.I., I.S.Curthoys, C.H.Markham**, *Acta Otolaryng.* 80:185-196(1975)
26   **Ezure,K., W. Graf**, *Neuroscience* 12:95-109 (1984)
27   **Baker,J., J. Goldberg, B.W. Peterson**, *J. Neurophysiology* 54:735-756(1985)
28   **Pellionisz, A.J.** In:*Adapt.Mech.of Gaze Cont* Berthoz A(ed)Elsevier 281-296(1985)
29   **Pellionisz, A.J.** In: *Cerebellar Funct.* Bloedel JR et al.(eds) Springer 201-229(1985)
30   **Pellionisz, A.J.** In:*Visuomotor Coordination* , (Lara, R, Arbib M. Eds)   pp.1-20(1983)
31   **Pellionisz, A.J., R. Llinas**, *Neuroscience* 16:245-274 (1985)
32   **Pellionisz, A.J, B W.Peterson** In:*Control of Head Movements*, Acad.Press(1986)
33   **Gielen, C.C.A.M., E.J. van Zuylen**, *Neuroscience* 17:527-539 (1986)

Supported by Grant NS 22999 from NINCDS

DIGITAL SIGNAL PROCESSOR ACCELERATORS FOR NEURAL NETWORK SIMULATIONS

P. Andrew Penz

Richard Wiggins

Texas Instruments, Dallas TX 75265

ABSTRACT

The artificial neural network is a parallel computer architecture invented three decades ago as an alternative to the serial, von Neumann machine. Neural networks do not lend themselves to implementation by traditional semiconductor hardware (resistors are required) nor to fast simulation by standard serial methods (multiplication operations dominate). Digital signal processors execute multiplications in hardware and can accelerate neural network simulations. Neural network simulations can be efficiently mapped onto an array of digital signal processors for further acceleration. The acceleration possible with one particular digital signal processor is presented, and a possible system design for a neural network workstation is outlined. The approach is consistent with standard semiconductor technology and can be expected to follow traditional semiconductor cost/functionality learning curves.

INTRODUCTION

Artificial neural networks or NN for short originated with the work of Rosenblatt[1], Widrow[2], and Steinbuch[3] in the late 1950's for the purpose of characterizing input vectors relative to vectors stored as a matrix. The input vector would represent a pattern to be recognized and the output format depended on the storage algorithm. Rosenblatt and Widrow used a local storage method. The resultant output was an ordered list of numbers which represented the degree of closeness between the input and the ordered stored states. The Steinbuch storage method was a non-local/distributed algorithm and produced vectors that had been stored. Both methods will be quickly reviewed to determine the computationally difficult parts of the algorithms

Let U^s be the M ($1 \leq s \leq M$) vectors to be stored, having components U_j^s ($1 \leq j \leq N$). The input vector to be characterized is $V^?$ where the superscript stands for the unknown state. The local algorithms store the states U in terms of a matrix T whose components are given by:

$$T_{sj} = U_j^s$$

Eq. 1

i.e. the sth row of the storage matrix was the sth state vector. One then performs a matrix multiplication between the storage matrix and the unknown vector to characterize the unknown vector component:

$$output_s = \sum_j T_{sj} V_j^? = \sum_j U_j^s V_j^?$$

Eq. 2

The output is then a list of the dot[inner] product between the stored vectors and the unknown vector. Given normalized vectors, the stored vector most parallel to the unknown vector would give the largest output list entry which would constitute identification by the system. The hope, but rarely the reality, was that the output vector would contain one large component which would uniquely characterize the input in terms of the stored states. Unfortunately the result often was a list with small differences in the components, thus making identification difficult. It should be noted that the storage algoritm is local in nature, one matrix component assigned to each vector component.

Fig 1. shows a possible implementation of the local storage NN via resistors which are inversely proportional to the matrix coupling strengths:

Eq. 3

$$R_{ij} = 1/T_{ij}$$

The resistor implementation is shown more for pedagogical reasons than for an indication of a practical design. With input voltages $V_j^?$ applied to the columns the currents resulting in the row operational amplifiers will be just the matrix multiplication indicated in Eq.2. Such a representation will be more intuitive to some than the matrix equations. The design leaves several things to be desired, however, from a practical VLSI point of view: programmable resistors do not exist, long bus lines are already the main problem in current dynamic random access memory design and steady current drain through resistors is contrary to the CMOS low power trend.

A variety of improvements[3-9] have been generated to increase the contrast in the output list of the local storage method and in response to a generic criticism by Minsky and Papert[10]. The important thing about these improvements from the point of view of this paper is that the computationally intensive part of these algorithms remains the matrix multiplications inherent in Fig. 1 and Eq. 2.

The second type of storage is the non-local or distributed storage and is based on the sum over outer products of the stored vectors:[11]

$$T_{ij} = \sum_s U_i^s U_j^s.$$

Eq. 4

Note that the outer product of two vectors results in a matrix and the inner product results in a scalar. The utility of this storage algorithm can be appreciated by considering the storage of orthonormal vectors. With this condition, it can easily be shown that the eigenvectors of the storage matrix include the vectors used in the storage algorithm (Eq.4). Note that the components of the storage matrix are now functions of data from each vector. This distributes the storage and produces fault tolerance by correlation. Traditionally the stored states are binary vectors in the distributed storage case but may be real number-vectors in the local storage case.

Fig.2 shows the use of the distributed storage algorithm in the same resistor representation as Eq. 3 with feedback between the rows and columns. The feedback and the threshold logic at the output have been shown by Anderson[12] and Hopfield[13] to produce stable states which are the digital stored states. The fact that such a network might have stable states can be inferred by considering an input vector that was specially constructed to be an eigenvector of the matrix. The output would then be parallel to the input in an N dimensional vector space, and the thresholding operation would act as a normalization. The fed back vector would be a clipped, pseudo eigenvector and could be expected to output a parallel version of the input. Thus the circuit tends to search for pseudo eigenvectors as stable states, although the nonlinear thresholding obviously complicates such a simplistic analysis. With the same caveat, the stored states are pseudo-eigen vectors of the storage matrix, and so the network searches for stored states. Such a "physical" line of reasoning can be used to interpret the Anderson/Hopfield result. Note that the output is now a stored vector, not a list of inner products between the unknown input and the stored states, as in the local storage case. In fact, the output in the distributed storage case is the stored vector closest in the Hamming metric (defined as the number of 1's different between two digital vectors) to the input state. Thus Fig.2 is a content addressable memory in the Hamming sense.

As with the local storage case, one can see from Fig.2 that the computationally intensive parts of the distributed storage algorithm will be matrix multiplications. The same comments about the problems with traditional methods of computing such algorithms therefore follow. The use of digital signal processors (DSP) to accelerate the matrix multiplication parts of the NN algorithms will now be discussed.

DSP CHIP AS A NN ACCELERATOR

The pervasiveness of digital semiconductor technology for computation and the limited applicability of traditional analog signal processors has led to the development of the DSP special purpose microprocessor[14]. The key DSP feature for this discussion is that DSP chips perform multiplication operations in dedicated hardware. The DSP dedicated hardware produces faster execution of matrix operations, and this feature has led to considerable commercial interest in such chips. For a comparison of execution rates for DSP vs. standard computers, consider the example of a TMS 32020 DSP[15] relative to three members of the DEC VAX family as shown in Table 1. The table was constructed for a 16x16 bit fixed point multiplication and double precision accumulation (TMS 32020 mode) working a 256 square component matrix multiplying a 256 component vector. The TMS 32020 executes each multiplication of this computation by recalling the matrix component from memory and performing a single multiply/accumulate instruction. The multiply inner loop takes 600nS and the vector computation takes a total of 600x256x256 = 39mS. The VAX operation was estimated by Wallace Anderson[16] assuming a three instruction loop, with no time for indexing.

TABLE 1 EXECUTION TIME COMPARISONS FOR ONE TMS 32020 VS VAX SYSTEMS

OPERATION	TMS 32020	VAX 8600	VAX 11/785	VAX 11/750
Compute One Vector $$\sum_{j=1}^{256} T_{ij} V_j$$	39mS	98mS	196mS	360mS
Performance Relative to TMS 32020	100%	40%	20%	10%

As expected, a single TMS 32020 executing a multiplication/accumulation operation in 200nS and associated dynamic random access memory chips can offer up to an order of magnitude of acceleration in matrix operations relative to VAX systems. One could either perform the non-matrix parts of the algorithm on another general purpose microprocessor and use the DSP for matrix operations alone or one could use the programmability of the TMS 32020 DSP to perform a limited amount of non-matrix operations.   The cost and size advantages of a microcomputer approach relative to a main frame approach are additional to the speed advantage. Since the DSP approach depends only on high volume VLSI chips, one would expect the method to remain advantageous as the performance of such chips is improved. Since DSP boards are commercially available for personal computers, one can use this approach in a pc environment to evaluate the NN applicability.

DSP ACCURACY vs. NEURAL NETWORK REQUIREMENTS

An important consideration is the accuracy requirements of the multiplication/accumulation operations in Eq.2, relative to the problem being attacked. In this paper the design criterion is for maximum rate operation, and there is the possibility of accuracy/throughput tradeoffs. Traditionally one can represent variables in floating point and sacrifice throughput, or one can represent numbers as integers for maximum machine speed - for a given processor word size. The integer method requires double precision accumulation and more care by the programmer to avoid overflow but is more consistent with this paper's design criterion. The TMS32020 DSP can represent inputs with + 32,268 to -32,268 resolution and has a dynamic range of + /- 2 10^9. For distributed storage NNs, the input vectors tend to be digital and the matrix values tend to be integers bounded by the maximum size of the matrix. Thus the TMS 32020 is directly applicable to distributed storage NNs. For local storage/look-up-table NNs, the inputs and the matrix elements can be real numbers. Here full attention to the overflow potential needs to be paid by the programmer. Tools such as the multiply/divide instructions using a 32 bit intermediate result in the Forth language will need to be employed (TMS 32020 software supports Forth79).[17] One also needs to remember that NNs basically are an attempt to use many simple processors working in parallel to solve large problems. It is inconsistent to expect the simple processors to handle high precision arithmetic and still be mass produced at very low cost.

DSP ARRAYS TO FURTHER ACCELERATE NN ALGORITHMS

One of the standard methods for speeding up a computation which is already running on a single processor is to attempt to distribute the computation among several similar processors. Texas Instruments has developed a parallel computer architecture consisting of many DSP modules sharing a common bus to provide the necessary computational power for advanced signal processing applications such as speech recognition.[18] The architecture for this DSP array is shown in Fig.3 and a picture of the completed board, called the Odyssey is shown in Fig.4. Each DSP module contains a TMS 32020, up to 16k bytes of program memory, and up to 128k bytes of data memory. Each processor has its own local bus for accessing program and data memory, and all processors can execute concurrently with fair arbitration protocol. Assuming that the memory is equally divided between data and program, the 64k words of data memory will store a 256 x 256 array of 16 bit numbers, e.g., T_{ij} for a 256 x 256 problem.

Each board is capable of 20 million arithmetic operations per second. The architecture was design for numerical problems which may be divided into several processing modules which may be executed concurrently and which require a limited data exchange between modules.

Given such a canonical parallel architecture as the Odyssey, it then must be shown that generic NN algorithms can be mapped efficiently onto the architecture. This demonstration is straight forward when one remembers that the part of the NN algorithm which one wants to accelerate is the matrix multiplications. Consider the 16 bit multiplication and 32 bit accumulation example above, but expand the size of the problem to a 1024x1024 matrix times a 1024 component vector. The approach is to subdivide the problem into sizes that each TMS 32020 module can handle. The formula for the matrix operation can be broken into 16 submatrices (each 256 x 256) by compartmentalizing both i and j indices into regions (1,256), (257,512), (513,768), (769,1024). For each of the four ranges ($1 \le i \le 256$), ($257 \le i \le 512$), ($513 \le i \le 768$) and ($769 \le i \le 1024$) the output is:

$$OUTPUT_i = \sum_{j=1}^{256} T_{ij} V_j^? + \sum_{j=257}^{512} T_{ij} V_j^? + \sum_{j=513}^{768} T_{ij} V_j^? + \sum_{j=769}^{1024} T_{ij} V_j^?$$

Eq. 5

One assigns the first DSP to the first sum, the second to the second, the third to the third, and the fourth to the fourth. In this fashion, four DSPs compute a partial sum for one component in .15mS, and a total sum can be computed from the partial sums in less than a microsecond. The point is that the summation of the partial sums is trivial relative to the time of a partial sum. Thus four TMS 32020s on a single Odyssey board can execute the 256 component computations for one of the four i ranges in 39mS. Putting four Odyssey boards to work on the problem simultaneously, one would pick four rows to be computed simultaneously (one from each grouping) and assign each individual row to one Odyssey to achieve a total time of 39mS execution. The results are compared in Table 2 for the Odyssey system vs the same VAX systems as in Table 1. Generally for an N x N problem, one requires $(N/256)^2$ TMS 32020 chips or $(N/512)^2$ Odyssey boards to execute the problem in 39mS.

TABLE 2

OPERATION	# ODYSSEY BOARDS	ODYSSEY TIME	VAX 8600	VAX 11/785	VAX 11/750
512 x 512 Matrix Times 512 Vector	1	39mS	392mS	784mS	1440mS
Performance Relative to DSP Strategy	1	100%	10%	5%	2.7%
1000 x 1000 Matrix Times 1000 Vector	4	39mS	1598mS	3136mS	5760mS
Performance Relative to DSP Strategy	4	100%	2.5%	1.2%	.7%

Thus an NxN matrix vector calculations can be handled in 39mS by $(N/256)^2$ TMS 32020s working on $(N/512)^2$ Odyssey boards. Of course, such an improvement can be achieved from any similar parallel arrangement of DSPs. The Odyssey representation is used only for a concrete example. One would use such a DSP array in conjunction with a host workstation which would carry out the non-matrix tasks.

CONCLUSIONS

The artificial neural network field has been briefly summarized to demonstrate that the execution of most neural network algorithms is computationally dominated by matrix multiplication operations. Such algorithms can be accelerated with a single DSP and further accelerated with a coarse array of DSPs. The approach takes advantage of standard semiconductor technology and so can be expected to increase in productivity and cost effectiveness with standard semiconductor learning experience. Model simulation using such an accelerator in a workstation environment can treat more realistic problems than have been possible up until now and so will assist in determining the utility of artificial NNs.

BIBLIOGRAPHY

1. F. Rosenblatt, "Principles of Neurodynamics," Spartan Books, New York (1962).
2. B.Widrow, in "Self-Organizing Systems 1962, ed. M.C.Yovits, G.T. Jacobi, and G.D.Goldstein, Spartan Books, Washington DC (1962) p.435.
3. T. Kohonen, "Self Organization and Associative Memory," Springer Verlag, Berlin (1984).
4. S. Grossberg, "Studies of Mind and Brain," Boston Studies in Phil. Sci., 70/, Reidel (1982).
5. L.Cooper, F. Liberman, and E. Dja, Bio. Cyber. 33/, 9 (1979).
6. G. Palm, "Neural Assemblies," Studies of Brain Function #7, Springer Verlag, Berlin (1982).
7. K. Fukushima and S. Miyake, Biol. Cyber. 50/, 377 (1984).
8. S.Amari and M.A. Arbib, "Competition and Cooperation in Neural Nets," Lecture Notes in Biomath #45, Springer Verlag, Berlin (1982).
9. R. Hecht-Nielsen, Proc. SPIE, 360/, 180 (1983).
10. M. Minsky and S. Papert, "Perceptrons," MIT Press, Cambridge, MA (1969).
11. K. Steinbuch and U.A.W. Piske, IEEE Trans. Elec. Computers, 12/, 846 (1963).
12. J.A. Anderson, J.W. Silverstein, S.A. Ritz and R. Jones, Psych. Rev. 84/, 413 (1977).
13. J. Hopfield, Proc. Nat. Acad. Sci. USA 79/, 2554 (1982); 81/, 3088 (1984).
14. S.Y. Kung, H.J. Whitehouse, T. Kailath, eds. "VLSI and Modern Signal Processing" Prentice-Hall, Englewood Cliffs, NJ, (1985).
15. TMS 32020 User's Guide, Texas Instruments, Incorporated, Houston, Texas (1985).
16. Wallace Anderson, private communication.
17. H. Harrison, Proc. Speech Tech. '86, New York, April 1986.
18. W. Gass, R. Tarrant and G. Doddington, ICASSP 1986 Proceedings, Tokyo, Japan, p. 2887-2890.

352

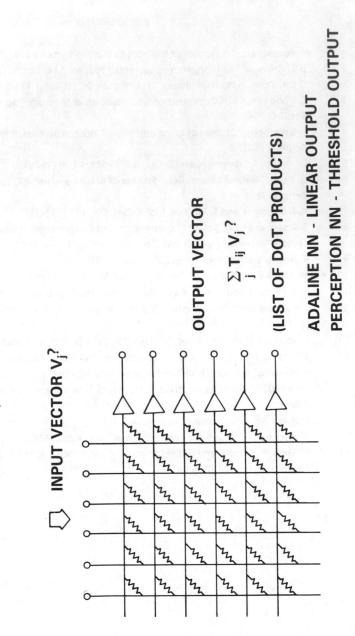

LOCAL STORAGE $1/R_{ij} = U^i_j$

INPUT VECTOR V^i_j?

OUTPUT VECTOR
$\sum_j T_{ij} V_j$?

(LIST OF DOT PRODUCTS)

ADALINE NN - LINEAR OUTPUT
PERCEPTION NN - THRESHOLD OUTPUT

Figure 1. Resistor representation of a local storage NN during execution.

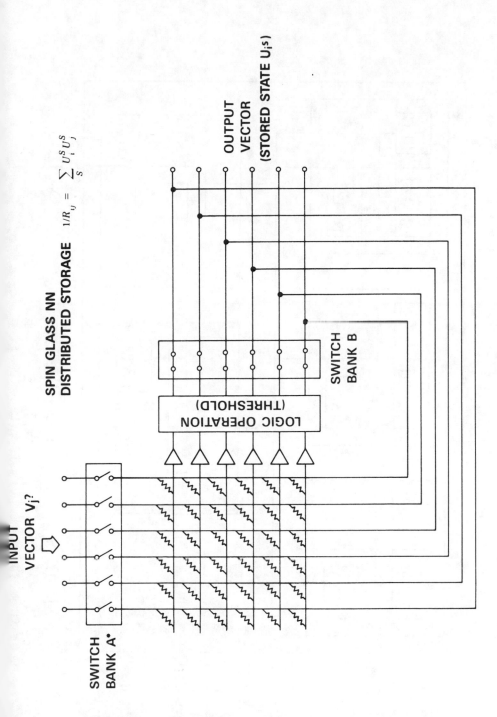

Figure 2. Resistor representation of a distributed storage NN during
execution. (Switch bank A open, bank B closed.)

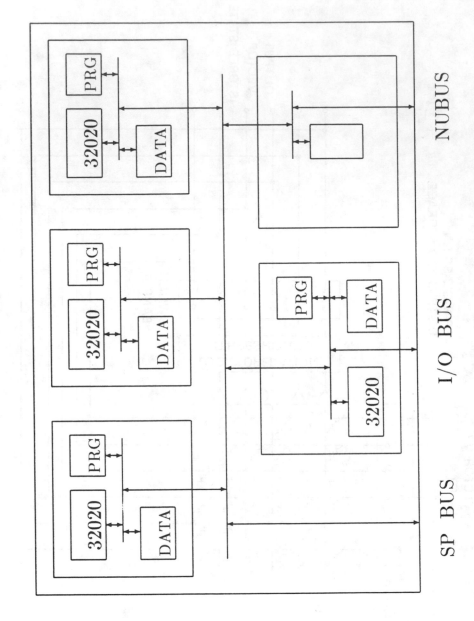

ODYSSEY ARCHITECTURE

Figure 3. Odyssey architecture: 4 TMS 32020 DSPs per board.

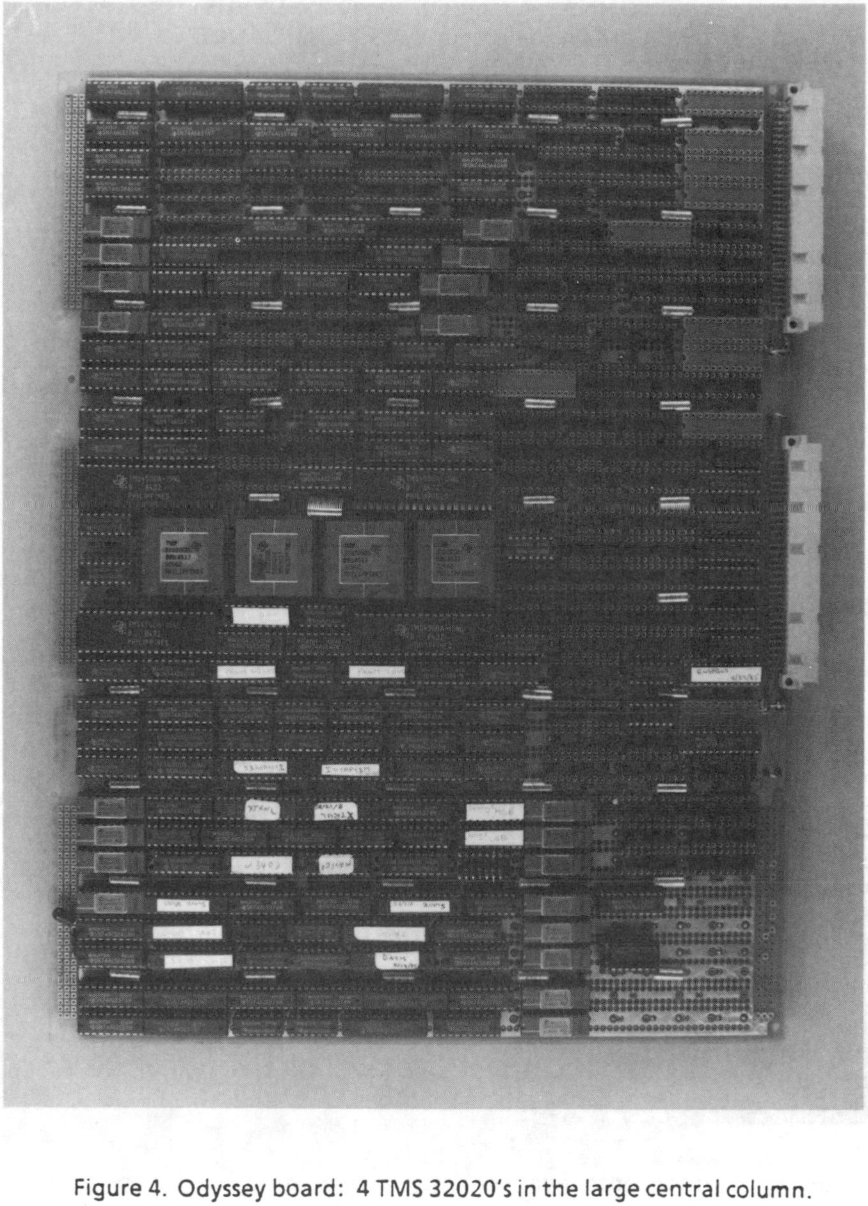

Figure 4. Odyssey board: 4 TMS 32020's in the large central column.

DESIGNING A NEURAL NETWORK SATISFYING A GIVEN SET OF CONSTRAINTS

L. PERSONNAZ, I. GUYON and G. DREYFUS

E.S.P.C.I., 10 rue Vauquelin, 75005 Paris FRANCE

INTRODUCTION

The learning rules which have been proposed for networks of Mc Culloch-Pitts formal neurons were designed to store informations as attractors of the dynamics of the system. Two approaches have been developed : several authors have studied the storage capacity of networks using Hebb's learning rule ; conversely, it has been shown that a generalization of Hebb's rule leads to the perfect storage of any number of information patterns (see Pef. 1 and references therein). The problem which is addressed in this paper goes beyond the stability of the stored informations : we show that it is possible to design a neural network satisfying a given set of constraints such as a given set of stable states and/or transitions and/or cycles. From an engineering point of view, this provides a powerful tool to perform new functions such as associations between informations and concepts. By formulating these dynamic behavior constraints in term of linear algebra, we derive an **"associating rule"** which imposes a given set of one-step transitions in state space, thus performing associations. Applications to character recognition are presented.

NEURAL NETWORK DESIGN: THE ASSOCIATING RULE.

We consider a fully connected network of n Mc Culloch-Pitts formal neurons operating in parallel with period τ , without sensory inputs. The thresholds will be taken equal to zero. The state of a neuron i at time t, $\sigma_i(t)$ is a binary variable, with value 1 or -1, which depends on the states of the other neurons at time $t-\tau$ in the following way:

$$\sigma_i(t) = \text{sgn}(v_i(t-\tau)) \qquad \text{if } v_i(t-\tau) \neq 0$$

$$\sigma_i(t) = \sigma_i(t-\tau) \qquad \text{if } v_i(t-\tau) = 0$$

where $v_i(t-\tau) = \sum_j C_{ij} \sigma_j(t-\tau)$ is the membrane potential of neuron i at time $t-\tau$.

Suppose that we want to impose p one-step transitions in state space defined by :

$$\vec{\sigma}^k \longrightarrow \vec{\sigma}'^k \qquad k = 1 \text{ to } p$$

The coupling coefficients must satisfy a system of n x p inequations :

$$\left(\sum_j C_{ij} \sigma_j^k \right) \sigma_i'^k > 0$$

but it is sufficient to solve the following linear system of n x p equations :

0094-243X/86/1510356-4$3.00 Copyright 1986 American Institute of Physics

$$\sum_j c_{ij} \; \sigma_j^k = \sigma_i'^k$$

Defining two matrices $\Sigma = \left[\vec{\sigma}^1, \dots, \vec{\sigma}^P\right]$ and $\Sigma' = \left[\vec{\sigma}'^1, \dots, \vec{\sigma}'^P\right]$, the above conditions can be written in matrix form :

$$C \Sigma = \Sigma' \quad .$$

This equation has an infinite number of solutions if $\Sigma' \Sigma^I \Sigma = \Sigma'$, where Σ^I is the pseudoinverse[2] of Σ . The most appropriate solution is :

$$C = \Sigma' \Sigma^I \quad (1)$$

since, if $\Sigma' \Sigma^I \Sigma \neq \Sigma'$, it gives the matrix which minimizes the euclidean norm of the error-matrix $C \Sigma - \Sigma'$.

Notice that an imposed transition such that $\vec{\sigma}'^m = \vec{\sigma}^m$ reduces to imposing the stability of the state $\vec{\sigma}^m$.

Relation (1) will be referenced in the following[2] as the **associating rule**. There exist several algorithms[2] to compute the coupling matrix C iteratively by introducing sequentially the transitions to be memorized, as usual in most learning processes. Since the learning rule is able to store one-step transitions, it can be used to impose sequences of transitions leading to a stable attractor state, cycles...

Example : handwritten character recognition.

In this example of application a set of handwritten numerals has been read by a solid state camera and the video signal digitized by an automatic threshold system ; the character are subsequently stored in a 32 x 32 memory. A ten bit field coding the class is added to each state to be learnt (this scheme has been used[3] for linear associative memories). After the memorization of the 44 transitions shown in figure 1.a. with the associating rule, 250 unknown patterns have been presented to the network ; approximately 80% of them are well recognized, the other are misclassified (with a wrong code) and only a small percentage are unrecognized (with a meaningless code). Some examples of application are shown in figure 1.b. These results can be interpreted as follows : the **associating rule** digs wide basins of attraction for each class of numerals and very few spurious attractors appear : the network almost always reaches one of the imposed attractors. This can be very suitable for some classification applications ; for handwritten character recognition, however, having more unrecognized than misclassified patterns is highly desirable ; in this respect, the learning process which is presented in the next paragraph is better. Finally it should be noticed that no special preprocessing, such as feature extraction, is required, and that the choice of the learnt examples has not yet been optimized.

NEURAL NETWORK DESIGN : THE PROJECTION RULE.

In the case where only stable states are to be stored, the **associating rule** (1) reduces to the **projection rule** :

$$C = \Sigma \Sigma^{I}$$

C always exists : it is the projection matrix into the subspace spanned by the set of states to be learnt. The **projection rule** guarantees the perfect storage of any set of vectors as stable states,
but the attractivity decreases hyperbolically with the number p of
stored vectors. The stored vectors have the lowest energy and it has
been shown that no cycles can occur.

The **projection rule**, as well as the **associating** rule, are
non local rules. Nevertheless, in the case of weakly correlated patterns, the **projection rule** can be approximated by a local lear-
ning rule : $\Delta C_{ij} = \frac{1}{n} (\sigma_i \sigma_j - v_i \sigma_j - \sigma_i v_j)$;

furthermore, if the last two terms are negligible, the latter relation reduces to Hebb's rule : $\Delta C_{ij} = (1/n) \sigma_i \sigma_j$.

Example : handwritten character recognition.

In order to compare with the previous example, we have stored the same 44 handwritten numerals, with their 10-bit code, using the **projection rule** (Figure 2.a). The same 250 unknown patterns have
been presented to the network, with the following results : 80% are
well recognized (with the right code), 10% are misclassified and 10%
are unrecognized. The latter result is remarkable since the number
of misclassified patterns is smaller, and the number of unrecognized
patterns is larger, than in the previous example. This is interesting as far as potential practical applications in pattern recognition are concerned. Note that no characters whatsoever are recognized if Hebb's rule is used instead of the **projection rule**.Some
examples of evolutions in state space are shown in Figure 2.b. One
should notice that some of the correctly recognized characters lead
to stable states which are not present in the training set, but do
look like the patterns to be recognized, and have the right code :
they are "thresholded" linear combinations of patterns with similar
shapes. Therefore, in this case, these "spurious" states are used to
our advantage. Some other spurious stable states are "thresholded"
linear combinations of patterns with very different shapes ; they
are also useful because they act as garbage collectors with a meaningless code. Thus, the evolutions in state space can be interpreted as follows : the **projection rule** generates many low energy
stable states at the bottom of the basins of attraction, which can
be used for pattern recognition, and also a lot of bumps at the top
of the energy barriers, which act as traps for unrecognizable states.

(1) L. Personnaz, I. Guyon, G. Dreyfus, J. de Phys. Lettres **46**,
L359 (1985).
L. Personnaz, I. Guyon, G. Dreyfus, "Disordered Systems and Biological Organization", p. 227, NATO ASI Series (Springer, 1986).
(2) A. Albert, "Regression and the Moore=Penrose Pseudoinverse" (Academic Press, 1972).
(3) T. Kohonen, "Self-Organization and Associative Memory" (Springer, 1984).

FIGURE 2.a.

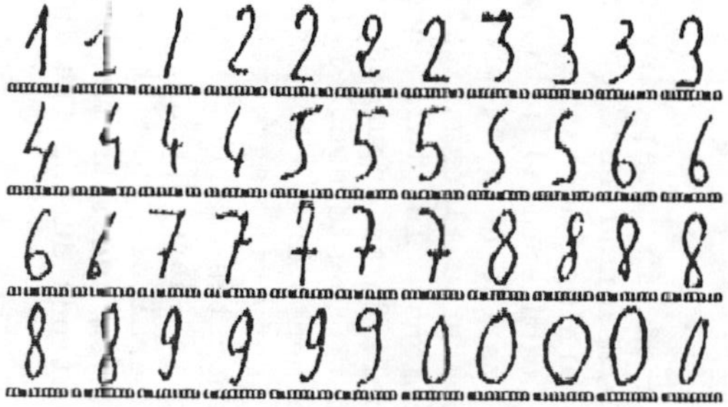

FIGURE 2.b.

80 %
RECOGNIZED

10% UNRECOGNIZED

10% MISCLASSIFIED

FIGURE 1.a.

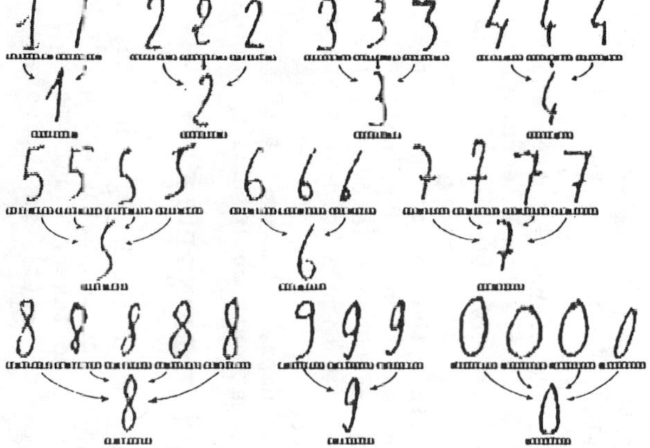

FIGURE 1.b.

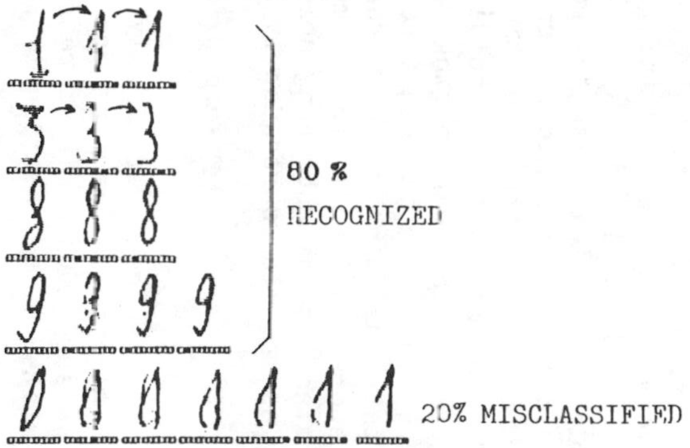

80 %
RECOGNIZED

20% MISCLASSIFIED

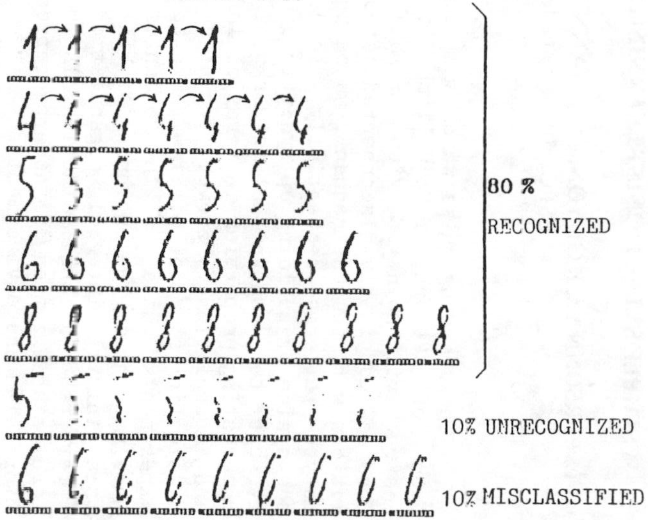

A SIMPLE SELECTIONIST LEARNING RULE FOR NEURAL NETWORKS

L. PERSONNAZ, I. GUYON, A. JOHANNET, G. DREYFUS and G. TOULOUSE

ESPCI, 10 rue Vauquelin, 75005 Paris

The learning rules which have been used in most investigations of networks of Mc Culloch-Pitts formal neurons stem from an instructivist point of view[1] : such learning mechanisms start from a completely disconnected network (tabula rasa) ; the interactions with the environment, which presents the network with prototype patterns sequentially, lead to a gradual organization of the system by building up synaptic connections. In this context, it has been shown that networks of simple formal neurons exhibit interesting properties in terms of information storage, retrieval and association. However, recent developments in biology advocate a selectionist point of view which seems more in agreement with experimental data[2] : the organism generates connections between neurons at the beginning of the brain development ; thus, this initial internal organization determines some "knowledge" (prerepresentations) ; the learning mechanism consists in a selective stabilization of these prerepresentations during the interactions with the environment.

Moreover most instructivist learning rules are questionable in two additional respects:

i- the learning mechanism leads to a symmetrical synaptic matrix.

ii- during the learning phase, sign reversals of the synaptic strengths are possible, which means that an excitatory synapse (with positive synaptic strength) might become an inhibitory synapse (with negative synaptic strength) ; such phenomena are very seldom observed in biological systems.

Following ideas developed in Reference 3, the above considerations lead us to propose a simple local learning rule which starts from a randomly connected network (tabula non rasa) and avoids the above mentioned pitfalls[4]. Moreover, its mechanism is selectionist : besides storing the prototype patterns, it erases the neighbouring prerepresentations and selectively preserves the prerepresentations which are uncorrelated to the stored patterns. In other words, the rule digs a hole in state space to store the informations, smooths out the edges of the hole by erasing the neighbouring holes and bumps, but does not affect the holes in which subsequently learnt patterns could fit.

A SELECTIONIST LEARNING RULE

We consider a fully connected network of n Mc Culloch Pitts formal neurons operating in parallel with period τ, without sensory inputs. The thresholds will be taken equal to zero. The state of a neuron i at time t, $\sigma_i(t)$, is a binary variable, with value 1 or –1, which depends on the states of the other neurons at time t– in the following way :

$$\sigma_i(t) = \text{sgn}(v_i(t-\tau)) \qquad \text{if } v_i(t-\tau) \neq 0$$

$$\sigma_i(t) = \sigma_i(t-\tau) \qquad \text{if } v_i(t-\tau) = 0$$

where $v_i(t-\tau) = \sum_j C_{ij} \; \sigma_j(t-\tau)$ is the membrane

potential of neuron i at time t–τ .

It has been shown[5] that the (n x n) synaptic matrix C can be computed conveniently so as to impose the stability of any set of prototype states $\vec{\sigma}^k$ by relation :

$$C = \Sigma \Sigma^I + B (I - \Sigma \Sigma^I)$$

where $\Sigma = [\vec{\sigma}^1, \ldots, \vec{\sigma}^p]$ and Σ^I is the pseudoinverse[6] of Σ . This rule can be put into an iterative form. The increment of matrix C in order to learn a kth pattern $\vec{\sigma}^k$ will be :

$$\Delta C(k) = (I - B) \frac{\vec{u}^k \vec{u}^{k^T}}{\|\vec{u}^k\|^2}$$

where u^k is the orthogonal projection into the supplementary subspace of the subspace spanned by the k–1 previously learnt vectors.

It can easily be shown that $C(k-1)\,\vec{u}^k = B\,\vec{u}^k$ where C(k–1) is the value of the synaptic matrix prior to learning the new pattern $\vec{\sigma}^k$. Therefore :

$$\Delta C(k) = (I - C(k-1)) \frac{\vec{u}^k \vec{u}^{k^T}}{\|\vec{u}^k\|^2}$$

and $C(0) = B$.

Hence, two opposite attitudes may be adopted : for simplicity, a zero initial matrix can be chosen, so that C is the orthogonal projection matrix into the subspace spanned by the prototype vectors[5] ; in this case, the learning rule is an instructive process starting from an initial "tabula rasa". On the contrary, if a non zero initial matrix is taken, learning starts from a system which already contains some knowledge called "prerepresentations", and consists in modifying the synaptic coefficients. It can be shown analytically, and has been checked by numerical simulations, that the learning rule operates by selection :

- If a pattern is very correlated to a prerepresentation, it is learnt without effort because the learning process consists in digging a basin of attraction in state space, in the vicinity of a preexisting hole ; the prerepresentation is erased and attracted by the learnt pattern so that the learning rule smooths out the edges of the basin.

- The prerepresentations which are uncorrelated to the learnt patterns are not altered and remain useful for further learning.

If the initial matrix B is symmetrical and its components taken at random, the resulting phase space structure can be viewed as a spin glass energy landscape[7], and the attractors are the bottoms of the valleys (static prerepresentations). However, even if matrix B is symmetrical, the synaptic matrix C after learning will not, in general, be symmetrical, which is one of the conditions for a learning model to be plausible from a biological standpoint. It can be noticed that the learning process described above allows to memorize and retrieve perfectly any set of patterns, but is not a local rule because all the synapses computed previously are involved in the modification of one synaptic coefficient. This last point is not realistic from a biological standpoint ; in the next paragraph, we shall study a local version of the rule which is suboptimal as far as the storage capacity is concerned.

COUPLING COEFFICIENTS VERSUS SYNAPSES

In our previous studies[5] on networks of formal neurons our aim was to find efficient learning rules from the storage capacity viewpoint. The projection rule has been proved to be the best in this respect, but, being a non local rule, it has no obvious biological meaning. However, it is very well known that, for almost uncorrelated patterns, the projection rule can be approximated by HEBB's rule which is local in nature. Following the same idea, we present, in this paragraph, a local version of our selectionist learning rule, which allows to memorize perfectly any set of orthogonal patterns, but is suboptimal for almost uncorrelated patterns. This provides a synaptic model where the synapse can learn by itself from locally available information only. Moreover, we show that the initial nature (excitatory or inhibitory) of the synapses remains unchanged during the learning phase.

If orthogonal patterns are to be learnt, then
$$\Sigma^{I} = (1/n)\,\Sigma^{T} \quad \text{or equivalently} \quad \vec{u}^{k} = \vec{\sigma}^{k};$$
therefore the increment of matrix C necessary to learn a new pattern $\vec{\sigma}^{k}$ becomes :

$$\Delta C(k) = (I - C(k-1))\frac{\vec{\sigma}^{k}\vec{\sigma}^{kT}}{n} \qquad \text{and} \quad C(0) = B$$

The variation of a synaptic coefficient is given by :

$$\Delta C_{ij}(k) = 1/n(\sigma_{i}^{k}\sigma_{j}^{k} - v_{i}\sigma_{j}^{k})$$
where $v_{i} = \sum_{r} C_{ir}(k-1)\,\sigma_{r}^{k}$;

v_i is the membrane potential of neuron i when the new pattern to be learnt is input to the network $C(k-1)$.

The first term corresponds to the classical HEBB's rule, giving a contribution of $\pm 1/n$. Let us assume that the initial synaptic matrix B is random with values $\pm 1/\sqrt{n}$ (for the membrane potential to remain finite for very large n) ; the second term is $O(n)$ and, since v_i is a graded variable, it provides a fine tuning of the synaptic coefficients taking into account the information learnt previously. Hence, starting from an initial synaptic coefficient with value $\pm 1/\sqrt{n}$, learning consists in small modifications around this average value. The number of patterns that can be stored without sign reversal of a synaptic coefficient has been evaluated : before reaching Proba(sign reversal) = 0.05, one can store p $\simeq$ n/7 uncorrelated patterns. This value may be compared to the limitations on the storage capacity introduced by the suboptimality of the local rule for non orthogonal patterns. The two limitations are of the same order of magnitude.

CONCLUSION

We have attempted to include biological constraints into a network of formal neurons. The two main ones were : first, biological synapses are either inhibitory or excitatory but, in most cases, their initial nature cannot change ; secondly, connections between neurons are not, in general, symmetrical. The outcome of this investigation is a learning rule which, besides satisfying the above constraints, turns out to be a very simple mechanism of learning by selection. Moreover, it is, just like Hebb's rule, a local rule whereby synapses are modified according to locally available information only. Such a model may help bridging the gaps between biological neural networks and networks of formal neurons.

REFERENCES

(1) W.A. Little, Math. Biosci. **19**, 101, (1974).
    J.J. Hopfield, Proc. Natl. Acad. Sci. USA **79**, 2554, (1982)
    P. Peretto, Biol. Cybern. **50**, 51, (1984).
    D.J. Amit, H. Gutfreund and H. Sompolinsky, Phys. Rev. A
    32, 1007, (1985) ; Phys. Rev. Lett. **55**, 1530, (1985).

(2) J.P. Changeux, T. Heidmann and P. Patte, in "The Biology of
    Learning", ed. by P. Marler and H. Terrace (Springer Verlag,
    1984).

(3) G. Toulouse, S. Dehaene and J.P. Changeux, Proc. Natl. Acad.
    Sci. USA, to be published.

(4) L. Personnaz, I. Guyon, G. Dreyfus and G. Toulouse, J. Stat.
    Phys. **43**, 411, (1986).

(5) L. Personnaz, I. Guyon and G. Dreyfus, J. Phys. Lett. **46**,
    L-359, (1985) ; Phys. Rev. A. (submitted for publication).

(6) A. Albert, in "Regression and the Moore-Penrose Pseudoinverse"
    (Academic Press, 1972).

(7) Heidelberg Colloquium on Spin Glasses, Lecture Notes in
    Physics, vol. 192 (Springer Verlag, 1983).

Analog Decoding Using Neural Networks

J. C. Platt

J. J. Hopfield*

California Institute of Technology, Pasadena, CA 91125

Abstract

This paper develops a particular error correction code which can be effectively decoded by a relatively simple neural network. In high noise situations, this code is comparable to that used at present in deep space communications. The neural decoder has $N!$ stable states with only N^2 neurons, and can quickly extract information from analog noise. This example illustrates the effectiveness of neural networks in solving real problems when the problem can be cast in such a fashion that it fits gracefully on the network.

1. Introduction

When information is to be transmitted or stored under noisy circumstances, error correction codes provide a means to retain the information faithfully. Error correction codes add redundancy to information so that the information survives the noise.

In this paper, background on error correction codes will be presented. Then, two codes amenable to neural network decoding will be shown. The neural decoders will be illustrated, then results of simulations of these decoders will be presented.

2. Error Correcting Codes

A desirable code is one which suppresses errors as well as possible when codes are finite, does not require an impossible quantity of storage, and can be decoded with a reasonable amount of computation If the system is such that either a bit error is made or it is not (e.g., a soft fail in ROM), this is the only coding problem. In the case of the transmission of weak signals in the presence of noise, however, the whole problem is truly an analog problem. A transmitted 1 or 0 actually corresponds to the level of an analog signal being at the level of a nominal 1 or a nominal zero. Without noise, 1's and 0's could be accurately recovered by a threshold circuit with its threshold level set halfway between the voltage level of a nominal 1 and a nominal 0. With additive Gaussian noise, however, there is a probability of the actual voltage when a 0 is sent being greater than threshold, or of a 1 being less than threshold. Thus, possible errors will be made in the identification of 1's and 0's.

If the signal is immediately digitized at the receiver, the distinction between almost certain bits and bits which are "iffy" is lost. With a Gaussian channel, this loss of information corresponds to a loss of approximately 2 dB, or 40 percent, of the signal power, compared to a scheme which keeps this information [1].

To make optimal use of the available signal energy, it is necessary to use a code which permits initial analog decoding, effectively making a decision about an entire analog code word at a time. This paper describes some such codes which can be initially decoded by "neural" networks. Any real encoding-decoding system would use in addition a standard digital encoding as well, and would follow the analog decoding by a further digital decoding procedure [2].

This problem is of interest for two reasons. First, it exemplifies the design of a network of N neurons with fixed connections, and having far more than N

---

* Also with Department of Molecular Biophysics, AT&T Bell Laboratories, Murray Hill, NJ 07974

stable states, to solve a real problem. (Having more than N stable states violates no information theorem, since the actual stable states are highly correlated, and thus contain much less than N bits of information per state). Second, decoding is a stylized example of the typical problem in perception. For example, given a visual scene, in 0.2 seconds, the eye will take in about 10^8 pixels, each with a few bits of intensity information. The real information in a visual glance is much less than 10^9 bits. The process of perception can be viewed as a contraction of this huge state space down into the much smaller space of essential information. While we have nothing to say about how to do the decoding in this extremely difficult case, vision clearly is "merely" a decoding problem, and seeing how to decode even in highly stylized and abstract codes should be of some use in thinking about the computational problem of perception.

3. Codes for Neural Networks

3a. 1 in N Code

A code that can be decoded simply by a neural network is a 1 in N code. We will consider using frequency-shift-keying (FSK), where the sender puts energy into N closely spaced frequencies. For a 1 in N code, the sender only puts energy into one of the frequencies at any time. This is close to optimal, since the 1s are much rarer than the 0s. In the decoding process, a "1" is assigned to the one of the N frequencies in which the largest signal is found.

This decoding process can be done by a "neural" circuit made as an N-flop, with $T_{ij} = -1, i \neq j$, with an appropriate threshold, and with the input from frequency j connected to neuron j. Thus, the input for the decoder is fed to the neural network in parallel.

The neurons in this circuit compete, the neuron with the largest input wins, and shuts off all of the other neurons. Thus, the circuit does the decoding exactly.

3b. Permutation Matrix Code

An interesting and more complex code can be based on a 1 in N position coding. Consider a codeword that is N symbols long. The symbols are chosen from an alphabet that has N symbols. Now, specify that each symbol in the alphabet will appear in the codeword exactly once. Thus, an N by N permutation matrix describes which symbol goes into which position in the codeword. In this matrix, there is only one 1 per row, and one 1 per column. Thus, the permutation matrix code can be viewed as an extension of a 1 in N code, where N old codewords are grouped into one new codeword.

There are $N!$ possible N by N permutation matrices, so the information in such a matrix is $\log_2 N!$ bits. Without the restriction to be a permutation matrix, the information would be $N \log_2 N$ bits.

Again, FSK is a practical implementation of the permutation matrix code. Using FSK, each symbol in the alphabet is a different frequency. N corresponding bandpass filters are listening to the channel. Thus, the output of bandpass i at time j at can formed into a matrix I_{ij}. This matrix nominally looks like a permutation matrix V_{ij}, and the decoding procedure is to find the permutation matrix closest to the input.

The decoding can be performed by extending the neural network that decoded the 1 in N code. To enforce one on per row and one on per column, there should be simultaneous mutual inhibition along the rows and columns.

Consider a typical code word: the identity matrix. If, because of noise, one of the off-diagonal inputs is large, two neurons on the diagonal will cooperate to shut off the neuron corresponding to the spurious input. Thus, the code should be immune to one symbol errors.

4. Network Implementations

In this section, various differential equations will be presented that decode the permutation matrix code. Let I_{ij} be the input for the neuron in the ith row and the jth column. Let V_{ij} be the state of that neuron.

4a. Linear Programming

Consider the problem of finding the closest permutation matrix to the input. Thus,

$$- \sum_{i,j} V_{ij} I_{ij} \tag{1}$$

should be minimized. However, the final output of the decoder should be digital, so there is an additional constraint that

$$V_{ij} \in \{0, 1\}. \tag{2}$$

(1) and (2) comprise an integer programming problem. Integer programming is usually quite difficult. However, the permutation matrix coding problem is equivalent to weighted bipartite graph matching (WBGM). The task in WBGM is to assign workers i to tasks j. Every worker should be assigned exactly one task. Also, there is an efficiency I_{ij} that each worker does a task. WBGM consists of finding a digital permutation matrix V that assigns workers to tasks, such that the function in (1) is minimized.

WBGM can be shown to be solved with the following constraints.

$$\sum_i V_{ij} \leq 1, \qquad \sum_j V_{ij} \leq 1, \qquad V_{ij} \geq 0 \tag{3}$$

Since the quantity that is to be minimized is linear, and the constraints are linear, WBGM can be solved by linear programming. The solution to the WBGM linear constraints can be shown to always lie on $V_{ij} \in \{0, 1\}$ [3].

Thus, to decode the permutation matrix, one can use the simplex algorithm or Karmarkar's algorithm. To speed it up even further, one can implement it in analog hardware that will be governed by neural network differential equations.

Tank and Hopfield has previously discussed the approximate solution of linear programming by neural networks [4]. Following that work, establish an energy function

$$E = - \sum_{i,j} V_{ij} I_{ij} + f \left(\sum_i V_{ij} - 1 \right) + f \left(\sum_j V_{ij} - 1 \right) + f \left(-V_{ij} \right). \tag{4}$$

where $f(x) = \exp(kx)/k$, where k is large.

The network will slide down this energy function. The first term is merely the quantity that needs to be minimized. The steep exponential terms enforce the constraints by creating walls in the state space. As the exponential gets steeper, the system performs better, but an infinite k is not physically realizable. Also, as N increases, k must increase, otherwise the system will hang up in the center of the space.

The equations of motion corresponding to this energy function are

$$\tau \dot{V}_{ij} = I_{ij} - f' \left(\sum_i V_{ij} - 1 \right) - f' \left(\sum_j V_{ij} - 1 \right) + f' \left(-V_{ij} \right). \tag{5}$$

where $f'(x) = \exp(kx)$.

Notice that the terms corresponding to the permutation matrix constraints are merely exponential row and column inhibition. Because the system has steep exponential walls, the equations are stiff, and need to be simulated with Gear's method [5].

When simulated, the network starts out in the center of the space, moves down the flat plane created by the first term until it slams into one of the walls. It then crawls along the wall until it hits more walls, and it finally comes to rest, usually at a vertex of the constraint hyperplanes. Occasionally, when then system doesn't have a strong drive towards one particular vertex, the system hangs up in the middle of the space, because the interior isn't perfectly linear, due to the exponential energy terms.

4b. Current Summing

Alternative equations of motion can be proposed that are closer to VLSI analog circuits, namely, using current summing instead of voltage averaging using resistors.

$$\tau \dot{V}_{ij} = A(V_{ij}) \left(I_{ij} + 2ce^{kV_{ij}} \right) - B(V_{ij}) \left(\sum_i ce^{kV_{ij}} + \sum_j ce^{kV_{ij}} \right) \qquad (6)$$

A is a function that shuts off excitation when $V_{ij} = 1$. For example,

$$A(V) = (1 - V)/(1 + \epsilon - V). \qquad (7)$$

B is a function that shuts off inhibition when $V_{ij} = 0$. Thus,

$$B(V) = V/(V + \epsilon). \qquad (8)$$

The exponentials reflect the behaviour of transistors beneath threshold. The A and B functions simulate transistors shutting off when the voltage across them is close to 0 [6].

Notice that the inhibition is a sum of exponentials, instead of an exponential of sums, as in §3a. Syntax is enforced much more strictly: at $N^2 = 64$, the system never broke syntax. However, these equations of motion do not correspond to any known energy function, since the excitation and the inhibition are shut off with different functions, and hence cannot be proven to converge to a fixed point.

5. Results

The codes presented in §3 were simulated with a Gaussian channel. Thus, Gaussian noise was added to the input of each neuron, and the various decoders were simulated.

To compare the neural code to other codes, a standard performance graph is plotted: the error probability per information bit versus the signal-to-noise power ratio. The error per bit, P_E, is computed from the word error probability P_W.

$$P_E = 1 - (1 - P_W)^{1/M}, \qquad (9)$$

where M is the number of information bits in a codeword.

E_b/N_0 is commonly used for signal-to-noise comparisons. E_b is the energy sent per information bit and N_0 is the noise power [7]. Now,

$$E_b = P/R, \qquad (10)$$

where P = average signal power and R = the information rate of the code, in bits per time step. If the signal sent down the channel is either X or 0, and the standard deviation of the noise is σ, and the time per symbol is T, then

$$P = X^2/T, \qquad N_0 = 2\sigma^2/T. \tag{11}$$

Let s = signal-to-noise ratio = X/σ. Then,

$$E_b/N_0 = s^2/2R. \tag{12}$$

Shannon's Source-Channel Coding Theorem states that one can transmit over a Gaussian channel with arbitrarily low error probability, as long as $E_b/N_0 > \ln 2$. In practice, E_b/N_0 of 3 to 5dB are used [8].

At low error rates, an approximate analytic theory of the error rate in the permutation code can be constructed. The essential idea is that at low error rates, the typical error can be thought of as the exchange of 1's between two particular rows. This is an effectively 2 x 2 permutation matrix plus noise problem, and can be solved exactly as an error function.

The probability of making such an error is then a combinatorial coefficient times this elementary probability. The low-error curve in the figure was constructed in this way.

P_E versus E_b/N_0 are plotted for various codes in figure 1.

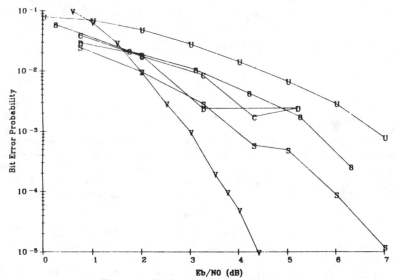

Figure 1. Performance of Various Codes

In figure 1, the 1 in 8 code is marked with an "8." The ideal performance of the 8 by 8 code is marked with an "S." A $K = 7, r = 0.5, Q = 3$ Viterbi code is marked with a "V." Uncoded phase-shift-keying is marked with a "U." Also shown are the results of the linear programming ("D") and current summing neural networks ("C") that implement the 8 by 8 code. The ideal 8 by 8 code comes close to the performance of the $K = 7$ Viterbi code that NASA uses as its analog code for Voyager [9]. The 1 in 8 code does somewhat worse, with the neural networks falling somewhere in between.

6. Conclusions

We have shown by example that efficient analog error-correcting codes can be designed in such a way that they can be efficiently decoded on neural networks. At high error rates, the 8 by 8 code is comparable to the $K = 7$, $r = 1/2$ Viterbi code which is at present in use in deep space communication. Below bit error probabilites of 10^{-2}, the 8 by 8 code gradually degrades in performance in comparison with that Viterbi code.

The 8 by 8 code could be improved a few tenths of a dB by transmitting a low power negative signal instead of a zero power in the bands which are transmitting nominal zeroes. With such tuning, the 8 by 8 code may have a noticable advantage over the Viterbi scheme in high noise situations.

The circuit for this code is very simple—it would require less than 200 transistors and 100 resistors on a chip to do the decoding, and the relevant analog signals could be carried to their appropriate locations by using analog CCD technology with 8 parallel bucket lines. Alternatively, a current summer uses 360 transistors. A decoding of 15 bits can readily be made in about 10^{-6} seconds. This combination of speed and simplicity exemplify the power of a neural circuit in a task which is well matched to the circuit. While a large digital chip can also be made to do this same task, it would take more time and chip area.

More complex codes can also be constructed which link the positions of 1's in a three dimensional structure of connections. These are interesting to explore, but require a circuit more like that used by Hopfield and Tank in the Traveling Salesman Problem [10] to solve them, since they are not reducible to linear programming.

This paper is supported, in part, by an AT&T Bell Laboratories Fellowship (JCP) and by NSF Grant PCM-8406049.

Thanks to R. J. McEliece, E. C. Posner, D. W. Tank, and Y. S. Abu-Mostafa for their helpful comments and suggestions.

7. References

1. R. J. McEliece, *The Theory of Information and Coding*, (Addison-Wesley, London, 1977), p 103.
2. J. H. Yuen, ed., *Deep Space Telecommunications Systems Engineering*, JPL Publication 82-76 (1982).
3. C. H. Papadimiriou, K. Steiglitz, *Combinatorial Optimization: Algorithms and Complexity*, (Prentice-Hall, Englewood Cliffs, NJ, 1982), p. 248.
4. D. W. Tank, J. J. Hopfield, *IEEE Cir. & Syst.*, (1986), in press.
5. C. W. Gear, *Numerical Initial value Problems in Ordinary Differential Equations*, (Prentice-Hall, Englewood Cliffs, NJ, 1971).
6. J. C. Platt, *Sequential Threshold Circuits*, Caltech Computer Science Technical Report, 5197:TR:85, (1985).
7. R. J. McEliece, *The Theory of Information and Coding*, (Addison-Wesley, London, 1977), p 116.
8. E. C. Posner, Personal Communication.
9. J. H. Yuen, ed., *Deep Space Telecommunications Systems Engineering*, JPL Publication 82-76 (1982), p 225.
10. J. J. Hopfield, D. W. Tank, *Biol. Cyber.*, v. 52, pp 141-152, (1985).

NONLINEAR DISCRIMINANT FUNCTIONS AND ASSOCIATIVE MEMORIES

Demetri Psaltis and Cheol Hoon Park
California Institute of Technology
Department of Electrical Engineering
Pasadena, CA, 91125

INTRODUCTION

A discriminant function maps a feature vector to either a one or a zero and in this manner each possible input pattern is mapped to one of two possible classes. An array of discriminant fucnctions, each implementing a distinct grouping of the same input vectors in two classes, can be thought of as a memory that associates each input feature vector to an associated output vector that is binary valued. Let $f(j)$ denote the input vector and let $g(i)$ be the output of the ith discriminant function. Data are stored in the array of discriminant functions by selecting how each of them classifies a set of specified input vectors, denoted by $f^{(m)}(j)$. In this manner each vector $f^{(m)}(j)$ is associated with a binary output vector $g^{(m)}(i)$.

Linear discriminant functions (LDF's) have received most of the attention in the literature because they are relatively easy to analyze and also easy to implement. Associative memories that can be thought of as arrays of LDF's have also been extensively investigated[1,3,6]. The output vector $g(i)$ in this case is obtained as follows:

$$g(i)=\text{sgn}\{\sum_{j}^{N} w_{ij}f(j)\} \qquad [1]$$

where $\text{sgn}\{x\}=1$ if $x>0$ and -1 otherwise. Many of the properties of associative memories of this type are derived directly from the corresponding properties of LDF's. For instance, the storage capacity of a LDF, defined as the maximum number M of vectors that can be separated in two classes in any specified manner, is approximately equal to N, where N is the number of element of the feature vector $f^{(m)}(i)$[2]. The number of associations that can be stored in a memory that consists of an array of LDF's is also upperbounded by N. Algorithms that have been developed for training LDF's can also be directly used for storing information in associative memories. A particularly simple method is the "outer-product" method in which the matrix of weights w_{ij} is calculated as the sum of the outer products of the associated vectors[1,3,4].

$$w_{ij} = \sum_{m}^{M} g^{(m)}(i) f^{(m)}(j). \qquad\qquad [2]$$

The outer product scheme is suboptimum in terms of storage capacity, i.e. the bound M=N cannot be achieved. In experimental simulations M is typically .1N-.15N. The theoretical estimate for the capacity of the outer product storage method is M<(N/4logN).

QUADRATIC ASSOCIATIVE MEMORY

We now consider the characteristics of associative memories that are constructed as arrays of discriminant functions other than LDF's. Specifically we consider discriminant functions that are calculated as polynomial expansions of the feature vector f(j):

$$g(i) = \text{sgn}\{w_o + \sum_{j}^{N} w_{ij} f(j) + \sum_{j}^{N} \sum_{k}^{N} w_{ijk} f(j) f(k) + \ldots\}. \qquad [3]$$

The motivation for looking at higher order discriminant functions is the potential for increased storage capacity due to the increase in the available degrees of freedom. The total number of nonredundant weights that can be specified in an rth order expansion is N+r choose r. For r=1 (LDF's) the number of degrees of freedom is N (ignoring the threshold variable w_o), whereas for r=2 (i.e. an expansion that includes the linear and quadratic terms) it increases to N'=(N+1)(N+2)/2. These additional degrees of freedom directly translate to additional storage capacity in an associative memory that is constructed as an array of quadratic discriminant functions.

In what follows we will concentrate on a specific form of quadratic associative memory that includes only the quadratic term in the expansion of equation 3. The next item is to determine methods for storing data in the weights of a quadratic associative memory. The most straightforward method is to consider the nonredundant patterns $f^{(m)}(j) f^{(m)}(k)$ as new feature vectors with dimensionality N' and then apply an array of LDF's to these patterns to construct an overall quadratic memory that associates the original vectors $f^{(m)}(j)$ to corresponding output vectors $g^{(m)}(i)$. This is shown schematically in Figure 1 which is suggestive of an optical implementation of this scheme. The outer product of the input vector f(j) with itself is first formed to create an N X N image of the new feature space. This image then illuminates the volume hologram which is pre-exposed to interconnect the N^2 points of the input image to N points at the output. The total number of weights that need to be specified in this case is N^3

372

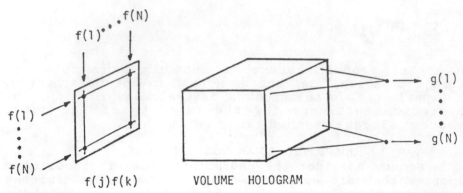

Figure 1. Holographic implementation of a quadratic associative memory.

which very conveniently matches the degrees of freedom provided by a volume hologram whose shape is a cube.

QUADRATIC ASSOCIATIVE MEMORIES WITH OUTER PRODUCT STORAGE

The weights w_{ijk} can be specified using any one of the algorithms that have been developed for training LDF's. For the quadratic associative memory, the outer product construction is of particular interest because as we will show it leads to a very simple implementation. For the quadratic memory, storing data as sum of outer products leads to the following expression:

$$w_{ijk} = \sum_m^M g^{(m)}(i) f^{(m)}(j) f^{(m)}(k) \qquad [4]$$

The memory is addressed by evaluating the following quadratic form:

$$g'(i) = \sum_j^N \sum_k^N w_{ijk} f(j) f(k)$$

$$= \sum_m^M [\sum_j^N f^{(m)}(j) f(j)]^2 g^{(m)}(i) \qquad [5]$$

The second form of the above equation was obtained by substituting equation 4 in 5 and interchanging the order in which the summations are performed. If we perform a similar substitution for the linear associative memory (equations 1 and 2), the result is identical to equation 5, with the exception that the summation over j is raised to the second power in the quadratic memory. This is shown schematically in Figure 2 for the general case of

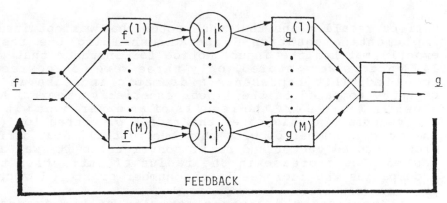

Figure 2. Associative memory constructed with polynomial
           disctiminant functions of order k and the
           outer product storage algorithm.

a kth-order memory, i.e. a memory that is constructed as
an array of discriminant functions that are the kth order
expansion of the input vectors. Memories of this type are
also discussed in reference 5.

Numerical simulations of the quadratic memory using
outer product storage were performed based on equations 4
and 5.   For an   autoassociative memory (i.e. $f^{(m)}=g^{(m)}$)
), feedback can be used in a fashion completely analogous
to a Hopfield style network. The simulations were done
for an autoassociative case with feedback. Binary valued
vectors (+1,-1) were randomly selected. Two sets of
experiments are displayed in Table 1. The first column in
Table 1 lists the number of 63 bit vectors that were
stored and the second column lists how many of these same
vectorswere correctly recalled in each case.

N=63			N=32		
Stored Vectors	Stable Vectors		Stored Vectors	Stable Vectors	Attraction Radius
60	60		8	8	9.0
70	70		16	16	8.3
80	79 (1)		24	24	8.3
90	89 (1)		32	32	8.0
100	98 (1)		40	39	7.0
110	108 (1)		48	45	5.0
120	119		56	49	4.8
130	128				
140	137				
150	140				

Table 1. Computer Simulations

Correct recall means that the vector that was obtained by implementing equation 5 and thresholding the result exactly matched the input. Notice for instance that when 140 vectors were stored, only three were not correctly recalled. This represents a dramatic increase in the capacity compared to the linear associative memory. Also shown in Table 1 are the results of an experiment with 32 bit vectors. In this case the radius of attraction (i.e. the average number of bits that the initial input differs from a stored vector and still converges to it) was also tested. The decrease in the radius of attraction that accompanies the increase in the number of stored vectors is very gradual.

The increased storage capacity of the quadratic memory compared to the linear is dramatic particularly in view of the minor modification (i.e. squaring) that one needs to do in order to convert a linear memory to quadratic. The number of nonredundant weights in equation 4 is approximately equal to $N^3/3!$ compared to $N^2/2$ for the linear memory. The available degrees of freedom of the system increase by a factor of $N/3$. The storage capacity increases by $N/3$ also as the following signal to noise ratio analysis demonstrates.

We assume that the vectors $f^{(m)}(j)$ are binary valued $(+1,-1)$ and

$$E\{f^{(m1)}(j1)f^{(m2)}(j2)\}=\delta_{m1,m2}\delta_{j1,j2}$$

where $\delta_{n,m}$ is the Kronecker delta function. Now let us take the input to the quadratic memory to be the m0-th stored vector. Then equation 5 can be written as follows:

$$g'(i)=N^2g^{(m0)}(i)+\sum_{m\neq m0}^{M}[\sum_{j}^{N}f^{(m)}(j)f^{(m0)}(j)]^2g^{(m)}(i)$$

The first term in the above expression is the desired recalled signal and it is equal to $E\{g'(i)\}$. The second term is interference with standard deviation approximately equal to $N(3M)^{1/2}$. The ratio of the expected value to this standard deviation is

$$SNR=N/(3M)^{1/2}.$$

A similar analysis for the linear memory yields a signal to noise ratio equal to $(N/M')^{1/2}$, where M' is the number of vectors stored in the linear memory. We obtain an estimate for the relative storage capacity of the two memories by equating the two SNR's. This yields M=M'N/3, an increase precisely equal to the increase in the number of degrees of freedom used. The computer simulations, even though limited, they are consistent with this result. For instance, for N=63 we expect an increase in storage capacity by a factor of approximately 20. If we take as 140 the number of vectors that were correctly stored in the above example, then we calculate the number of vectors that can be stored in the linear memory to be 7, which is about right.

DISCUSSION

The outer product associative memory leads to a simple implementation for the quadratic case, involving only square law nonlinearity. This is particularly significant in terms of an optical implementation[7] where this can be easily accomplished by detecting the intensity of a coherent wave. The added capacity that is available in the quadratic memory can be used for storing more data or it can be combined with the shift invariant property of an optical correlator to yield a shift invariant associative memory[8].

The storage capacity can be increased further if we consider polynomial discriminant functions of higher order. We have shown that if we consider expansions up to the Nth order of binary feature vectors, then any associative mapping is possible. This means that each of the possible 2^N input vectors can be arbitarily mapped to any chosen binary output vector. The number of nonredundant terms in such an expansion is 2^N confirming again that the available degrees of freedom determine the storage capacity.

ACKNOWLEDGEMENTS

We thank E.G. Paek and J.Hong for their help and many helpful discussions.
This work is supported by the Defense Advanced Research Projects Agency.

REFERENCES

[1] J. A. Anderson, Math. Biosci. 14, 197-220 (1972)
[2] T. M. Cover, IEEE Trans. Elec. Comp. EC-14, 326-334 (1965)
[3] J. J. Hopfield, Proc. Natl. Acad. Sci. USA 79, 2554-2558 (1982)
[4] T. Kohonen, Self Organization and Associative Memory (Springer-Verlag, N.Y.,1984)
[5] Y. C. Lee, G. Doolen, H. H. Chen, G. Z. Sun, T. Maxwell, H. Y. Lee, L. Giles, this volume.
[6] K. Nakano, IEEE Trans. Syst. Man and Cyber. SMC-2, 3 (1972)
[7] D. Psaltis and N. Farhat, Opt. Lett. 10, 98-100 (1985)
[8] D. Psaltis, J. Hong and S. Venkatesh, 625-27, SPIE, Jan 86, LA, California.

Topology Conserving Mappings for Learning Motor Tasks

H. Ritter and K. Schulten
Physik-Department
Technische Universität München, D-8046 Garching

Abstract

Topology conserving mappings play an important role for biological processing of sensory input. We suggest that principles found capable of establishing such maps can also be applied to organize the learning of motor tasks. As an example we consider the task of learning to balance a pole.

Introduction

In this contribution we like to suggest that principles found capable of establishing topology conserving mappings between sensory input and cortical brain areas can also organize the learning of motor tasks. This is demonstrated for an example, the teaching of a robot device to balance a pole. Our study is motivated by the observation that the aquisition of many higher motor skills requires a period of consciously controlled generation of the respective movements until their execution becomes automatic. We model this behaviour by a control system which consists of three components: i) a controller C with output f^C capable of solving the control task at hand and acting as a teacher for ii) an initially unorganized array A of units y , which modifies the controller's output f^C to a value f and passes it to iii) the system S to be controlled (Fig.1).

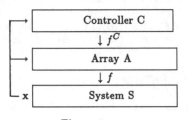

Figure 1.

Both controller C and array A receive information about the current state x of the system S and the task of the array shall be to learn gradually to take over the function of the controller by establishing a topology conserving mapping between the states of the system and suitable control responses. For this purpose we employ a modified form of an algorithm which originally goes back to Kohonen[1,2,3,4] and which we have investigated recently in the context of somatotopic mappings[5].

The Model

The evolution of the system dynamics and the topology conserving mapping will be described in discrete timesteps $t_n = n\Delta t$. The array A will play the most important role in the following. It serves two purposes: First it monitors at each time step t_n the state x_n of the system S associating a particular unit $y^* \in A$ with the current state. Second, it associates responses $f^A(y)$ with each of its units y. The response $f^A(y^*)$ of the unit y^* selected at time t_n constitutes then the force which controls the system S until one timestep has elapsed and a new system state x_{n+1} is obtained.

Let X denote the state space of the system S to be controlled, A the array of units $y \in A$ and F the space of admissible forces to control S. With each unit $y \in A$ we associate at time t_n two vectors $\phi_n(y) \in X$ and $f_n^A(y) \in F$. The force for a given state $x \in X$ shall then be given by $f^A(y^*)$, where y^* is that particular unit for which $\phi_n(y^*)$ is closest to x, i.e. $|\phi_n(y^*) - x| = \min_{y \in A} |\phi_n(y) - x|$. This prescription specifies a (discretized) map Φ_n between states $x \in X$ and control responses $f \in F$ at time t_n. The initial map Φ_0 is chosen arbitrarily, e.g. is random. This map shall now be gradually transformed into a map sending state vectors x into adequate control actions f.

The desired final map should fulfill two demands: (1) Each of the units y shall be devoted to a small subregion of X and to a response $f^A(y)$, such that neighbouring units in the array belong to neighbouring subregions in X and to similiar responses f^A. (2) The resolution of the discretized map shall be fine in those regions of the state space X which are often realized by S and may be coarser in those regions which are less frequently assumed by the system. These are just the demands met by algorithms capable of establishing topology conserving maps between a sensory source and a cortical brain area[1-5]. The role of sensory signals for such maps is played in the present application by the sequence of pairs (x_n, f^C) produced by the time evolution of the state of S and the respective action of the "teacher" C. However, for a good mapping the dimension of X should not exceed the dimension of A. Although this is not so severe a restriction for a technical application, it is a difficulty in the biological case, where one would expect A to correspond to a two dimensional neural sheet thus allowing only a mapping of the two most relevant degrees of freedom. This problem seems to be overcome in the biological organism by using several suitably interacting neural sheets or by compressing higher dimensional spaces into 2-dimensional sheets at the price of discontinuities.

The goal of our algorithm is the gradual refinement of the map Φ_n from an initial random choice to a state where Φ_n can take over the control of the system S. For this purpose we suggest the following refinement procedure for Φ_n at each time step n:

1) Search for unit y^* with

$$\|\phi_n(y^*) - x_n\| = \min_{y \in A} \|\phi_n(y) - x_n\| \tag{1}$$

where x_n is the state of S at time t_n.

2) Update Φ_n via

$$\begin{aligned}
\phi_{n+1}(y) &= \phi_n(y) + h(y - y^*, t_n) \cdot (x_n - \phi_n(y)) \\
f^A_{n+1}(y) &= f^A_n(y) + h(y - y^*, t_n) \cdot (f^C - f^A_n(y))
\end{aligned} \tag{2}$$

where f^C is the output of the controller C for state x_n and $h(y, t)$ is a function of Gaussian type centered at zero in its first argument and of width and amplitude decreasing with increasing second argument t.

3) Act upon S with a control force

$$f = \alpha(t_{n+1}) f^A_{n+1}(y^*) + (1 - \alpha(t_{n+1})) f^C \tag{3}$$

until the next time step. Here $\alpha(t)$ is a function which gradually increases from $\alpha = 0$ at t=0 to $\alpha = 1$ at the end of the learning.

Steps 1) and 2) have been shown to lead to a topology conserving map if the pairs (x_n, f^C) can be considered as a series of independent stationary random variables[1,2,4]. However here the developing map itself feeds back onto the source of its input by controlling the time evolution of S. This can modify the precise dynamics of the evolution of the map, but our simulations suggest that the properties of this process to converge to the topology conserving map remain preserved.

The advantage of the above algorithm over a control rule given by a fixed table of values ϕ and corresponding forces f^A lies in its capability to distribute the pairs (ϕ, f^A) over the space $X \otimes F$ with regard to the density of control actions required by the given motor task. This adaptation occurs automatically in the course of learning without the need for prior knowledge. Furthermore the spreading of the local adjustments into the immediate neighbourhood of a selected unit y^* in step 2 of the above refinement procedure brings about the topology conserving property of the resulting map and can be considered as a rudimentary form of generalization which facilitates convergency. The values f^A need not necessarily specify directly a control force. They can as well be input parameters of a lower level control law which serves to calculate the proper response actions from the parameter f^A and the system state x. From the view of the algorithm this amounts to replacing the original system S by a new system S', which is the concatenation of this control law and S.

In the following we will show the results of the algorithm for the case of learning to balance a pole. Here the system S represents a massless rod with two point masses 1 and m attached to its upper and lower end, respectively. The rod's motion is restricted to a vertical plane with the mass m confined to glide along the horizontal x-axis and gravity pointing downward along the negative z-axis. θ denotes the counterclockwise angle subtended by the rod and the positive x-axis and balancing can be achieved by exerting a horizontal force f upon the mass m (Fig.2). The pole motion was simulated in time steps of 0.1s.

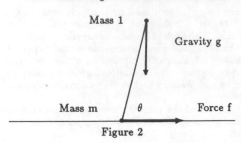

Figure 2

The state vector of the system S was represented by the last two successive pole inclinations, i.e. $x_n = (\theta_n, \theta_{n-1})$. The force delivered by the controller was given by

$$f^C = const \cdot (1.4 \cos \theta - 0.1 \dot{\theta}). \tag{4}$$

A two dimensional square array of 25x25 units was assumed for A. Learning was achieved by a series of trials, each trial starting with an initial value for θ drawn randomly from the interval $[60^o, 120^o]$ and lasting until either θ had left the interval $[0^o, 90^o]$ or 60 timesteps had elapsed.

Figures 3–6 below show the development of the map Φ_n during the simulation. Each of the left diagrams shows the distribution of the units y in the state space X of pairs of successive pole inclinations θ_n, θ_{n-1}. Each unit y is depicted at the location in X given by its associated vector $\phi_n(y)$. In order to illustrate the emerging topology conserving character of Φ_n we have connected those points $\phi_n(y^{(1)}) = (\theta_n^{(1)}, \theta_{n-1}^{(1)})$ and $\phi_n(y^{(2)}) = (\theta_n^{(2)}, \theta_{n-1}^{(2)})$ for which $y^{(1)}$ and $y^{(2)}$ are neighbouring units in A. The increased order of these connections in going from Figure 3 (initial map Φ_0) to Figure 6 (final map Φ_n, n=10 000) attests to the emerging topology conserving property of the $\phi_n(y)$.

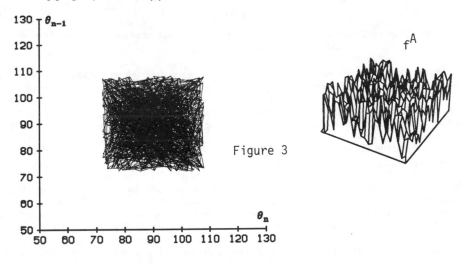

Figure 3

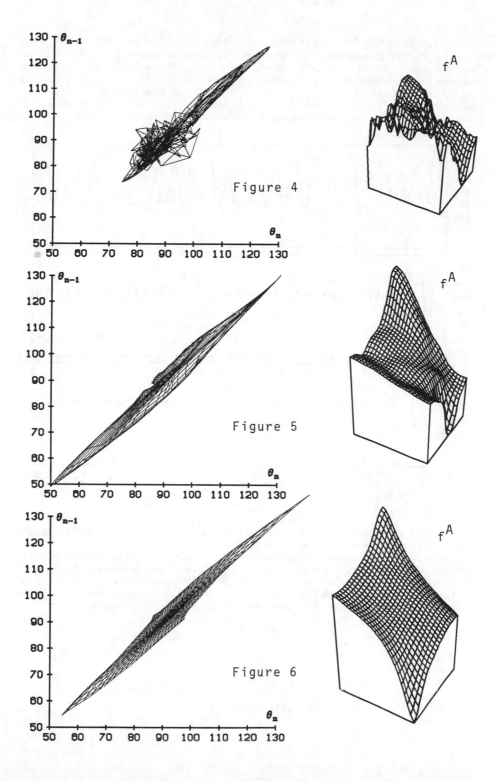

Figure 4

Figure 5

Figure 6

380

The right diagrams in Figs. 3–6 show the value of the force $f^A(y)$ as a mesh surface above the array A. Figure 3 presents the initial state of Φ. The values for both $\phi(y)$ and $f^A(y)$ for each unit y were chosen randomly from fixed intervals. The following figures 4 and 5 show the gradual development of the map after 100 and 1000 timesteps respectively. The units rapidly begin to "tune in" along the line $\theta_n = \theta_{n-1}$, which reflects the fact that successive pole inclinations θ usually differ only by a small amount and which therefore constitutes the essential region of X for the problem. Finally Fig.6 shows the asymptotic state reached after 10 000 time steps.

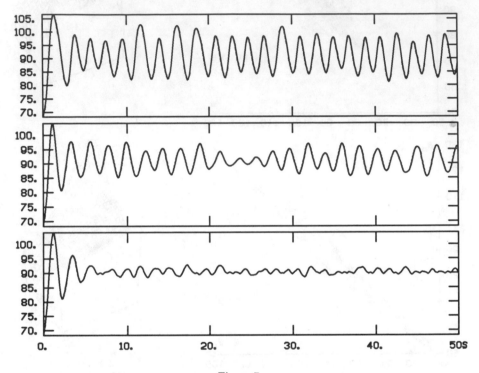

Figure 7

A smooth mapping has evolved and the controller now may be disposed of without significantly affecting the balancing performance of the system.

Actually the array is capable of performing the balancing task on its own considerably earlier, but at a correspondingly reduced level of performance. This is shown in Fig.7, which compares the time evolution of the inclination θ for a pole initially inclined at an angle of $\theta = 70°$ under the control of Φ after 1000, 5000 and 10 000 time steps. Each diagram shows $\theta(t)$ for the first 50 seconds after the release of the pole with the controller C being disabled.

REFERENCES

1. T.Kohonen, Biol.Cybern. 43, 59-69 (1982).
2. T.Kohonen, Biol.Cybern. 44, 135-140 (1982).
3. T.Kohonen, Proc. 6th Int. Conf. on Pattern Recognition 114-128 (1982).
4. T.Kohonen, Self-Organization and Associative Memory (Springer, Berlin 1984).
5. H.Ritter, K.Schulten, Biol.Cybern 54 (1986) (to appear).

AN ARTIFICIAL NEURAL NETWORK INTEGRATED CIRCUIT
BASED ON MNOS/CCD PRINCIPLES[+]

J. P. Sage, K. Thompson, and R. S. Withers
Lincoln Laboratory, Massachusetts Institute of Technology
Lexington, MA 02173.

ABSTRACT

This paper describes the design principles for an implementation of an artificial neural network (ANN) in the form of a silicon integrated circuit based on charge-coupled device (CCD) and metal-nitride-oxide-semiconductor (MNOS) technologies. The significant features of this design are: (1) the synaptic coupling strengths stored in the MNOS devices can take on continuous, analog values and (2) the synaptic weights can be reprogrammed at any time under electrical control and in response to conditions in the network. These features should make possible ANNs that are capable of dynamic, in situ learning.

INTRODUCTION

In this paper we describe the basic design principles behind a silicon integrated circuit implementation of artificial neural networks (ANN). These principles, which can be used to implement a wide range of ANN types, are illustrated by a test chip now in fabrication that realizes a simple network of the Hopfield type [1]. The design is based on two concepts: charge-coupled device (CCD) charge packets to represent analog information transmitted through a synapse and metal-nitride-oxide-semiconductor (MNOS) device structures to store electrically changeable nonvolatile analog synaptic weighting values.

CCD technology is well developed. It is the dominant technology used to make solid-state visible imagers and is also used in high-speed analog signal processing devices for functions such as matched filtering and correlation [2]. MNOS technology is also not new. MNOS electrically erasable programmable digital read-only memories (EEPROMs) have been produced commercially, and at Lincoln Laboratory we have demonstrated a technique for storing analog information in MNOS structures in an accurate, well-controlled way [3].

The design of our artificial neural network incorporates the following significant and unique features:

(1) Electrical Programmability: The synaptic coupling strengths are not built into the device at the time the chip is fabricated, nor are they or programmed into the device in a single irreversible operation. New values for the weights can be established under electrical control and stored at any time. This capability is essential if artificial neural circuits are to perform dynamic, in situ learning.

(2) Analog Storage: The synaptic coupling strengths are not limited to binary or ternary values but can take on continuous values over a wide dynamic range (equivalent to 4 to 8 bits). Although some neural network algorithms can work with binary weights, the information storage density and the performance are superior with analog weights. Some algorithms cannot function effectively

---

[+] This work is supported by the Department of the Air Force.

382

unless analog weights are available. In particular, adaptive learning generally requires incremental adjustment of the weights.

ARCHITECTURE

The essential features which characterize what is called a neural net are: (1) a relatively large number of relatively simple computational elements and (2) a very high degree of direct interconnection between these computing elements. By analogy with biological neural networks, we will refer to the computing elements as neurons, the signal lines that carry output from one neuron to other neurons as axons, and the site at which one neuron receives and processes input from the axon of another neuron as a synapse.

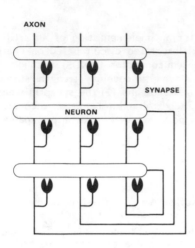

Fig. 1. Neural network schematic.

Our integrated circuit test chip implements a network of the form shown in Fig. 1, in which there is complete pair-wise interconnection between the neurons. The ovals represent the neurons; the vertical signal lines, the axons; and the tulip-shaped patterns, the synapses. The axons are designed to carry binary information from the neurons, which, in turn, make their decisions by summing linearly the analog-weighted inputs arriving via the synapses and comparing that sum to a threshold value.

STRUCTURE AND OPERATION

There are many ways to use CCD/MNOS techniques to implement artificial neural networks. We will present here one very simple implementation that illustrates the basic principles. First we will describe the overall physical realization, and then we will cover the details of its operation.

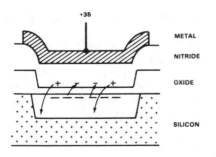

Fig. 2. MNOS structure.

The substrate material is p-type silicon. The neurons consist of electrically conductive n-type diffusions which, during a readout operation, receive charge packets from the synapses using CCD charge-transfer techniques. The axons are conductive lines that carry electrical signals to the synapses and enable or disable the formation of charge packets there. The size of the charge packets, when their formation is enabled by the axon, is controlled by the state of the MNOS synapse.

An MNOS device is shown in cross section in Fig. 2. It is similar to the pervasive MOS (metal-oxide-semiconductor) device except that (1) the main gate insulator is silicon nitride and (2) the silicon oxide insulating layer is so thin (25 Å) that, at gate voltages of approximately +35 V or -35 V, electrons and holes can move by quantum-mechanical tunneling between the underlying silicon and long-lifetime traps in the nitride layer. Fig. 2 shows a high positive voltage on the gate causing a net negative shift in the charge stored in the traps. When the gate voltage is kept below 10 V, as in normal operation, virtually no tunneling occurs, and the trapped charge is essentially permanent. The effect of the trapped charge is to make the voltage on the gate appear to be higher or lower than it actually is as far as carriers in the silicon are concerned. This apparent modulation of the gate voltage is used to control the size of the synaptically transfered charge packets.

We do not have the space here to explain in detail the operational procedures for changing the synaptic weights by adjusting the amounts of trapped nitride charge in the MNOS devices. We will simply state that CCD charge-metering techniques are used to produce appropriate charge packets in the silicon and that the charge in these packets is transferred by tunneling into the nitride traps. The circuit in our test chip is able to perform the Hebbian calculation $V_i \cdot V_j$, where V_i is the binary state of neuron i and V_j the binary state of neuron j for a state pattern V to be learned. We expect the chip to be incrementally programmable so that additional states can be stored at any time without having to erase and reprogram the entire memory.

We will now explain how a simplified version of the circuit is operated to perform one cycle of a Hopfield-type readout. This readout, in its simplest form, requires the following mathematical operation:

$$N_i = \text{sign} \left[\sum_j S_{ij} \cdot A_j \right]$$

where N_i is the next state of neuron i, A_j is the current state of axon j (i.e., the present output from neuron j), and the S_{ij} are the synaptic weighting coefficients.

Fig. 3 shows a plan view of one synaptic site in an MNOS/CCD neural network. This site bridges neuron i, which is connected to all synapses in the row, and axon j, which is connected to all synapses in the column. Neuron i receives charge which reaches diffusion D_{out} across control gate G_{out} from synapse S_{ij}. All G_{out} gates on the chip are connected to a single control line G, and all synapse bias gates S_{ij} are connected to a single line S. Charge needed to form a packet under gate S_{ij} is injected from diffusion D_{in} across control gate G_{in}. All diffusions D_{in} on the chip are driven from control line D. The input control gate G_{in} is connected to axon j. If the voltage on A_j is high (active axon, $A_j=1$), charge is allowed to flow from

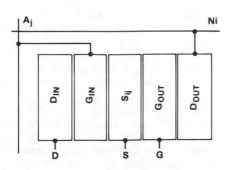

Fig. 3. Synaptic cell in network.

D_{in} to S_{ij}, where a precise quantity is metered out for later transfer to D_{out}; if the voltage on A_j is low (inactive axon, $A_j=0$), charge flow from D_{in} is blocked, and N_i will receive no charge from this synapse. Line N_i accumulates the charge from all synapses connected to it, and a sense circuit compares this charge to a threshold value to determine the state of A_i for the next cycle of the circuit.

Fig. 4 shows the readout operation in detail. A cross section of the circuit is shown at the top of the figure. Charges drawn in the insulator under gate S are meant to represent the trapped charge stored in the nitride layer. The rest of the figure shows potential diagrams and charge densities at various steps in the readout operation. Line A shows the initial condition. D_{out} (N_i) has been biased very positive (deep potential well), gate G has been turned off to isolate the synapse region from the output diffusion, and the input diffusion D_{in} is held at a high positive voltage.

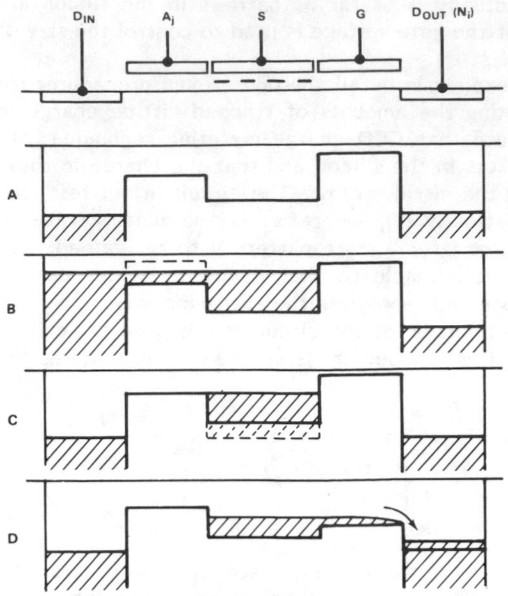

The readout cycle begins (line B) when the voltage on D_{in} is pulsed low, raising the charge level and filling the synaptic region with charge. The dashed line under A_j shows the potential for an axon that is inactive. In that case no charge flows into the region under gate S. After the pulse on D_{in} is finished (line C), any excess charge in the synaptic region flows back to D_{in}, leaving a neatly metered packet of charge under the MNOS gate S. The depth of the well, and hence the amount of charge in the packet, depends on the empty-well potential under gate S. Although all synapse gates have the same applied voltage, if the trapped nitride

Fig. 4. CCD clocking sequence during readout.

charge is more positive, then the effective gate bias is more positive, the well will be deeper, and the charge packet will be larger. This is indicated by the dashed potential curve. The final step (line D) is to transfer the charge packet to the neuron (D_{out}). First, the diffusion D_{out} is disconnected from its bias source so that its potential is floating. Next, by turning gate G on and then (not shown) turning gate S off, any charge under the MNOS gate will flow onto the neuron, causing its potential to decrease. This change is sensed by a high-impedance detector.

The simple readout scheme described here assumes axon logical values of 0 and +1 and cannot handle alternative values of -1 and +1. Nor does it account for bipolar values of the synaptic weighting coefficients. The actual circuit we are building on our test chip achieves these functions using a more complex unit cell with differential inputs and outputs.

CONCLUSIONS

We have described how CCD and MNOS techniques can be used to implement an artificial neural network with analog synaptic weights that can be reprogrammed under electrical control. New states of the neural system evolve synchronously at the CCD clocking rate, which can easily be in excess of 10 MHz. Reprogramming of the MNOS synapses in response to conditions within the network can be used to implement in situ adaptation and learning algorithms.

REFERENCES

1. J. J. Hopfield, Proc. Natl. Acad. Sci. USA, 79 (Biophysics), 2554 (1982).
2. S. C. Munroe, IEEE MILCOM Conf. Proc., (1982).
3. R. S. Withers, R. W. Ralston, and E. Stern, IEEE Elec. Dev. Let., EDL-1, 42 (1980).

FORGETTING AS A WAY TO IMPROVE NEURAL-NET BEHAVIOR

Richard J. Sasiela

M.I.T. Lincoln Laboratory, Lexington, Massachusetts 02173-0073

ABSTRACT

The output of neural nets of the Hopfield type favors some states more than others. Spurious states which are not in the learned set are also produced. It has been shown that by exciting the input of the net with random states and then "unlearning" the output state that the frequency of occurrence of the desired states becomes more uniform and that of spurious states is reduced. This has been proposed as one of the purposes of dreams.

A mathematical model for the behavior of neural nets is presented. The eigenvalues of the matrix of neural connections are examined when unlearning takes place. It is shown that output states with the higher eigenvalues will occur more frequently. The unlearning algorithm preferentially reduces the larger eigenvalues, thereby, making the eigenvalues of the learned states closer in size. This tends to equalize the frequency of occurrence of the learned states. Spurious states have their eigenvalues reduced to less than unity and, thereby, decay in the iteration process of recall. In a "learning" mode the desired states can be reenforced and become more accessible.

INTRODUCTION

Hopfield, Feinstein, and Palmer[1] conducted computer experiments in which they trained a neural net to store certain patterns. They found that if a random pattern was presented to the input, the output would be one of the learned states 60% of the time and a spurious states 40% of the time. They then used the output state from the random inputs to unlearn that state. They found after several hundred iterations of this unlearning process that the frequency of occurrence of the spurious states had been reduced to less than 10%. Too many iterations of the unlearning process caused the trained memory to deteriorate.

Crick[2] has proposed that in REM (rapid eye movement) sleep the memory is being excited by a random input and the output state is used to unlearn it. The purpose of this process is to consolidate the more recent memories into the associative memory of the brain by eliminating spurious states and making the other memories more equally accessible.

This process is examined mathematically to gain some insight into this unlearning process. The information contained in the system is stored in the matrix values which represent the connection strength between neurons. It is found that the size of the matrix eigenvalues are very important in determining the frequency of occurrence of a given output state. Unlearning has the property of reducing the eigenvalues of the matrix. Since accessibility is related to the size of the eigenvalues, the states associated with the higher eigenvalues will be reduced more quickly than those with smaller eigenvalues. Thus eigenvalues of stored states will converge to each other, thus making those states more equally accessible.

The many spurious states, which each occur infrequently, have eigenvalues which are smaller than those of the desired states. Reducing these eigenvalues further in the unlearning process reduces them to values under unity so that these states decay in the iterative process of recall. The net effect is that the learned states even though they grow more slowly in the iterative process will be more accessible.

The unlearning process, since it reduces all eigenvalues will eventually reduce those of the desired states to less than unity so that they will no longer be accessible. This explains why the unlearning process is eventually unstable - too much unlearning destroys the desired memory.

In the reenforcing process, the output of the memory is used with the opposite sign of that in the unlearning process with the net effect that the eigenvalues are increased in size. If a certain input appears more frequently than any other, then it will have the largest eigenvalue and will, therefore, appear with greater frequency on the output if a random input is applied. This means that frequently presented inputs will be recognized with more noise or with less information than other inputs. This behavior is desirable and has physical correlates.

EFFECT OF UNLEARNING ON THE MEMORY MATRIX

In all theories of associative memories, in which the information is stored in the connection strengths between system elements, one can recall the entire stored memory vector from a partial or noisy input vector. In order to do this categorization process, one must invoke some nonlinear process to decide which of the stored vectors the input is most likely to be. In a simple minded way one could just take a dot product of the input vector with each of the stored vectors and then select the largest output as the correct one. One can establish a threshold which if it is not exceeded results in a "not known" category. This process is done in a more roundabout way in most associative net models. The reason for this is that the mechanisms are thought to better model the behavior of biological systems and also to better fit into a framework in which learning can take place.

Anderson[3] in his "brain in the box" model performs this non-linear selection process by iterating the input through the connection matrix and then feeding it back to its input. Some components of the output vector will grow in time and some will decay. The maximum size of components of the output vector is clamped to a certain value, and after a sufficient number of iterations the output vector is in a corner in n-dimensional space, hence the name "brain in the box" model.

In Anderson's model the initial growth is linear with the non-linear clamping process causing the output to go to a corner of the box in which the linear trajectory was passing close to. Therefore, one can obtain insight by examining the behavior of the system in the linear regime. The detailed examination of the nonlinear behavior is much more difficult analytically and it will not be done here.

To build an associative net which works well, i.e., gives the desired output when the input is degraded by noise or deletions of some of the vector components, requires one to construct an appropriate matrix. Anderson constructs this matrix A by making it a sum of matrices, each of which is the outer product of one of the desired output vectors, g_i, with the noise-free input vector f_i. The matrix is equal to

$$A = \sum_i g_i f_i^T \quad . \tag{1}$$

In the Hopfield formulation the output state g_i is equal to the input state.

It is easy to see that this matrix would produce the desired output if all the f vectors were orthonormal, then the output vector is

$$g = Af \quad . \tag{2}$$

Normally, the set of input vectors will not be orthonormal and the output will not be the desired output vector. Anderson provides a means to train the matrix in that he modifies this initial starting matrix by using the Widrow-Hoff iteration scheme. The set of inputs is applied sequentially and after each application the A matrix is modified by ΔA which is equal to

$$\Delta A = \eta(g - Af) f^T \tag{3}$$

where η is a small number.

After many iterations the resulting A matrix has the desired properties.

In the unlearning algorithm, a random input is applied and after the output has settled to g, the A matrix is changed by

$$\Delta A = -\eta \ gg^T \quad . \tag{4}$$

These memories have been shown to work best when the number of stored states is less than 15% of the dimensionality of the matrix.

We will examine the effects of unlearning by contructing a matrix which would have the ideal properties and then using this to show the effects of unlearning on the Hopfield matrix. Consider M linearly independent input vectors, f_i, with N components. Form M orthonormal vectors x_1, x_2,...,x_M by a linear transformation on $\{f_i\}$

$$x_i = Hf_i \quad . \tag{5}$$

Form the orthogonal matrix, S, where the M vectors of x are the first M columns and the remaining N-M columns are a set of orthonormal vectors which complete the basis set. An A matrix will be contructed from S and its properties will be examined. Consider

$$A = S \Lambda S^T \tag{6}$$

where Λ is a diagonal matrix. Booause of the construction of S its transpose is equal to its inverse, i. e.

$$S^{-1} = S^T \quad . \tag{7}$$

The elements of Λ are the eigenvalues of the matrix A. Using A in an iteration process in the linear regime results in

$$g = A^R f = S\Lambda^R S^T f \quad . \tag{8}$$

If the eigenvalues of the last N-M eigenvalues are chosen to be zero, then the projection of the output vector onto any of these vectors will be zero after one iteration. If the eigenvalues of the learned vectors are equal, then the eigenvector which had the largest projection on the input vector will give the largest output. The Λ matrix with these properties is equal to

$$\Lambda = \begin{bmatrix} a & & & & \\ & a & & & \\ & & a & & \\ & & & \cdot & \\ & & & & o \end{bmatrix} \tag{9}$$

and the matrix A is semi-positive definite.

The initial matrix set up by Hopfield is not equal to this ideal matrix. The eigenvalues of the learned states can differ in size, and the eigenvalues of the spurious states can be greater than unity, although, they will be less than the eigenvalues of the desired states.

Consider the result of transforming a random vector input with the matrix. Assume that the output is an eigenvector x_s. The argument can be presented even if the output is not an eigenvector, but it is more complicated. The new A matrix becomes

$$A+\Delta A=A-\eta x_s x_s^T=SAS^{-1} -[00..0,x_s,0..0]\begin{bmatrix} o & & & & \\ & o & & & \\ & & \cdot & & \\ & & & \eta & \\ & & & & \cdot \\ & & & & & \cdot \\ & & & & & & o \end{bmatrix}\begin{bmatrix} o \\ o \\ \cdot \\ x_s \\ \cdot \\ \cdot \\ o \end{bmatrix} \quad (10)$$

This can be written as

$$A + \Delta A = SAS^{-1} -S\begin{bmatrix} o & & & & \\ & o & & & \\ & & \cdot & & \\ & & & \eta & \\ & & & \cdot & \\ & & & & \cdot \\ & & & & & o \end{bmatrix}S^{-1}= S\begin{bmatrix} \lambda_1 & & & & \\ & \lambda_2 & & & \\ & & \cdot & & \\ & & & \lambda_s-\eta & \\ & & & & \cdot \\ & & & & & \cdot \\ & & & & & \lambda_N \end{bmatrix}S^{-1} \quad . (11)$$

Therefore, the unlearning process reduces the eigenvalue of the vector. If the eigenvalue were slightly greater than unity then this process could reduce it to be less than unity and it would decay in the iteration process given in Equation 8. For the desired output vectors, the ones with the largest eigenvalues will occur most frequently with a random input and their eigenvalues will be reduced the largest number of times. In this fashion the eigenvalues will approach each other and the learned vectors will tend to occur more equally in frequency with a random input.

Note that one is always reducing all eigenvalues and eventually the eigenvalues of the desired states will be reduced to less than unity and the memory of these vectors will be destroyed.

In the learning mode, Equation 4 is used with the opposite sign and the occurrence of a vector will increase the eigenvalue. Inputs that occurred most frequently would have the largest eigenvalues and they would capture input vectors over the largest volume of the N-dimensional space. After a sufficient learning period, if the most learned input vector is presented with errors or deletions there may still be enough of a projection along the noise-free learned vector that the large eigenvalue causes this vector to swamp the others and it would be the output vector. This type of reenforcement is desirable and it is observed in biological systems.

CONCLUSIONS

A method has been shown to construct a connection matrix which has the desirable properties of producing a learned output vector which is the closest to an input vector and of suppressing output vectors which are not in the learned set. It was shown that if one starts with a connection matrix like that of Hopfield that the unlearning process will tend to equalize the frequency of the desired outputs and suppress spurious outputs.

This matrix also has the property that in the learning mode the more often a vector is presented to the system the greater the number of errors or deletions allowed in the input vector while still producing the correct output vector.

REFERENCES

1. J. J. Hopfield, D. L. Feinstein, & R. G. Palmer, "Unlearning" has a stabilizing effect in collective memories, Nature Vol. 304, 158-159 (14 July 1983).
2. F. Crick & G. Mitchison, The function of dream sleep, Nature Vol. 304, 111-114 (14 July 1983).
3. J. A. Anderson, Cognitive and phychological computation with neural models, IEEE Transactions On Systems, Man, And Cybernetics, Vol. SMC-13, No. 5, 799-815 (September/October 1983).

This work is sponsored by the Department of the Air Force.

"The views expressed are those of the author and do not reflect the official policy or position of the U.S. government."

PARALLEL ANALOG NEURAL NETWORKS FOR
TREE SEARCHING

Janet Saylor and David G. Stork
Department of Physics & Program in Neuroscience
Clark University, Worcester, MA 01610

ABSTRACT

We have modeled parallel analog neural networks designed such that their evolution toward final states is equivalent to finding optimal (or nearly optimal) paths through decision trees. This work extends that done on the Traveling Salesman Problem (**TSP**)[1] and sheds light on the conditions under which analog neural networks can and cannot find solutions to discrete optimization problems. Neural networks show considerable specificity in finding optimal solutions for tree searches; in the cases when a final state does represent a syntactically correct path, that path will be the best path 70 - 90% of the time – even for trees with up to two thousand nodes. However, it appears that except for trivial networks lacking the ability to "think globally," there exists no general network architecture that can strictly insure the convergence a state that represents a single, continuous, unambiguous path. In fact, we find that for roughly 15% of trees with six generations, 40% of trees with eight generations, and 70% of trees with ten generations, networks evolve to "broken paths," i.e., combinations of the beginning of one and the end of another path through a tree.

Tree searches illustrate well neural dynamics because tree structures make the effects of competition and positive feedback apparent. We have found that 1) convergence times for networks with up to 2000 neurons are very rapid, depend on the gain of neurons and magnitude of neural connections but not on the number of generations or branching factor of a tree, 2) all neurons along a "winning" path turn on exponentially with the same exponent, and 3) the general computational mechanism of these networks appears to be the pruning of a tree from the outer branches inward, as chain reactions of neurons being quenched tend to propagate along possible paths.

THE OPTIMIZATION PROBLEM AND METHOD

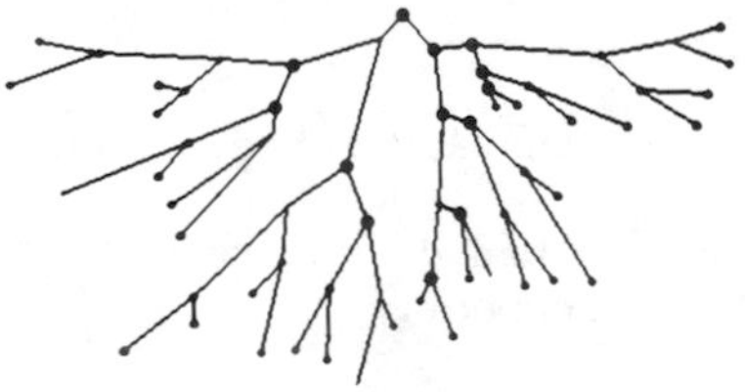

Figure 1: A decision tree with branching factor **n** = 2

A decision tree (Cayley tree, or simply, tree [2]) is a set of nodes arranged in generations with parents and "decendents" (Figure 1). The "root node" (at the top) is connected to **n** decendent nodes; we call **n** the "branching factor." Each node is in turn connected to **n** neurons in the next generation, and so on. The length of the lines linking the nodes represents the "cost function" – the cost for traveling between the two connected nodes. (In all cases, our costs were chosen randomly in a finite bound.) Our version of the tree searching problem is to find a path through any tree linking the root node to an (unspecified) end node that minimizes the total cost function (the sum of the cost functions between the nodes in the final route). In this respect, tree searching is a discrete optimization problem like the **TSP**. However, while in the **TSP** every city is visited and the goal is to find the order that minimizes the total cost (i.e., total pathlength), in tree searching the order in which subsets of nodes (cities) can be visited is fixed and the goal is to find the right <u>subset</u> to visit.

Using the neural network equations of Grossberg[3] and the general approach of Hopfield and Tank[1], we map the above problem onto a network of **N** interconnected identical nonlinear amplifiers – " neurons" – each of which obey:

$$\partial u_i / \partial t = \sum_{j \neq i} T_{ij} V_j - u_i \left(\sum_{j \neq i} |T_{ij}| + 1/R_0 \right) \qquad (1)$$

Here V_i is the output voltage for the ith neuron, u_i its input voltage, R_0 the inherent resistance of each neuron (assumed equal for all neurons). The inherent capacitances are also assumed equal, and are set to unity without loss of generality. The connectivity matrix, T_{ij}, represents the interconnections between the output of neuron **i** and its input to neuron **j**:

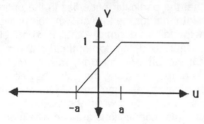

Figure 2: Input output characteristic, g(u), for each nonlinear neuron. The gain of each neuron is 1/2**a**.

Positive elements in T_{ij} represent excitatory interactions, while negative elements represent inhibitory interactions; we allow no self-interaction terms, i.e., $T_{ii} = 0$. Each neuron has a strictly saturating nonlinear input-output characteristic function, **g**, which could be of a general sigmoidal form; however, for simplicity in digital simulation we used the function shown in Figure 2. The gain of each neuron is 1/2**a**; here chosen to be 5 x 10^3. We map each of the **N** nodes in a tree (including the root node) to a different neuron – there is a one to one correspondence between nodes and neurons. The system evolves to a minimum of the Lyapunov "energy" function, **E**, given by[1]:

$$E = -\frac{1}{2} \sum_{j \neq i} T_{ij} V_i V_j - \sum_i \left\{ \int^{V_i} g^{-1}(V) \, dV \right\} \qquad (2)$$

The connection matrix should, ideally, be chosen such that all local energy minima correspond to stable states in which the neurons along a single valid path are on (i.e., saturated high) and all other neurons are off. Also, the matrix should be chosen such that the deepest energy wells – those with the largest basins of attraction – correspond to the shortest paths, so that the network will favor to <u>shortest</u> path over merely <u>short</u> paths. We briefly investigated several T_{ij} strategies before analyzing extensively the simplest: symmetric inhibitory connections between all members in the same generation – $T_{ij} = -B_G(1- \delta_{ij})$ for i and j members of the same generation, **G** – analogous to the inhibitory matrix elements used by Hopfield and Tank to require that in a stable state there be one and only one neuron on in every row. Our "data" terms of T_{ij} were porportional to a constant minus the length (cost) of the path segment between connected neurons **i** and **j**, and zero for unconnected neurons. These excitatory connections tended to force neurons in a short path (i.e., connected by small costs) to saturate high, while the inhibitory connections helped to insure that the final stable state could be interpreted as a meaningful route. The matrix just described does not completely satisfy the criteria established above: in addition to the desired energy minima, this matrix yields many spurious local energy minima which correspond to combinations of pieces of different paths, as we shall see below.

DYNAMICS OF THE NETWORK

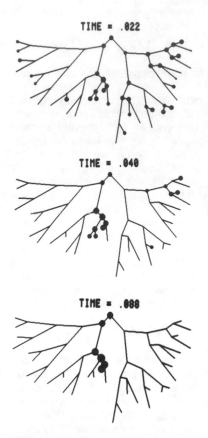

TIME = .022

TIME = .040

TIME = .068

Figures 3-5: Time evolution of a network finding an optimal path from a symmetric starting state

Figures 3-5 show the evolution of a network having the type of connectivity matrix described above. The starting state was one in which the output voltages of all neurons were equal. The size of a circular dot at any node is a function of the voltage of the corresponding neuron. We can think of this voltage as representing the tentative truth value for the proposition that the particular node lies in the optimal path. Note that for the tree shown, the network indeed found the optimal – that is shortest – route. Neurons along non-optimal paths have been turned off one by one; this quenching tends to propagate along the paths having long segments (large cost functions). More precisely, a neuron is turned off when the inputs from its positive "data" connections are overwhelmed by the inhibitory input from other neurons in the same generation. We have proven analytically and verified in simulations that the voltage on any neuron can be written as the sum of three terms: 1) a constant, **a**, related to the inverse gain, 2) a "short-term," exponential, ϵ_G, whose time constant is determined by the inhibitory connections in its generation and 3) a "long-term" fanning out component, ϵ_i, whose specific evolution is determined by the particular excitatory connections of neuron i. The output voltage, $V_i(t)$, the sum of a, $\epsilon_G(t)$ and $\epsilon_i(t)$, is shown in Figure 6, for all neurons in the second to last generation of a six generational tree (G=5): the voltages on the neurons start at zero, and quickly evolve toward **a**, after which they split (as determined by the data terms) until ultimately one neuron dominates the generation, and increases exponentially until saturation. (Had we used a sigmoidal amplifier characterisitcs, this increase would be even faster than exponential.) We find experimentally that for a syntactically correct, "winning" path, all neurons along that path are governed by the same exponent, even though their particular excitatory terms differ.

From starting states of all neurons having the same voltage, our networks converge to saturated, stable states in times that are approximately the same for all trees, independent of the branching factor or number of generations. For instance, in our model with a neural gain of

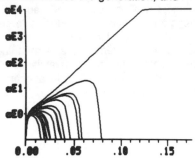

Figure 6: Output voltages, $V_i(t)$, for all neurons in the fifth generation of a tree

5×10^3, trees with 100 neurons converge to stable states in only 0.1 to 0.2 RC time constants, while trees with 2000 neurons converge to stable states in 0.2 to 0.8 RCs – only two to four times slower. In our model, effective neural resistances, and thereby the effective neural time constants, vary from generation to generation according to:

$$1/(\text{time const.}) = 1/R_i = \sum_{i \neq j} |T_{ij}| + 1/R_o \qquad (3)$$

As a result, we arbitrarily chose as our unit of time the RC for a neuron that has 100 inhibitory connections. One reason that the convergence times for our networks are shorter than those observed by Hopfield and Tank[1] for the **TSP** is that in our model neurons are ideal amplifiers with constant gain over the whole region in between -**a** and +**a** and with unphysical discontinuities in gain at those two input voltages. A real device with an average gain of 5×10^3 would have flatter input-output curve at small voltages, where much of the evolution of the system takes place (effective gain = 200 when $V_i = 0.01$). Therefore, our model converges at least ten times faster than would a real system with a sigmoidal g(u), even though the qualitative dynamics of competition is virtually identical for these two gain functions.

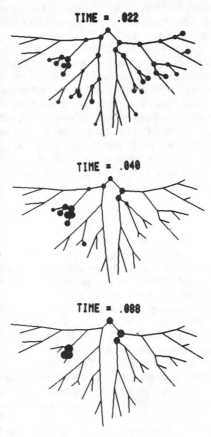

TIME = .022

TIME = .040

TIME = .088

Our networks for tree searching do not always evolve to syntactically meaningful states. Figures 7-9 show the evolution of a network to a "broken path." For the lack of a suitable inhibitory structure, our network architecture does not implement explicitly the requirement that a final state represent a single, complete and connected path. A network seems to evolve and find its final state through the propagation of chain reactions of neurons turning off or on along paths through the tree. When, along the same route, a chain reaction of neurons being quenched meets a chain reaction of neurons turned on, the result is generally such a broken path. Additional inhibitory inter-generational connections do not alleviate this problem effectively, but only result in the skipping of one generation between the end of one "partial path" and the beginning of the other.

However, there exists at least one "quick fix" for the problem of broken paths. If the inhibitory connections throughout the system sufficiently are sufficiently low, all significant broken pieces will develop into full paths, with virtual certainty that the best path will be one of the two or three chosen. This approach is reasonable for finding solutions for tree searches because in trees any superposition of paths is seperable. In contrast, in the case of an optimization problem

Figures 7-9: Time evolution of a tree finding a "broken path" from a symmetric starting state

with a combinatorial solution space, such as the **TSP**, two separate final states can not be retrieved after being combined into a single state.

In tree searching, when a state does represent a syntactically correct path, that path will be the optimal (shortest) path 70 - 90% of the time — even for large trees with up to two thousand nodes. Often a network will show considerable specificity in choosing the shortest path over those which are only a few percent longer. On the other hand, sometimes a network will evolve to a state which represents the tenth or twentieth best path through the tree — or worse. In these cases, we have found that there is almost always a recognizable and revealing reason for the failure such as: 1) If two of the shortest paths both have their shortest segments between the same generations, then their "splitting the prizes," may let a third path win overall — a path that may be much longer but has short segments (low cost functions) between other, less competetive generations. 2) If we start with the root neuron (node) saturated high and all other neurons off, then the chain reaction started by that root neuron will tend to propagate in a top-down, "greedy," non-parallel, "non-globally thinking" manner. At each decision point, the neuron along the shortest segment will be turned on the most with no look ahead to the next generation. The result is a path which may or may not be the shortest[2]. We observe the same top-down evolution in cases when, because of competition, neurons near the bottom end of path may, during the early evolution of the tree, have their output suppressed compared to the neurons nearer to the root node. In this case, top-down evolution may start from the middle generations of the optimal path. 3) There exists second-order feedback between nodes (siblings) in the same family through the (shared) parent node. This interaction will tend to promote a path in a particular subtree with many other short paths, over an optimal path which is buried in a subtree with many long paths. This effect can in part, but not completely, be canceled by letting the inhibitions between "sister " neurons, (i.e., those with the same parent) be a function of the two step chain of positive matrix elements between them. This problem of constructive and destructive interference between neurons along related paths increases in severity with the branching factor of a tree, and is a limiting probem for trees with a branching factor greater than four.

CONCLUSIONS

Our purpose in examining tree searches was to study the dynamics of neural networks for computation in a different and hopefully clearer context than the **TSP**. In this respect our efforts have been rewarding despite the fact that because of "broken paths" the success of the algorithm itself deteriorates with tree size. We have seen that the evolution of neural voltages in a tree, includes: a short term uniform convergence toward the voltage, **a**; a relative fanning out of voltages in a generation; and a pruning of the tree, as the behavior becomes exponential. The convergence times for our networks are rapid and almost completely independent of the size of the network. The long term behavior of all neurons along a given "winning" path seems to be governed a single exponent, even though their excitory connections differ. And, the computational mechanism of these networks seems to involve chain reactions along paths of neurons being quenched and saturated.

The **TSP** and tree searching are closely related in that all paths in both systems have the form of chains of mutually excitatory neurons. (In **TSP** the chains are closed loops.) As a result, we feel that it will be interesting to learn more about how our understanding of the dynamics of networks for tree searching pertains to the dynamics of networks for combinatorial optimization problems. In particular, it may be possible that there is a fundamental difference between networks with one dimension of inhibitions (the rows in tree searching) and two (the rows and columns in **TSP**).

In order to break the symmetry that arises from the 2N degeneracy in the possible representation of any tour through **N** cities, Hopfield and Tank[1] began their simulations from random starting states. The success of their algorithm depended on the probability that any starting state will be in the basin of attraction of one of best solution. We have approached tree searching somewhat differently. Instead of considering whether the optimal path through a tree has a broad basin of attraction (in fact we know that the energy landscape for our network, has many tightly packed local minima), we focused on the dynamics of these networks, which deterministically find optimal solutions for small trees (<256 neurons) from symmetric starting states. In the future, we hope to understand these dynamics better, compare them to those for the **TSP** and investigate their accuracy for different (i.e., random or non-symmetric) starting states[4].

REFERENCES

[1] Hopfield, John J. and David W. Tank "'Neural' Computation of Decisions in Optimization Problems" Biol. Cyber. **52** pp. 141-52 (1985)

[2] Winston, Patrick Henry **Artificial Intelligence** (2nd ed.) Addison-Wesley Publishing Co. (1984)

[3] Grossberg, Stephen "On the Development of Feature Detectors in the Visual Cortex with Applications to Learning and Reaction-Diffusion Systems" Biol. Cyber. **21** pp. 145-59 (1976)

[4] Saylor, Janet and David G. Stork "Neural Networks for Tree Searching: Dynamics, Accuracy and Symmetries" (in preparation)

HIGHER-ORDER BOLTZMANN MACHINES

Terrence J. Sejnowski

Johns Hopkins University, Baltimore, MD 21218

ABSTRACT

The Boltzmann machine is a nonlinear network of stochastic binary processing units that interact pairwise through symmetric connection strengths. In a third-order Boltzmann machine, triples of units interact through symmetric conjunctive interactions. The Boltzmann learning algorithm is generalized to higher-order interactions. The rate of learning for internal representations in a higher-order Boltzmann machine should be much faster than for a second-order Boltzmann machine based on pairwise interactions.

INTRODUCTION

Thousands of hours of practice are required by humans to become experts in domains such as chess, mathematics and physics[1]. Learning in these domains requires the mastery of a large number of highly interrelated ideas, and a deep understanding requires generalization as well as memorization. There are two traditions in the literature on learning in neural network models. One class of models is based on the problem of content-addressable memory and emphasizes a fast, one-shot form of learning. The second class of models uses slow, incremental learning, which requires many repetitions of examples. It is difficult in humans to study fast and slow learning in isolation. In some amnesics, however, the long-term retention of facts is severely impaired, but the slow acquisition of skills, including cognitive skills, is spared[2]. Thus, it is possible that separate memory mechanisms are used to implement fast learning and slow learning.

Long practice is required to become an expert, but expert performance is swift and difficult to analyze; with more practice there is faster performance[1]. Why is slow learning so slow? One possibility is that the expert develops internal representations that allow fast parallel searches for solutions to problems in the task domain, in contrast to a novice who must apply knowledge piecemeal. An internal representation is a mental model of the task domain; that is, internal degrees of freedom between the sensory inputs and motor outputs that efficiently encode the variables relevant to the solution of the problem. This approach can be made more precise by specifying neural network models and showing how they incorporate internal representations.

LEARNING IN NETWORK MODELS

Network models of fast learning include linear correlation-matrix models[3,4,5,6] and the more recent nonlinear autoassociative models[7,8,9,10]. These models use the Hebb learning rule to store information that can be retrieved by the completion of partially specified input patterns. New patterns are stored by imposing the pattern on the network and altering the connection strengths between the pairs of units that are above threshold. The information that is stored therefore concerns the correlations, or second-order relationships between the components of the pattern. The internal model is built from correlations.

Network models of slow learning include the perceptron[11] and adaline[12]. These networks can classify input patterns given only examples of inputs and desired outputs. The connection strengths are changed incrementally during the training and the network gradually converges to a set of weights that solves the problem if such as set of weights exists. Unfortunately, there are many difficult problems that cannot be solved with these networks, such as the prediction of parity[13]. The perceptron and adaline are limited because they have only one layer of modifiable connection strengths and can only implement linear discriminant functions. Higher-order problems like parity cannot be solved by storing the desired patterns using the class of content-addressable algorithms based on the Hebb learning rule. These models are limited because the metric of similarity is based on Hamming distance and only correlations can be used to access patterns.

The first network model to demonstrably learn to solve higher-order problems was the Boltzmann machine, which overcame the limitations of previous network models by introducing hidden units[14,15,16]. Hidden units are added to the network to mediate between the input and output units; they provide the extra internal degrees of freedom needed to form internal representations. The Boltzmann learning algorithm incrementally modifies internal connections in the network to build higher-order pattern detectors. The hidden units can be recruited to form internal representations for any problem; however, the learning may require an extremely large number of training examples and can be excessively slow. One way to speed up the learning is to use hidden units that have higher-order interactions with other units.

THIRD-ORDER BOLTZMANN MACHINES

Consider a Boltzmann machine with a cubic global energy function:

$$E = -\frac{1}{3}\sum_i\sum_j\sum_k w_{ijk}\, s_i\, s_j\, s_k$$

where s_i is the state of the i th binary unit and w_{ijk} is a weight between triples of units. This type of interaction generalizes the pairwise interactions in Hopfield networks[10] and Boltzmann machines, which contribute a quadratic term to the energy. Fig. 1 shows an interpretation of the cubic term as conjunctive synapses. Each unit in the network updates its binary state asynchronously with probability

$$p_i = \frac{1}{1 + e^{-\Delta E_i/T}}$$

where T is a parameter analagous to the temperature and the total input to the ith unit is given by

$$\Delta E_i = \sum_j \sum_k w_{ijk}\, s_j\, s_k$$

If w_{ijk} is symmetric on all pairs of indices

$$w_{ijk} = w_{jik} = w_{ikj} = w_{kji}$$

then the energy of the network is nonincreasing. It can be shown that in equilibrium the probabilities of global states P_α follow a Boltzmann distribution

$$\frac{P_\alpha}{P_\beta} = e^{-\frac{E_\alpha}{E_\beta}}$$

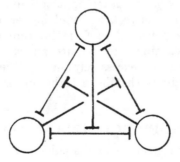

Fig. 1. Third-order interactions between three units. In the diagram the lines between units represent reciprocal interactions that are activated only when the third unit is in the *on* state. The third unit acts presynaptically to conjunctively control the pairwise interactions.

There are two forms of the Boltzmann learning algorithm, one for networks with inputs and outputs treated identically, and a second for networks where the input units are always clamped[15]. The former learning algorithm will be generalized for third-order interactions. The learning metric on weight space remains the same:

$$G = \sum_{\alpha} P_{\alpha} \log \frac{P_{\alpha}}{P_{\alpha}'}$$

where P_{α} is the probability of a global state with both the inputs and outputs clamped, and P_{α}' is the probability of a global state when the network is allowed to run freely. It can be shown that the gradient of G is given by

$$\frac{\partial G}{\partial w_{ijk}} = -\frac{1}{T}(p_{ijk} - p_{ijk}')$$

where p_{ijk} is the ensemble average probability of three units all being in the *on* state when the input and output units are clamped, and p_{ijk}' is the corresponding probability when the network is running freely. To minimize G, it is sufficient to measure the time averaged triple co-occurence probabilities when the network is in equilibrium under the two conditions and to change each weight according to

$$\Delta w_{ijk} = \epsilon \, (p_{ijk} - p_{ijk}')$$

where ϵ scales the size of each weight change.

HIGHER-ORDER BOLTZMANN MACHINES

Define the energy of a k-th order Boltzmann machine as

$$E = -\frac{1}{k}\sum_{\gamma_1}\sum_{\gamma_2}\cdots\sum_{\gamma_k} w_{\gamma_1\gamma_2\cdots\gamma_k} s_{\gamma_1} s_{\gamma_2} \cdots s_{\gamma_k}$$

where $w_{\gamma_1\gamma_2\cdots\gamma_k}$ is a k-dimensional weight matrix symmetric on all pairs of indices. The G matrix can be minimized by gradient descent:

$$\Delta w_{\gamma_1\gamma_2\cdots\gamma_k} = \epsilon \, (p_{\gamma_1\gamma_2\cdots\gamma_k} - p_{\gamma_1\gamma_2\cdots\gamma_k}')$$

where $p_{\gamma_1\gamma_2\cdots\gamma_k}$ is the probability of the k-tuple co-occurence of the $(s_{\gamma_1}, s_{\gamma_2}, \cdots s_{\gamma_k})$ when the inputs and outputs are clamped, and $p_{\gamma_1\gamma_2\cdots\gamma_k}'$ is the corresponding probability when the network is freely running.

In general, the energy for a Boltzmann machine is the sum over all orders of interaction and the learning algorithm is a linear combination of terms from each order. This is a Markov random field with polynomial interactions[17].

DISCUSSION

Conjunctive synapses such as those studied here can be used to model multiplicative relationships[18]. In a third-order Boltzmann machine the conjunctive interactions must be symmetric between all three pairs of units in a triple. This configuration has been used to implement shape recognition using mappings from a retinal-based frame of reference to object-based frames of reference[19,20]. In principle, these mappings could be learned by a sufficiently large number of hidden units with only pairwise interactions, but in practice the number of units and time required would be prohibitive. Learning this mappings using third order interactions occurs much more quickly.

Higher-order interactions have recently been introduced into content-addressable networks with fast learning[21,22]. The storage capacity of these networks is much larger than networks with only pairwise connections, but the number of connections is also much larger. Another advantage of higher-order interactions is the possibility of storing higher-order predicates[13]. However, these networks remain limited in their ability to generalize because they can only memorize the stored patterns; without hidden units they cannot generate new internal representations.

One of the serious problems with all higher-order schemes is the proliferation of connections, which tend to be the most expensive part of an implementation. A network of n units would require $O(n^k)$ connections to implement all interactions of order k. For example, consider the problem of learning mirror symmetries[16]. Random-dot patterns are generated with a mirror symmetry along one of several axes in an $N \times N$ grid. The task is to learn to classify new patterns given only examples of correctly classified mirror-symmetric patterns. A Boltzmann machine with pairwise interactions and 12 hidden units between the input and output layer can learn to classify patterns in about 50,000 trials. Using third-order interactions between the input and output layer would require $O(N^4)$ connections, most of which would be superfluous since only $O(N^2)$ of these connections carry any information relevant to the solution of the problem. Thus, learning may be faster but the price in connections may be prohibitive.

Whether a higher-order Boltzmann machine is of practical value depends on the tradeoff between the increased number of connections and the decreased learning time. At present it is not known how learning in Boltzmann machines scales with the size and difficulty of a problem, but it should be possible to simulate higher-order Boltzmann machines for small problems and compare them with conventional second-order Boltzmann machines. Other incremental learning algorithms, such as backpropagation[23] can also be generalized to higher-order units.

REFERENCES

1. D. A. Norman, Learning and Memory (W. H. Freeman, New York, 1982).

2. L. R. Squire & N. J. Cohen, In: J. L. McGaugh, N. W. Weinberger & G. Lynch (Eds.) Neurobiology of Learning and Memory (Guilford Press: New York, 1984)

3. K. Steinbuch, Kybernetik 1, 36-45 (1961)

4. J. A. Anderson, Math. Biosci. 14, 197-220 (1972)

5. T. Kohonen, IEEE Trans. C-21, 353 (1972)

6. H. C. Longuet-Higgins, Nature 217, 104 (1968)

7. J. A. Anderson & M. C. Mozer, In: G. E. Hinton & J. A. Anderson (Eds.), Parallel Models of Associative Memory (Erlbaum Associates, Hillsdale, N. J., 1981).

8. T. Kohonen, Self-Organization and Associative Memory (Springer-Verlag, New York, 1984).

9. T. J. Sejnowski, In: G. E. Hinton & J. A. Anderson (Eds.), Parallel Models of Associative Memory (Erlbaum Associates, Hillsdale, N. J., 1981).

10. J. J. Hopfield, Proceedings of the National Academy of Sciences USA 79, 2554-2558 (1982).

11. Rosenblatt, F., Principles of Neurodynamics, (Spartan, New York, 1959).

12. B. Widrow, In: M. C. Yovits, G. T. Jacobi & G. D. Goldstein (Eds.), Self-Organizing Systems 1962 (Spartan Books, Washington, D. C., 1962)

13. M. Minsky, & S. Papert, Perceptrons, (MIT Press, Cambridge, 1969).

14. G. E. Hinton, & T. J. Sejnowski, Proceedings of the IEEE Computer Society Conference on Computer Vision & Pattern Recognition, Washington, D. C., 448-453. (1983).

15. D. H. Ackley, G. E. Hinton, & T. J. Sejnowski, Cognitive Science 9, 147-169 (1985).

16. T. J. Sejnowski, P. K. Kienker & G. E. Hinton, Physica D (in press).

17. S. Geman, & D. Geman, IEEE Transactions on Pattern Analysis and Machine Intelligence 6, 721-741 (1984).

18. J. A. Feldman & D. H. Ballard, Cognitive Sci. 9, 205-254 (1983)

19. G. E. Hinton, Proc. 7th Int. Joint Conf. Artif. Intel. 1088-1096 (Kauffman, Los Altos, CA, 1981).

20. G. E. Hinton & K. J. Lang, Proc. 9th International Joint Conf. Artif. Intel. 252-259 (Kauffman, Los Altos, CA, 1985).

21. D. Psaltis, Proc. Snowbird Meeting on Neural Networks for Computing (1986)

22. Y. C. Lee, G. Doolen, H. H. Chen, G. Z. Sun, T. Maxwell, H. Y. Lee & L. Giles, Physica D (in press).

23. D. A. Rumelhart & G. E. Hinton, In: D. Rumelhart & J. McClelland (Eds.), Parallel Distributed Processing (MIT Press, Cambridge, 1986).

FIRING RESPONSE OF A NEURAL MODEL WITH THRESHOLD MODULATION AND NEURAL DYNAMICS

Paolo Sibani

Nordita, Blegdamsvej 17, DK-2100 Copenhagen Ø, Denmark

ABSTRACT

The attractors of the dynamics of a neural network are usually determined by the learned patterns via a Hebb-like learning rule or one of its modifications. Here we explore the possibility of constructing 'innate' attractors by a simple mechanism of threshold modulation in a model of neuronal firing.

INTRODUCTION

In the theory of categorization and associative recall, nonlinearities are important[1,2]. They provide attractors for the network dynamics which act as categorizers, a task which linear systems, which are good noise filters[3], cannot perform.

There are obviously many possible choices of nonlinear dynamics. One approach has exploited an analogy between neural networks and spin systems, resulting in a statistical mechanical description[4,5,6]. The dynamics of the network is stochastic, and the firing rate of the neuron is considered as analogous to an average magnetization in response to a local field. In this way, one is naturally led to the concept of temperature, which describes the statistics of the fluctuations of the network, and enters as a parameter in the relationship between magnetization and the local field.

A conceptually different approach[7] represents the state of the network at 'time' n by a vector X_n, of the same dimension as the number of neurons. Each coordinate of the vector is interpreted as the state of a particular neuron. The system is updated synchronously in the following way: first the input signal to the ith neuron is calculated by a weighted sum of the outputs from the other neurons, as the ith coordinate of the vector

$$Y_n = M X_n. \tag{1a}$$

Here M is a 'synaptic' matrix. The firing rates at time n + 1 are then computed as

$$X_{n+1} = \mathcal{C}(Y_n), \tag{1b}$$

where $\mathcal{C}$ is a nonlinear function which acts coordinatewise and which limits the neural response within finite bounds. In this scheme, the nonlinearity of the dynamics can be traced back to the property of a single neuron, and can in principle be understood in terms of neural models.

In this paper, we show how threshold modulation in a class of neural models influences the form of $\mathcal{C}$ – the transfer function of the neuron – and speculate on the possible consequences for the attractors of a network dynamics of the type given by eqs. (1a) and (1b).

'Integrate and fire' models of the neuron are the simplest ones available in the literature[8,9]. The neural membrane is considered as a condenser shunted by a resistance – describing membrane conductivity – and loaded by a current – resulting from the outputs from the other neurons. The state variable x(t) is the voltage across the membrane. It can increase up to a threshold T, whereafter the neuron fires and x is reset to, say, zero. In the Fohlmeister model[10] which we consider below, the membrane conductivity γ is also a time-dependent quantity, and the time evolution is given by the equations

$$\frac{dx}{dt} = -\gamma x + I \tag{2}$$

$$\frac{d\gamma}{dt} = -C\gamma + Dx \tag{3}$$

for $x(t) < T$. Here, I is the input current, and C and D are constants. After each firing, x is reset to zero, while γ is assumed to vary continuously. Other initial conditions can also be used[11], but are not necessary for our purposes. The transfer function is the relation between the input current I and the firing rate, or by a trivial rescaling, between input current and average output current.

In spite of the highly nonlinear character of the model, the transfer function of the above model is perfectly linear, and moreover shows no sign of saturation. Introducing a 'dead' period after each firing, in which the current is very ineffective in charging the membrane ($\gamma = \infty$), gives saturation and a transfer function similar to the piecewise linear sigmoid used in the BSB model. A quite different result is obtained by naively solving eqs. (2) and (3) on a digital computer by discretizing the time with a small steplength Δt. As shown in fig. 1, the system saturates at a frequency $\omega_1 = 1/\Delta t$. In addition, the transfer function develops plateaus at frequencies $\omega_n = \omega_1/n$, $n = 1, 2, 3, \ldots$. This is the well-known phenomenon of phase locking. The same effect can be obtained on an analog computer[11], by alternating periods of length Δt where the threshold is very (infinitely) high, to much shorter periods where the threshold is T. This means that the threshold devices samples the values of the state variable x(t) at a given frequency ω_1, which of course is the the maximal firing frequency of the model. Exactly the same thing happens, albeit implicitly, in numerical integration on a digital computer. This type of threshold modulation give only 1/n phase locking. A sinusoidal modulation gives all the substeps at frequencies $(m/n)\omega_1$, where $m < n$.

The behaviour of the system is summarized by the following observations:
(i)   for low I there is a linear regime, with an extension along the I axis controlled by the value of T relative to a = $\sqrt{IC/D}$, which is the asymptotic value of x(t) for $t \to \infty$ when the firing condition is removed from the equations.
(ii)   Changes in C and D which keep C/D constant do not change the result. There are therefore three free parameters, C/D, T and Δt, which can be reduced to two by a rescaling of the time axis.
(iii) A line drawn through the midpoints of the steps is only slightly concave.

Threshold modulation has previously been suggested[12] as a possibility in excitable tissue, and can arise by a quite generic negative feedback mechanism acting on the threshold value. It does not seem to contradict experimental evidence. In fact, recent experiments on the visual response of behaving monkeys to a sequence of Walsh stimuli[13] show interesting temporal patterns in the spike sequences from a single neuron, which could also arise by a phase locking mechanism. The device can be built into electronic circuitry, it is very simple and gives rise to a complicated behaviour. Therefore it is worth looking at the consequences of a staircase-like transfer function on the network dynamics given by eq. (1). By adding a bias, the response can be shifted so that the upper and lower limits are symmetrically located with respect to zero. The phase space of the system is then a zero-centered hypercube. The parameters can also be chosen so that the average slope is close to one, and many steps intersect the x = y line. Finally, we let the matrix M of eq. (1a) have the form

$$M = 1 + \epsilon A, \qquad (4)$$

where ϵ is a scalar. This implies that the 'old' pattern is not completely erased by the updating. We think of A, the feedback matrix, as having positive eigenvalues.

When $\epsilon = 0$, eq. (1) is just an iteration of $\mathcal{C}$ in each coordinate separately, and all the intersections of the steps with the x = y line are attractive fixpoints, with a basin of attraction corresponding to the step width. Turning the interaction on, (some of) the coordinates of the state vector will tend to grow bigger, due to the effect of A. However, as long as they stay within a step width, the 'cut' operation will bring them back to their previous value. This means that, as long as ϵ is small enough, there is the possibility of 'local' saturation effect to a firing rate which is not the highest possible rate. We conjecture that the attractor morphology can be changed by varying the parameter ϵ, and go from a situation where there are many attractors, each with a small basin, to a situation where only broad categories survive. The parameter ϵ could then be thought of as an attention parameter.

In this approach, the possible stable attractors of the network dynamics arise as a combination of 'genetic' effects, i.e. the size and number of steps in the transfer function, with the 'experience' of the network, which can be encoded in the synaptic matrix M. This can be done by giving a pattern as input, letting the system relax to the corresponding attractor and then learning the attractor by modifying the synoptic matrix according to a Hebb or Woodroff-Hoff rule. Numerical work along these lines is in progress.

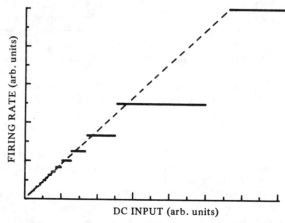

Fig. 1. Firing rate versus d.c. input current for the Fohlmeister model. The last step has infinite width, and corresponds to complete saturation. The curve was obtained by solving eqs. (2) and (3) on a digital computer with a discrete step length $\Delta t = 1/1000$. The other parameters were c = 0.01, d = 0.1, and T = 0.1.

ACKNOWLDEGEMENTS

I would like to thank Preben Alstrøm for many enlightening discussions of phase locking and Mogens Levinsen for the set-up of the neural model on an analog computer.

REFERENCES

1. J.J. Hopfield, Proc. Natl. Acad. Sci. USA, 79, 2554 (1982).
2. J.J. Hopfield, Proc. Natl. Acad. Sci. USA, 81, 3088 (1984).
3. T. Kohonen, Self-Organization and Associated Memory (Springer-Verlag, Berlin, 1984).
4. P. Peretto, Biol. Cybern. 50, 51 (1984).
5. D.J. Amit, H. Gutfreund and H. Sompolinski, Phys. Rev. A32, 1007 (1985).
6. D.J. Amit, H. Gutfreund and H. Sompolinski, Phys. Rev. Lett. 55, 1431 (1985).
7. J.A. Anderson, J.W. Silverstein, S.A. Ritz and R.S. Jones, Psychol. Rev. 84, 413 (1977).
8. R. Fitz Hugh, Bull. Math. Biophysics 17, 257 (1955).
9. J.P. Keener, F.C. Hoppensteadt and J. Rinzel, SIAM J. Appl. Math. 41, 503 (1981).
10. J.F. Fohlmeister, Kybernetik 13, 104 (1973).
11. M. Levinson, private communication.
12. L. Glass and M.C. Mackey, J. Math. Biol. 7, 339 (1979).
13. B.J. Richmond, L.M. Optican, M. Podell and H. Spitzer, Preprint, Lab. of Neuropsychology, NIMH, Bethesda, MD 20892.

VLSI Architectures for Implementation of Neural Networks

Massimo A. Sivilotti, Michael R. Emerling and Carver A. Mead[1]

California Institute of Technology, Pasadena CA 91125

April 15, 1986

Introduction

A large scale collective system implementing a specific model for associative memory was described by Hopfield [1]. A circuit model for this operation is illustrated in Figure 1, and consists of three major components. A collection of active gain elements (called amplifiers or "neurons") with gain function $V = g(v)$ are connected by a passive interconnect matrix which provides unidirectional excitatory or inhibitory connections ("synapses") between the output of one neuron and the input to another. The strength of this interconnection is given by the conductance $G_{ij} = G_0 T_{ij}$. The requirements placed on the gain function $g(v)$ are not very severe [2], and easily met by VLSI-realizable amplifiers. The third circuit element is the capacitances that determine the time evolution of the system, and are modelled as lumped capacitances.

This formulation leads to the equations of motion shown in Figure 2, and to a Liapunov energy function which determines the dynamics of the system, and predicts the location of stable states (memories) in the case of a symmetric matrix T.

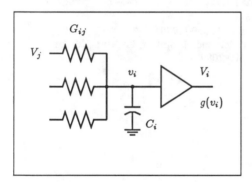

$$C \cdot \frac{dv_i}{dt} = \sum_{j=1}^{N} G_{ij}(\pm V_j - v_i)$$
$$= \sum_{j=1}^{N} G_{ij}(\pm g(v_j) - v_i)$$

$$E = -\frac{1}{2}\sum_i \sum_j G_{ij} V_i V_j$$
$$+ \sum_i \left(\sum_j G_{ij}\right) \int_0^{V_i} g_i^{-1}(V)dV$$

$$\frac{\partial E}{\partial t} \leq 0 \quad \text{and} \quad \frac{\partial E}{\partial t} = 0 \Rightarrow \frac{\partial V_i}{\partial t} = 0$$

Figure 1: Circuit model for Hopfield system   Figure 2: Differential Equations of Motion

VLSI Restrictions

Since collective systems exhibit interesting global properties as a consequence of having large numbers of individually simple elements, implementation with very large scale integration (VLSI) circuit technology appears very suitable. There are, however, a number of technology-dependent limitations that are introduced by such a choice.

Cost

The principal cost measure in VLSI is *area*. Even with the use of die-stitching techniques, there exists a physical limit on the maximum area a circuit can occupy. Also, the off-chip environment is quite different from the internal circuit, for electrical reasons. This fact, coupled

[1]This research was supported by the System Development Foundation

with the fundamental restriction on I/O pads, makes it desirable to integrate an *entire* system on a single chip.

Analog electronics, by exploiting the intrinsic physics of native devices, generally occupy less area per function than an implementation using a digital abstraction. For example, an analog differential-input multiplier may require as few as 8 transistors to perform the relatively complex calculation $(y = k(x_{1+} - x_{1-})(x_{2+} - x_{2-}))$. Furthermore, there is none of the overhead associated with mapping what is essentially a continuous problem into a discrete-time (sampled digital) system.

Power

A common complaint about analog computing elements is that their power consumption is high, due to a desire for maximum linearity at high operating speeds, and because discrete (off-chip) components present relatively highly capacitive loads. In a VLSI context, power dissipation must be limited to a few watts (for conventional packaging technologies). However, collective circuits implemented entirely on one die have no requirements to drive external loads, do not have to be particularly fast, and value symmetry much more highly than linearity.

It is important to note that the Hopfield circuit model exhibits non-zero power dissipation even after convergence is reached, and the computation is nominally terminated. For systems of several hundred amplifiers, it is not possible to build interconnect matrices of tens of thousands of resistors without explicitly limiting the power consumption of the amplifiers. A commonly suggested alternative, computation by current summing, is even more impractical, as the number of (power dissipating) current injectors that must be controlled scales with the number of synapses.

The approach we have taken to permit the implementation of large arrays is to limit the current consumption of the amplifiers, guaranteed by keeping most of the MOS devices in a subthreshold regime of operation [3]. For sufficiently low gate voltages (less than the so-called "threshold voltage", below which the digital abstraction of transistor operation classifies the transistor as "off"), the drain current is exponential in the gate voltage. This behavior is exactly the same (and indeed, the physics are identical) as bipolar transistors exhibit throughout their operating range, with the additional benefit that MOS transistors draw no gate current.

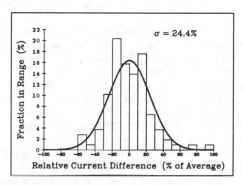

Figure 3: Full wafer current variation in 3x3 micron MOS transistors

Parameter Variation

An additional complication is introduced in the case of very small devices, where statistical or systematic doping variations can affect their transfer characteristics by significant amounts. These variations are particularly evident in the case of fabrication lines intended for digital chips (which are relatively insensitive to such variation). If an analog design methodology is used that requires currents to be precisely matched or subtracted, it is unlikely that sufficient accuracy can be obtained with single small transistors. The degree of variation to be expected is illustrated in Figure 3 [4], which shows the drain currents of identically biased MOS transistors from a typical MOSIS [5] digital process. When differences between *adjacent* transistor currents are taken (Figure 4), the variation in relative current differences is substantial, and indicates (in this case) a fairly random process, as opposed to some longer-range (die-scale) systematic variation. These variations can be minimized by using larger transistors (Figure 5), or by relying on statistical numbers of transistors to participate in a computation.

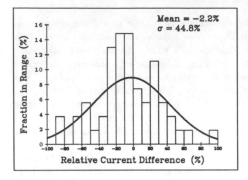

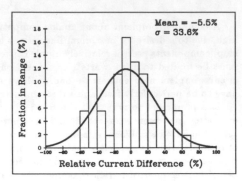

Figure 4: Variation in adjacent $3x3\mu$ MOST   Figure 5: Variation in adjacent $24x24\mu$ MOST

Furthermore, it is clear that a computational scheme must be designed that is robust against such variation, and that displays a high tolerance to noise. Such claims are commonly made of collective systems; they must be carefully examined, however, in light of the actual implementation.

VLSI Implementations

The two principal implementation issues that have yet to be addressed are the requirement for negative valued resistive elements for inhibitory connections, and the desire for symmetrical large signal behaviors to maximize noise margins. Together, these considerations led to the adoption of amplifiers with differential input stages, and complementary outputs. The symmetry of the resulting dual rail signal representation makes any element with sufficient gain adequate for the amplifier. For the first implementation, we picked the simplest element possible, the nMOS inverter. There is a controllable interconnection between the inverters, which can be used to enforce to varying degrees the complementarity of the outputs. Figure 6 shows the schematic of the amplifier, with the cross-couple network implemented as two cross-connected NAND gates. Input pass transistors, marked as gated on ϕ_1 are used to disconnect the inverters from the matrix, allowing the state of the system to be latched dynamically.

Figure 7 shows test data illustrating the transfer function of the actual amplifier. The output of the amplifier is plotted as a function of its input; all data are normalized to voltage rails of -1 to $+1$, and the differential nature of the inputs has been explicitly incorporated into the plot. As expected, the plot is symmetrical.

The other major component of the design is the interconnection element (Figure 8). Each T_{ij} has 4 pass transistors, operating in their ohmic (resistive) regime, and is capable of 3 interconnection strengths $(-1, 0, +1)$. The ϕ_2 line is used to selectively disconnect the matrix, and, in conjunction with the ϕ_1 line, is used to single-step through the chip's operation.

Nothing more would be required, if it were not for the desire to dynamically change the

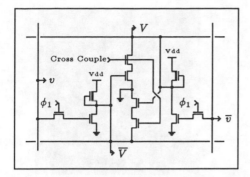

Figure 6: nMOS amplifier schematic

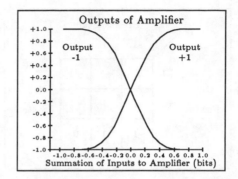

Figure 7: Amplifier transfer function

programming of the matrix. For the nMOS prototype, we opted for programmability by a generalized outer product scheme in which any two vectors may be multiplied (Figure 9). Each matrix element is stored at the corresponding interconnection element. The component of the cross product is generated at each T_{ij} site (by a simple AND pass network), and is added to the previous T_{ij} value, which is then stored in the tri-stable memory element.

This addition operation is truncating (addition table in Figure 10), it is *not* associative in the algebraic sense, but symmetry is preserved (if only symmetrical matrices are added). Simulation had indicated that the clipped T_{ij} matrix would not greatly impact the capacity of the system as a whole; this observation has been verified experimentally.

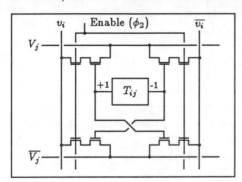

Figure 8: T_{ij} Element

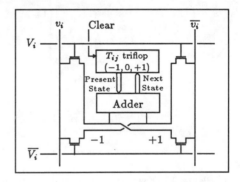

Figure 9: T_{ij} Block Diagram

The entire programmable T_{ij} element contained 41 transistors. At any time, only *four* of these (the dual-rail interconnect transistors in the corners of Figure 8) participate in the association process; the rest of the circuitry is dedicated to providing programmability.

Testing the nMOS prototype

A design containing 22 active elements, and a full interconnect matrix (462 elements) was fabricated using MOSIS' 4μm feature size nMOS technology. The entire project measured 6700μm by 5700μm, and required 53 I/O pads. The completed ASSOCMEM chip is shown in Figure 11.

The chip worked immediately, with 3 memories being routinely programmable. More rarely, 4 memories were possible, for carefully chosen (i.e. nearly orthogonal) vectors. These capacities

412

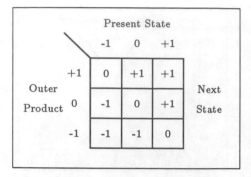

	Present State		
	-1	0	+1
+1	0	+1	+1
Outer Product 0	-1	0	+1
-1	-1	-1	0

Next State

Figure 10: Truncating adder table

Figure 11: The ASSOCMEM

do not include the complements of the desired memories, which are themselves stable due to the symmetry of the system.

With two memories in the system, if the starting state is any closer to either of the memories, the system was found to converge directly into that stable state. Figure 12 plots the probability of falling into each of two randomly chosen memories, as a function of the Hamming distance from the initial state to the memories.

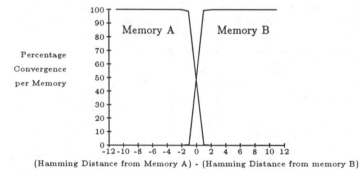

(Hamming Distance from Memory A) - (Hamming Distance from memory B)

Figure 12: Plot of Convergence Probability

Figure 13 shows an incremental association in progress. By driving the ϕ_1 and ϕ_2 lines with a 2-phase non-overlapping clock, the inputs and outputs of the amplifiers are selectively connected to the matrix. Thus, the association can be halted, and a discrete notion of time is introduced. In the example shown, the initial state is intermediate between two memories. Within one cycle, the common bits between the two memories become active. The system is now at a metastable point, and takes 2 cycles to evolve away from it, finally settling in one of the memories.

It is worth mentioning that when run in continuous mode, we were never able to detect an association still in progress after the $50\mu s$ that our instrumentation required to switch from driving to monitoring the I/O lines.

Finally, Figure 14 deals with the issue of fault tolerance. This chip had over 40% of its T_{ij} elements unprogrammable, yet could still associate to one of 2 stored states. Of the 10 chips returned by MOSIS, all had one particular element malfunction, 2 had no other elements fail, and the others ranged from 2% (10) to 12% (54) bad, excluding the chip in Figure 14. All could store at least 2 memories.

```
         Memory 1 = 0011110000111100001111

         Memory 2 = 0000000000111111111111

     Starting State = 0000000000000000001111


     Iteration 0:  V = 0000000000000000001111

     Iteration 1:  V = 0000000000111100001111

     Iteration 2:  V = 0000000000111100001111

     Iteration 3:  V = 0001010000111100001111

     Iteration 4:  V = 0011110000111100001111
```

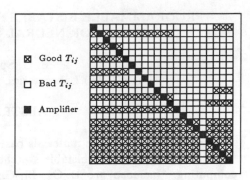

Figure 13: Incremental association   Figure 14: T_{ij} yield

Conclusions

Collective systems exhibit many appealing properties, including robustness and fault tolerance, an ability to deal with ill-posed problems and noisy data, which conventional digital architectures do not. Thus, neural computation suggests a design methodology that may permit the application of VLSI to difficult perception problems, such as vision and audition, with fault tolerance permitting wafer-scale processing surfaces.

Research is proceeding on alternative design techniques. The associative memory design has been transferred to subthreshold CMOS, and a 289 neuron chip has been fabricated. Arranged as a 17x17 pixel array, a subset of the ASCII character set has been programmed as memories. Testing is proceeding.

Finally, as feature sizes continue to shrink to the $1\mu m$ range, the design of chips containing 1000 fully-interconnected neurons becomes feasible. Such chips would represent viable tools to assist in the modelling of neural networks, by providing an efficient computational implementation.

References

[1] J. Hopfield. Neural networks and physical systems with emergent collective computational abilities. *Proceedings of the National Academy of Sciences USA*, 79:2554–2558, April 1982.

[2] J. Hopfield. Neurons with graded response have collective properties like those of two-state neurons. *Proceeding of the National Academy of Sciences USA*, 81:3088–3092, May 1984.

[3] C. Mead and M. Maher. A charge-controlled model for submicron MOS. In *Proceedings of the Colorado Microelectronics Conference*, May 1986.

[4] J. E. Tanner. *Integrated Optical Motion Detection*. PhD thesis, California Institute of Technology, 1986. In preparation.

[5] D. Cohen and G. Lewicki. MOSIS – the ARPA silicon broker. In *Proceedings from the Second Caltech Conference on VLSI*, pages 29–44, California Institute of Technology, Pasadena, CA, 1981.

PROGRAMMABLE BISTABLE SWITCHES AND RESISTORS FOR NEURAL NETWORKS

E. G. Spencer

AT&T Bell Laboratories, Murray Hill NJ 07974

ABSTRACT

A system of crystalline materials based on bismuth oxide is being evaluated for use as programmable bistable switches and resistors in neural networks for computing. Included are Bi_2O_3, $Bi_{12}GeO_{20}$ $Bi_{12}SiO_2$, and $Bi_4Ge_3O_{12}$. Both switching and electrical programming of resistors have been observed with voltages as low as 50 mV and with currents in the nanoampere range. Generally the resistances are of the order of 10^6 to $10^9\,\Omega$ in the OFF position, switching to the order of 10^4 to $10^5\,\Omega$ in the ON position. Present efforts are directed toward improving the metallurgy of the cross stripe conductors and the quality of the films in order to be able to fabricate large scale arrays in which all of the elements have the same operating characteristics with resistances in the $R=10^8\,\Omega$ to $10^{11}\,\Omega$ range. Also, computer controlled programming and measurement methods are being implemented which will duplicate the exact operating conditions in neural network analog computation.

INTRODUCTION: NEURAL NETWORK ARCHITECTURE

Neural networks for computing are being considered which consist of very large scale arrays of programmable bistable resistors or switches. The operation to be performed or the program to be computed will be written into the network array by setting each elemental switch in the ON position or by setting each elemental resistor to a specific value as determined by the mathematical equations and procedures of the problem to be solved. The dc biases and the computational signal strengths are set to be below the values required to activate the switches or to change the values of the resistors.

Programming is to occur, well defined, with a single pulse or with a specific series of pulses of the order of two or three times the operating bias or signal voltage. It is necessary that the switch or resistor remain in a stable state, even with all the biases turned off, until a reversing pulse is applied specifically, to return the material to its high resistance value.

PROGRAMMABLE BISTABLE SWITCHES AND RESISTORS

In these investigations the attempt is being made to employ an electric field to create microscopic size conducting paths approaching a few atomic spacings in cross section. Thus a single pulse in creating one or more conducting paths would increase the conductance of the element by a definite step. Each succeeding pulse would increase the conductance by a similar amount. It should be possible to write, or program, any desired value. Conversely reverse pulses would tend to return the element, in steps to its original more insulating state. A much stronger pulse would create a sufficient number of conducting paths to interact cooperatively and cause the element to switch to a final R(ON) value.

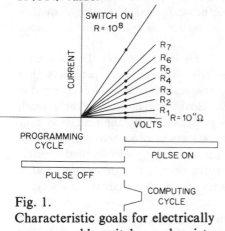

Fig. 1.
Characteristic goals for electrically programmable switches and resistors.

Some design goals[1] representing functions required of bistable switches and resistors are illustrated in Figure 1. The element in its initial state is to have a resistance of $R = 10^{11}\,\Omega$. A bias voltage of 100 mV is applied as in the lowest line. Application of a single pulse of 2V creates a weak microscopic size conducting path in the material and the resistance changes to a value represented by the line labeled $R1$. Successive single pulses strengthen the conducting path and change the resistance in steps to any desired value. The resistors are ohmic for signal or computing voltages below the critical value of ~1V.

Computational problems, as in linear programming, may require that the individual elements be set to specific values of resistance which may range over many orders of magnitude.[2] Other kinds of computations operate in a binary switching mode in which all of the elements initially are set to either of two specific values such as R(OFF) $= 10^{11}\,\Omega$ and R(ON) $= 10^8\,\Omega$. Voltage pulses of reversed polarity will tend to interrupt the microscopic size conducting paths and return the resistor, in steps, to the more insulating state. While Figure 1 represents design goals based on computer operation requirements, Figure 2 is intended to present the same goals with regards to materials development. A sufficiently strong reversing pulse will return the resistor to the OFF state.

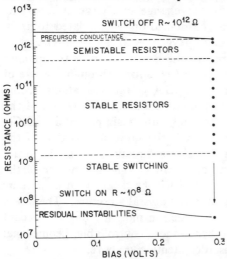

Fig. 2.   Materials characteristics

BISMUTH OXIDE SYSTEM OF MATERIALS

Investigations are being carried out on a system of materials based on bismuth sesquioxide. These include Bi_2O_3, $Bi_{12}GeO_{20}$, $Bi_{12}SiO_{20}$ and $Bi_4Ge_3O_{12}$. This system was considered for development initially because of the remarkable and unique crystallographic, optical, electrooptical, and ultrasonic properties[3] and because of the subsequent technological developments over the past two decades.[4] The materials melt congruently, and were found to grow readily from the melt.[5] For example almost perfect crystals of $Bi_4Ge_3O_{12}$ as large as 0.5 meter long and 0.5 meter in diameter are routinely produced for nuclear scintillation counters.[6] It was to be expected that high quality crystalline films could be made by vapor deposition as well as by ion beam milling or by sputtering.

In stoichiometric proportions $Bi_{12}GeO_{20}$ is an insulator and the measured resistivity is on the order of 10^{10} to 10^{13} Ω-cm. The physical processes that produce the finite conductivity necessary for the resistors and switches are due to an off stoichiometry oxygen defect, that is, vacant oxygen sites in the crystal. By controlling the number of vacancies, it is possible to vary the resistivity over many orders of magnitude. The conduction process involves moving electrons, along a line of oxygen vacancies in directions defined by the applied electric field. The oxygen vacancies can exist as positive or negative or neutral charge states. Since the charge states of the vacancy are modified there is always a shift, however small, in the positions of the ions. A strain field also is produced. When the electric field is removed, the electrons will return to positions that allow the strain to relax to its original value. The element will return to its higher insulating state.

In order to retain the conducting path there must be a residual strain pattern, weak enough for the electric field to overcome but strong enough to retain, or lock in, the strain produced by the change in charge state of the oxygen vacancy. This residual strain may be due to disorder related to grain boundaries, size and shape of crystallites or to oxygen vacancies themselves.

However, another reason for the interest in the Bi_2O_3 system is the manner in which high temperature crystal structures are stabilized through the use of Ge, Si or Ti additions.[7] For example in $Bi_{12}GeO_{20}$ the Ge atoms, being relatively few in number, are positioned at the corners and in the center of the unit cell. Thus they may be considered to constitute a simple cubic (CsCl) type super lattice in the Bi_2O_3 structure which is the same size as that of the unit cell, 10.5Å.[8] This super lattice then will give a directionality to the strain field pattern at each lattice site of an oxygen vacancy. The magnitude of the effect will be different for the different ions and will be determined in part by the cohesive energy of the particular species in the crystal structure. The strain field associated with the superlattice and the electron motion along the vacant oxygen sites will display a more uniform, hence more controllable, geometrical pattern than the strain field due to the disorder mentioned above.

Investigations are being carried out on test structures consisting of a longitudinal conducting stripe over which the oxide film is deposited. On top there are 20 conducting cross stripes thus forming 20 resistors or switching elements. Initial work was done using gold conductors 100 μm wide and films of $Bi_{12}GeO_{20}$ ~3000Å thick. Dc voltages of 1 to 5 volts cause the elements to switch ON and OFF repetitively with currents in the mA range. Usually upon first applying the voltage a series of small instabilities were observed at voltages much smaller than that required for switching. It was concluded that switching in the milliampere range was made up of a large number of switching events in the microampere or nanoampere range. Subsequent measurements were made using conductors 1 μm wide with the expectation of using the lower currents of the individual switching events to program resistors of specific values. The resistance of the element is given by $R=\rho\,t/a$, where ρ is in Ω-cm, t is the thickness and a is the cross sectional area. For example, for films that are 3000Å thick, $R=3\times10^3\rho$. This means that the resistivity will be $\rho=3.3\times10^7\,\Omega$-cm for the element to have $R(\text{OFF})=10^{11}\,\Omega$. Thus even in the OFF position the high resistivity required is achieved with a value of conductivity that is sufficient to establish definite microscopic conducting paths.

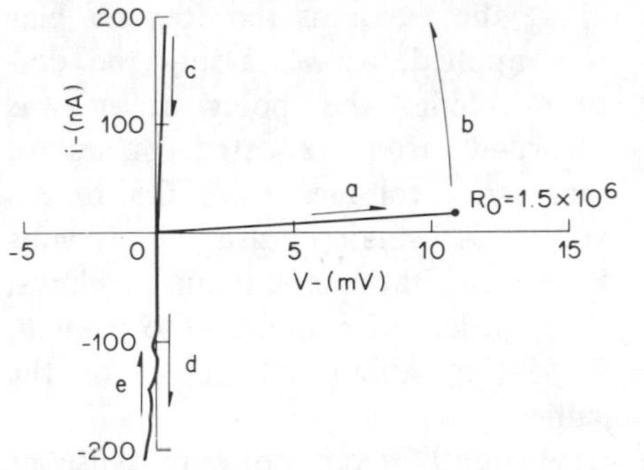

Fig. 3. $Bi_{12}GeO_{20}$ switching by applying a $+0.2$V pulse at a 10 mV bias.

EXPERIMENTAL

Data are shown in the figures that illustrate, a) switching using positive and negative dc biases, b) switching using a fixed bias and voltage pulses, and c) voltage pulse programming to produce a resistor of a specific value. All of the elements consist of films of the oxides ~3000Å thick with conducting leads 1 μm wide.

An x-y recorder tracing of an i-V curve is shown in Figure 3 for an element made using a film of $Bi_{12}GeO_{20}$ and gold cross stripes. With an applied 0.2 volt pulse, 10 μsec wide, and using a 10 mV bias, switching occurred from a resistance of $R=1.5\times10^6\,\Omega$ to $R=2\times10^3\,\Omega$, a ratio of 10^3 times. The element was returned to its high resistance value by applying a negative dc bias of about 2 or 3 mV.

After repeated cycling of this element for various tests, the data of Figure 4 were recorded which illustrate the ability to program bistable resistors of specific values. The straight lines are tracings from the x-y recorder as the 10 mV bias was turned on and later turned off.

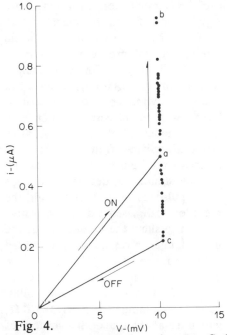

Fig. 4. Programmable resistor using $Bi_{12}GeO_{20}$.

Positive pulses, of 0.15V, 10 μsec in width, were applied at intervals of 10 seconds. The recorder was operated in the dot-print mode to record the increase in conductance after each pulse from a to b. Subsequently the polarity of the pulses was reversed and the decrease in conductance was recorded from b to c. The 75 successive reverse pulses reduced the bias current by approximately 10 nA per pulse. The pen was lowered to record the current as the bias was reduced to zero.

A pulse ON — pulse OFF switching mode is illustrated in Figure 5 using a similar film of crystalline bismuth sesquioxide, Bi_2O_3. The recorder trace lies

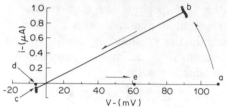

Fig. 5. Bi_2O_3 switching
$R = 10^8\,\Omega$ to $R = 10^5\,\Omega$

along the x-axis as the 100 mV bias was applied, at a. Using the dot-print mode, the point at a was recorded after repeated pulses of increasing voltages from 0.5 to 3.5 volts. A single pulse of 4 volts increased the conductance almost three orders of magnitude to point b. Repeating about 20 pulses of the same

magnitude caused the conductance to vary slightly from pulse to pulse as indicated by the printed dots before assuming a fixed value at b.

The pen was lowered and the bias was reduced to zero then fixed at -5 mV. Ohmic resistivity was recorded after three pulses of -3 V at c. The fourth pulse caused the element to switch to the original more insulating state at d, which was maintained at e, with the application of a 50 mV bias.

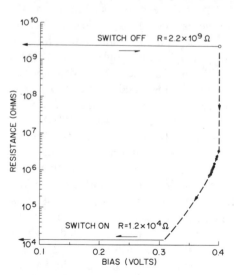

Fig. 6. Programmed resistors
and switching using $Bi_4Ge_3O_{12}$.

Complete switching while attempting to program resistors is shown in Figure 6. The test element was a film of $Bi_4Ge_3O_{12}$ using 1 μm wide molybdenum cross stripes. There was no measurable change in the resistance of $R=2.2\times10^9\,\Omega$ after repeated pulsing from 0.5V to 3V. Subsequently pulses of 3.5V caused the resistance to decrease in steps from pulse to pulse. Although the recorder scale used was not sensitive enough to show the fine details of the first changes in resistance, the printed dots indicate that the resistance changed in steps to a value of $R=5\times10^5\,\Omega$, a factor of 4000 times, before switching to $R=1.2\times10^4\,\Omega$, which is a change of five orders of magnitude.

DISCUSSION

Although the ability to program bistable switches and resistors of specific values using films based on the bismuth sesquioxide system has been demonstrated, both practical and basic physical problems remain to be worked out. The goal of operating in the nanoampere or picoampere range with resistances in the 10^8 to $10^{11}\ \Omega$ range appears to be reasonable. However, the techniques for reproducing these identical operating parameters for every element of a large scale array have not been developed. A part of the problem is in the present methods of measurements. In order to be able to display the fine scale variations in the data over the required four or five orders of magnitude, a computer controlled method is being set up. In addition to recording data precisely, the pulses and dc biases can be adjusted automatically after each programming pulse so as to avoid injury to the delicate element. It is especially important that the conditions for testing most probably will be exactly the conditions that will be used in operation of a neural network for computing. When these measurements are in operation the subtle differences among the different material parameters and fabricating methods can be studied in more detail.

ACKNOWLEDGEMENTS: The contributions of R. D. Yadvich, for the lithographic work, and M. E. Melczer, for many discussions, are greatfully appreciated.

REFERENCES

1. J. J. Hopfield and D. W. Tank, Biological Cybernetics 52, 378 (1985).
   J. J. Hopfield, Proc. Natl. Acad. USA, 81, 3088 (1984).
2. D. W. Tank and J. J. Hopfield, IEEE, Circuits and Systems (1986).
3. E. G. Spencer, P. V. Lenzo, A. A. Ballman, Proc. IEEE 55, 2074 (1967).
4. Some representative recent investigations. Y. H. Ja, Optics Comm. 44, 24 (1982). P. K. Grewal and M. J. Lea, J. Phys. C: Solid State Phys., 16, 247 (1983). M. G. Mitera, Optics Comm., 50, 79 (1984). P. Aubourg, J. P. Huignard, M. Hareng and R. A. Mullen, Appl. Optics 21, 3706 (1984). A. A. Ballman, H. Brown, P. K. Tien and R. J. Martin Jour. Cry. Growth 20, 251 (1973).
5. A. A. Ballman, Jour. of Cryst. Growth 1, 37 (1966).
6. O. H. Nestor and C. Y. Huang, IEEE Trans. on Nucl. Sci. 22, 68 (1975). L. A. H. van Hoof and W. J. Bartels, Mat. Res. Bull. 20, 79 (1985).
7. E. M. Levin and R. S. Roth, J. Res. Natl. Bur. Std. 68A, 197 (1964). G. Gattow and D. Schutze, Z. Anorg. Allgem.Chem. 328, 44 (1964).
8. S. C. Abrahams, P. B. Jamison and J. L. Bernstein, J. Chem. Phys., 47, 4034 (1967).

FAST SIMULATED ANNEALING*

Harold Szu

Naval Research Laboratory, Code 5709, Washington, D.C. 20375-5000

Abstract

Simulated annealing is a stochastic strategy for searching the ground state. A fast simulated annealing (FSA) is a semi-local search and consists of occasional long jumps. The cooling schedule of FSA algorithm is inversely linear in time which is fast compared with the classical simulated annealing (CSA) which is strictly a local search and requires the cooling schedule to be inversely proportional to the logarithmic function of time. A general D dimensional Cauchy probability for generating the state is given. Proofs for both FSA and CSA are sketched. A double potential well is used to numerically illustrate both schemes.

INTRODUCTION

When the classical energy/cost function $C(\bar{x})$ has a single minimum, the conventional method can provide the unique ground state, and any method of gradient descent can approach the minimum. However, when $C(\bar{x})$ has multiple extrema, a nonconvex optimization technique that allows tunnelling and variable sampling and accepting hill-climbing for escaping from local minima is required. To illustrate the concept, we first consider a serial processing. If a ball is rolling over a hilly terrain inside a box, one must shake the box gently enough in the vertical direction of perturbations that the ball cannot climb up the global minimum valley and sufficiently vigorously along the horizontal direction of sampling to escape from local minimum valleys. Thus, a strategy of variable perturbations is needed. We secondly consider a concurrent parallel processing. A molten solid having random thermal energy must be gradually cooled down in order to reach the (globally minimum energy) crystalline state. Thus, a thermal random noise is useful when it is gradually quenched. These algorithms may be called simulated annealing, or Monte Carlo method when a constant noise temperature is assumed as first proposed by Metropolis et al.[1] for computer simulation of hard-disc phase transitions. Recently, Kirkpatrick et al.[2] in classical systems and Ceperley and Alder[3] in quantum systems have investigated a general and powerful computing technique for changing noise temperature and sampling grid sizes. A necessary and sufficient condition for the convergence to the global minimum has been proven in 1984 by Geman and Geman[4] for the classical simulated annealing (CSA) based on a strictly local sampling. It is required that the time schedule of changing the fluctuation variance, described in terms of the artificial cooling temperature $T_a(t)$, which could be different from the true thermodynamic temperature T, is inversely proportional to a logarithmic function of time given a sufficient high initial temperature T_0.

$$T_a(t)/T_0 = 1/\log(1+t). \tag{1}$$

Such an artificial temperature cooling schedule is too slow to be practical. Instead, for arbitrary $T_o \neq 0$ the FSA has

$$T_c(t)/T_0 = 1/(1+t) \tag{2}$$

D-DIMENSIONAL CAUCHY PROBABILITY

Basically, the algorithm has three parts (1) States are generated with a probability density that has a Gaussian-like peak and Lorentzian wings that imply occasional long jumps

---

*NRL Invention Patent Case

among local sampling. (2) The canonical ensemble for a state acceptance probability allows occasional hill-climbing among descents. (3) An artificial cooling temperature enters both (1) and (2) as a control parameter of noise. FSA turns out to be better than any algorithm based on any bounded variance distribution, which is equivalent to the Gaussian diffusion process by the central limiting theorem. Starting in a random state, at each time step a new state is generated according to the generating probability. If this new state has lower cost it becomes the new state of the machine. If it has higher cost it becomes the new state with the probability determined by the acceptance function. Otherwise the old state is retained. Both the acceptance and generating functions vary according to the cooling schedule. When $T_a = 0$, it is a gradient descent method. Since the diffusion process used for the strictly local strategy is artificial, it can be replaced with a semi-local search with an occasional long jump among local diffusions described by a Lorentzian distribution defined in the D dimension as follows

$$g_c(\vec{x}) = (2\pi)^{-D} \int_{-\infty}^{\infty} \ldots \int_{-\infty}^{\infty} d\vec{k} \exp(-i\vec{k}\vec{x}) \exp(-c|\vec{k}|) = c/[(\vec{x}^2 + c^2)]^{(D+1)/2} \quad (3)$$

which has the Cauchy characteristic function

$$\chi(\vec{k}) = \exp(-c|\vec{k}|) \quad (4)$$

The parameter c is the temperature parameter $T_c(t)$ which decreases according to a cooling schedule to be determined. The Cauchy distribution implies an occasional long jump among essentially local sampling over the phase space. This proper tradeoff between local search and semi-global search allows a fast annealing schedule.

PROOFS OF COOLING SCHEDULES FOR FSA AND CSA

One of the most significant consequences of such a trade off observation is that we are able to prove generally the cooling schedule to be inversely proportional to the time, rather than to the logarithmic function of time[4]. Since a rigorous theorem based on a stochastic Markovian chain will be published elsewhere[6], we shall compare (CSA) with (FSA) and sketch the essential proofs for both cooling schedules in the arbitrary D-dimensional vector space. In FSA we separate the state-generating from the state-visiting, while the actual visit is decided by the hill-climbing acceptance criterion based on the canonical ensemble of a specific Hamiltonian. FSA demands the state-generating to be infinite often in time (i.o.t.), but CSA requires the state visiting to be i.o.t. Let the state-generating probability at the cooling temperature $T_c(t)$ at the time t and within a neighborhood be (bounded below by) $\geq g_t$. Then the probability of not generating a state in the neighborhood is obviously (bounded above by) $\leq [1 - g_t]$. To insure a globally optimal solution for all temperatures, a state in an arbitrary neighborhood must be able to be degenerated i.o.t., which does not, however, imply the ergodicity that requires actual visits i.o.t. To prove that a specific cooling schedule maintains the state-generation i.o.t., it is easier to prove the *negation* of the *converse*, i.e. the *impossibility* of *never* generating a state in the neighborhood after an arbitrary time t_0, namely such a negation probability vanished

$$\prod_{t=t_0}^{\infty} [1 - g_t] = 0 \quad (5)$$

Taking logarithm of (5) and Taylor expansion (noting that $\log 0 = -\infty$, $\log(1 - g_t) \approx - g_t$), to prove (5) is equivalent to prove (6),

$$\sum_{t=t_0}^{\infty} g_t = \infty \quad (6)$$

We can now verify those cooling schedules satisfying Eq. (6) in the D-dimension neighborhood for an arbitrary size $|\Delta \vec{x}_0|$ and t_0.

(i) Bounded variance type CSA: there exists an initial T_0 and for $t > 0$

$$T_a(t) = T_0/\log(t) \tag{7}$$

$$g_t \approx \exp\left[-\frac{|\Delta\bar{x}_0|^2}{T_a(t)}\right] T_a(t)^{-D/2} \tag{8}$$

$$\sum_{t=t_0}^{\infty} g_t \geq \exp\left(-\log(t)\right) = \sum_{t=t_0}^{\infty} \frac{1}{t} = \infty \tag{9}$$

(ii) Unbounded variance type *GSA*: For arbitrary $T_0 > 0$

$$T_c(t) = T_0/t \tag{10}$$

$$g_t \approx \frac{T_c(t)}{\left[T_c^2(t) + |\Delta\bar{x}_0|^2\right]^{(D+1)/2}} \approx \frac{T_0}{t\,|\Delta\bar{x}_0|^{D+1}} \tag{11}$$

$$\sum_{t=t_0}^{\infty} g_t \approx \frac{T_0}{|\Delta\bar{x}_0|^{D+1}} \sum_{t=t_0}^{\infty} \frac{1}{t} = \infty \tag{12}$$

So any neighborhood is visited i.o.t. and the cooling schedule algorithm is admissible. The advantage of using Cauchy distribution in D-dimensions is that the ability to take advantage of locality is preserved, but the presence of a small number of very long jumps allows faster escape from local minima. As a result we can be much less cautious in our cooling. In fact, we can cool as fast as $T_c(t) = T_0/t$ for any $T_0 > 0$. Because the rate of convergence of the annealing algorithm is bounded by the temperature, this means that the algorithm can converge much faster.

EXAMPLE OF NONCONVEX OPTIMIZATIONS

In order to illustrate both FSA and CSA we choose a one dimensional simple double well potential as the classical energy

$$C(x) = x^4 - 16x^2 + 5x \tag{13}$$

as illustrated in Fig. 1. In order to appreciate the analogy with the transition probability in quantum mechanical we plot both the normal distribution and the Cauchy/Lorentzian distribution over the shallower valley representing a trapping in the valley. While the wing of Lorentzian probability has reached the other, deeper, valley, the normal distribution has negligible value there and thus has less chance to escape. A higher temperature implies a faster sampling in a much more "coarse grained" fashion. As the temperature is gradually reduced, the Cauchy machine searches through the state space with more refined sampling. An artificial control temperature within the search state space is the state-space-search generating temperature T_C, which is different to a thermodynamic temperature along the energy/cost function in a canonical ensemble. For simplicity, we let the two be proportional or equal to each other $T_C = T$ without causing any confusion.

We apply stochastic optimization to the simple cost function (13). Apart from the automatic learning aspects, Boltzmann machines may be characterized by (i) bounded generating probability density (thermal diffusion), e.g., Gaussian

$$G(x) \cong \exp\left(-x^2/T_a(t)^2\right) \tag{14}$$

(ii) an inversely logarithmic update cooling schedule Eq. (1) and (iii) the canonical hill-climbing acceptance probability (putting the Boltzmann constant $K_B = 1$, i.e. $C = H/K_B$)

$$\exp\left(-C_{t+1}/T_a\right)/[\exp{-C_{t+1}/T_a}) + \exp\left(-C_t/T_a\right)] = (1 + \exp\left(\Delta C/T_a(t)\right))^{-1}. \tag{15}$$

Fig. 1 — Gaussian probability density versus Cauchy probability density plotted over a simplified model of cost functions

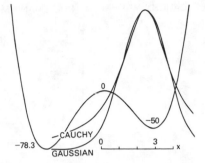

Where $\Delta C = C_{t+1} - C_t$ is the increase of cost incurred by a transition. The resulting cost at each time step shows the validity of the inversely logarithmic cooling schedule (1) (also plotted as the dotted line in Fig. 2.). The energy axis is the vertical axis which shows the first minimum, the zero, and the second minimum at the level of the horizontal axis as visited by several thousand trials.

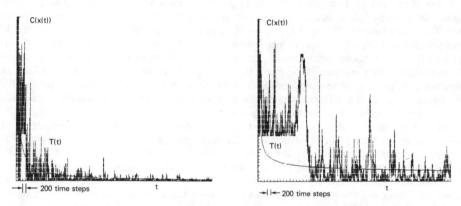

Fig. 2 — Actual cost $C(x(t))$ and cooling schedule $T(t)$ used in Boltzmann and Cauchy Machine are plotted against time steps $t = 1$ to 12000, where a tic mark is 200 time steps

Then we define the Cauchy Machine which replaces (i) the generating probability (14) with the Cauchy/Lorentzian distribution:

$$G_C(x) = T(t)/(T(t)^2 + x^2) \qquad (16)$$

Then, (ii) the update cooling schedule may be inversely linear in time, Eq. (2). For the sake of comparison, we use the identical hill-climbing acceptance probability (15) except $T_a(t)$ is replaced by $T_c(t)$. The success of the simulation shown in the right hand side of Fig. 3 supports our universal theorem of convergence for Cauchy machines, namely that the process finds the optimum with the cooling schedule Eq. (2). We shall first plot both the free-space random walk with displacements having the normal distribution together with the Cauchy random walk, and the actual random walk within the potential walls. It is evident that there is no bound on the variance of fluctuations of the Cauchy distribution. This provides us the opportunity of occasionally sampling the state space from one extreme to the other. Obviously, the Cauchy Machine is better in reaching and staying in the global minimum (shown in the left) as compared with the corresponding tracks generated by the Boltzmann machine (in the right).

424

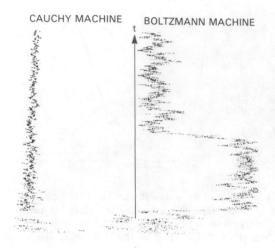

CAUCHY MACHINE   BOLTZMANN MACHINE

Fig. 3 — Comparison between Cauchy Machine search with Boltzmann Machine search (shown in right) within the identical cost constraint. The vertical axis is time steps and the horizontal is the displacement x.

We now turn to a conceptually simple but computationally complex application in a higher dimensional phase space. Given 100 points which have been randomly scattered from five lines (the ground truth), the problem is that to discover a best fit of those 100 points with five lines. This class of perception of random dot problem may be called "*unlabeled* (unknown correspondences between lines and points) mixture of densities" and to rediscover those labels and means is known as "*unsupervised* learning" in Computer Vision. In nonstochastic version, it is computationally complex or NP-difficult because there are 10^{10} possible ways to assign 100 random points to 5 groups, $(100)^5$ for a (dumb) exhaustive search. Furthermore, there exists ambiguous results and the unique solution cannot be guaranteed by the conventional gradient descent methodologies, which have no ability of de-trapping from local minima associated with each ambiguous result. A computationally complex NP-complete problem is the traveling salesman[2] problem which may be heuristically solved by the stochastic method of simulated annealing algorithm, and which would guarantee a unique solution if appropriate cooling schedules have been followed. The present example of "unsupervised learning" is solved using the standard Maximum likelihood formalism and FSA, and the result is presented.[5]

Basically, the Cauchy distribution helps us to preserve a local search and an occasional long jump in speed up the state-generation at an artificial noise temperature T_c which is conveniently separated from the thermodynamic temperature T used in physical distribution function $\exp(-H/k_B T)$ for the occasional hill climbing acceptance criterion of those generated states. Such a computational saving is comparable to what Tukey and Cooley did to the N^2 operations needed for 2D DFT with the observation of the harmonic pairing/butterfly giving the $N \log N$ operations needed for FFT. The computational saving of FSA when compared with CSA is similar to that FFT which revolutionized signal processing compared with DFT. Thus we can call it fast simulated annealing (FSA) which we hope should significantly broaden the applicability of simulated annealing to neural network computing[7] and physics problems.

References

1.   N. Metropolis, A.W. Rosenbluth, M.N. Rosenbluth, A.H. Teller, and E. Teller, *J. Chem. Phys.* **21**, 1087-1092, June 1953.
2.   S. Kirkpatrick, C.D. Gelatt, M.P. Vecchi, *Science* **220**, 671-680, 13 May 1983.
3.   D. Ceperley and B. Alder, Science **231**, 555-560, 7 Feb. 1986.

4. S. Geman and D. Geman, *IEEE Trans, Patt, Anan. Mach. Int.,* **PAMI-6** (No. 6), 721-741, Nov. 1984.

5. H.H. Szu and R.L. Hartley, "Simulated Annealing with Cauchy Probability" submitted to Optics Letter.

6. R.L. Hartley and H.H. Szu, "Generalized Simulated Annealing," submitted for publication.

7. H.H. Szu, "Neural Network Models for Computing," to appear in Applied Optics.

BINARY SYNAPTIC CONNECTIONS BASED ON
MEMORY SWITCHING IN a-Si:H

A.P. Thakoor, J.L. Lamb, A. Moopenn, and John Lambe

Jet Propulsion Laboratory
California Institute of Technology
Pasadena, CA 91109

ABSTRACT

Nonvolatile, associative, electronic memory based on neural network models promises high ($\sim 10^9$ bits/cm^2) information storage density since the information would be stored in a matrix of simple two terminal, passive interconnections. Device considerations dictate that such connections should be weak. For example, connections with $\sim 10^6 \Omega$ resistance are quite adequate in a 1000 x 1000 matrix ($\sim$250K bit ROM). Irreversible memory switching in hydrogenated amorphous silicon (a-Si:H) thin films is studied as a candidate mechanism for a prototype binary PROM matrix. The memory switching in a-Si:H is current induced and requires very low energy ($\sim$1 nanojoule for 1 μm^2 area of a 0.2 μm thick film) to switch from $\sim 10^{10} \Omega$ to $\sim 10^5 \Omega$. Resistivity-tailored, amorphous Ge$_{1-x}$M$_x$ (M=Al, Cu) provides the (synaptic) ballast resistor in the microswitch. A novel side saddle test structure exhibits a potential for a very high density binary connection matrix.

INTRODUCTION

An electronic, programmable binary memory matrix[1] with 32 neurons and 1024 interconnections based on Hopfield's neural network model[2], has been built in our laboratory. It has demonstrated a variety of model features including content-addressability, intrinsic fault tolerance, and memory retrieval in one machine cycle ($\sim$10 μsec). Furthermore, we have also shown[3] that the errors appearing due to the false memory states in such a matrix can be eliminated up to a significant proportion. This is accomplished by selecting an appropriate dilute coding for the information to be stored, employing an optimized local inhibition scheme, and making additional asymmetrical connections in the same binary matrix. Our results suggest that one can indeed have virtually error-free information retrieval from a binary memory matrix, with storage efficiency of up to $\sim$25% (total storage of $\sim 0.25 N^2$ bits in an N x N matrix) before the proportion of errors becomes unmanageable.

Nonvolatile, associative, electronic memory based on a binary connection matrix promises very high information storage density since the information would be stored in a dense array of "memory cells", each consisting of a simple two terminal, passive synaptic connection. The array of connections would primarily consist of two sets of crossed thin film wires with the connections made at some of its nodes. The synaptic "connections" in such a matrix need to be unusually "weak" (in the range of $\sim 10^6 \Omega$), primarily to limit the current in the wires to a reasonable density (to avoid electromigration) and to limit the overall power dissipation. Recently, Jackel et.al[4] have fabricated dense neural networks in thin film form using amorphous silicon based synapses patterned by e-beam lithography during the matrix fabrication. However, if an electronically switchable, resistive component with memory could be incorporated at each intersection, such a matrix could become an associative PROM, with a further possibility of E^2PROM if the memory switch could be made reversible. A microswitch at each node of the matrix would thus contain two components: a switchable memory element and a ballast resistor element. The ballast (synaptic) resistor would not only provide the required measured "connection strength" at each "ON" node, but would also control the energy delivered to the switching element during the "writing" process. In this paper, we report on the thin film structures consisting of a hydrogenated amorphous silicon element, which exhibits memory switching, and a stable synaptic (resistive) element of amorphous $Ge_{1-x}M_x$ (M=Al, Cu). Various configurations of the connections and the projected storage density of the matrix are discussed.

EXPERIMENTAL DETAILS

Hydrogenated amorphous silicon (a-Si:H) thin films were deposited onto glass substrates by reactive rf magnetron sputtering of a silicon target in a mixture of high (99.999%) purity argon and hydrogen (partial pressures 9 mTorr and 2 mTorr, respectively). Amorphous $Ge_{1-x}M_x$ films were deposited by thermal evaporation of presynthesized alloys of the desired compositions. Variation in the metal contents (x ranging from 0.06 to 0.10) in the alloy provided a convenient parameter to tailor the resistivity of the composite films. These films did not crystallize up to $\sim 250°C$ and their room temperature resistivity values showed little change on heat treatment. This was one of the most important selection criteria for device stability since moderate heat treatment (up to $\sim 150°C$) was unavoidable in the subsequent lithographic processing sequence. Sputter deposited Ni and Cr were used for the top and bottom thin film wires, respectively. Polyimide served as the dielectric between the wires where they crossed directly over each other. All patterning of the microswitch test structures with a nominal feature size of 10 μm was done by conventional photolithography and wet etching techniques. Switching characteristics of the as-deposited a-Si:H as well as the patterned test structures were studied using a microprober and a pulse generator.

RESULTS AND DISCUSSION

A variety of thin film materials have been discovered over the years, which exhibit electronic or field-induced (thermal, electrochemical, etc.) memory switching. However, the energy required to accomplish the switching is the most important factor which

428

determines the material suitability for any specific application. In the present case, a high ballast (synaptic) resistance, ~$10^6\Omega$, required in series with the switching material to guarantee the desired "ON" resistance of the node when switched, becomes an effective bottleneck which limits the energy available to the switching material. The switching should therefore be accomplished with very low energy. The microprober test circuit used for the switching studies is shown schematically in Fig. 1 (a). A current limiting resistance of $10^6\Omega$ was placed in the circuit, physically very close to the probing tip. It was extremely important to minimize the value of the hidden capacitance (c) in the test circuit to avoid misleading measurements of the energy actually consumed by

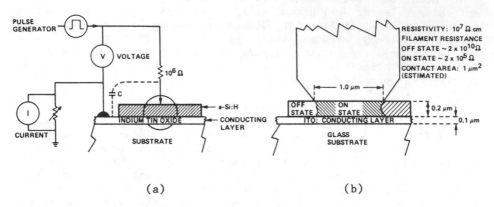

(a)   (b)

Fig. 1. (a) Microprober test circuit for the evaluation of switching behavior in a-Si:H films (b) OFF/ON resistance values for a 0.2 μm thick a-Si:H film.

switching material. Figure 1 (b) summarizes the switching characteristics of a 0.2 μm thick a-Si:H film deposited onto a glass substrate coated with a conducting oxide layer. The room temperature resistivity of the as-deposited a-Si:H film was ~$10^7\Omega$ cm. An approximately 1 μm^2 area of such a film is switched from ~$2\times10^{10}\Omega$ to ~$2\times10^5\Omega$ with the application of a voltage pulse of ~ 1μsec duration applied through a $10^6\Omega$ ballast resistor, thereby delivering up to ~1 nanojoule energy.

Figure 2 shows current-voltage (I-V) characteristics of memory switching, in a-Si:H thin film with a ballast resistance of $10^6\Omega$ in series. Regions 1,2, and 3 represent OFF, switching (one way only), and ON states, respectively. The electrical current-induced formation of a low resistivity, crystalline filament of silicon is believed to be the cause of memory switching a-Si:H. The detailed study of the switching mechanism will be published elsewhere[5].

Fig. 2 Current versus voltage (point-by-point I-V) characteristics of switching in a-Si:H, in series with a $10^6\Omega$ ballast resistor. Reversible regions (1) and (3) correspond to the OFF (before) and ON (after) states, respectively. Region (2) shows the actual irreversible change.

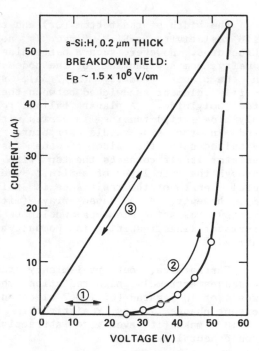

A planar and a side saddle structure, each consisting of a-Si:H switching element and a-Ge:Al/Cu resistor is shown in Fig. 3. The metal concentration in the germanium alloy is adjusted (in the range of 6 to 10 at %)to obtain a suitable resistivity value (1 to 100Ω cm) leading to a tailored resistance of ~$10^6\Omega$ in each geometry.

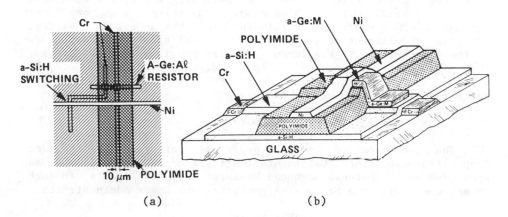

(a)   (b)

Fig. 3 Microswitch test structures containing a switching element (a-Si:H) and a ballast resistor (a-Ge:Al) in (a) planar configuration and (b) side-saddle geometry (not to scale).

The width of the bottom (Cr) and top (Ni) wires is 10 μm. The high resistivity a-Si:H layer is not patterned at all in these structures. However, such a patterning may be necessary in the high density, compact packing of the nodes to avoid crosstalk problems. In the planar geometry (Fig. 3a), switching takes place at the a-Si:H element sandwiched between the two wires of Cr and Ni, away from polyimide. A planar ballast resistor (a-Ge:Al, under the polyimide strip) remains in series with the switch. On the other hand, in the side-saddle structure (Fig. 3b), the resistor is deposited onto the side of the insulator (polyimide) and the resistor itself connects the top wire (Ni) to the bottom wire (Cr) through the thin layer of a-Si:H, where the switching takes place. Total length of the resistance in this case is ~25 μm (width ~10 μm), however, in a dense array (micron or even submicron line widths) resistive elements would be only on the side of the insulator lines resulting in compact nodes and a very high density matrix.

Furthermore, our preliminary results indicate that a-Si:H possesses extremely high radiation resistance. A total dose in excess of 10 megarad (of 1 MeV electrons) caused no more than a few percent change in its resistivity. High radiation resistance of a-Si:H makes it even more attractive for dense, rad-hard, electronic neural networks.

CONCLUSIONS

Hydrogenated amorphous silicon (a-Si:H) offers itself as an ideal candidate material for binary synaptic connections in an electrically programmable neural ROM. It exhibits excellent memory switching behavior with extremely low switching energy (~1 nanojoule for 1 μm^2 area, 0.2 μm thickness). Switching is accomplished through a $10^6 \Omega$ synaptic, thin film (a-Ge:Al or a-Ge:CU) resistor, the prime requirement of the matrix. A novel side-saddle structure of the microswitch in submicron geometry promises very high (over $10^9/cm^2$) connection density. High radiation resistance of the a-Si:H switch makes it even more attractive for high density, rad-hard, electronic neural networks.

ACKNOWLEDGEMENT

The work reported in this paper was performed by the Jet Propulsion Laboratory, California Institute of Technology, and was sponsored by the Defense Advanced Research Projects Agency, through an agreement with the National Aeronautics and Space Administration.

REFERENCES

1. John Lambe, A. Moopenn, and A.P. Thakoor; Proc. AIAA/ACM/NASA/-IEEE computers in Aerospace V. Conference, Long Beach, CA; Oct. 1985; p. 160.

2. J.J. Hopfield, Proc. Natl. Acad. Sci. USA, 79, 2554 (1982); 81, 3088 (1984).

3. A. Moopenn, S.K. Khanna, John Lambe, and A. P. Thakoor; See: "Error Correction and Asymmetry in a Binary Memory Matrix" elsewhere in this volume.

4. L. D. Jackel, R. E. Howard, H. P. Graf, B. Straughn, and J. S. Denker; J. Vac. Sci. Technology B (4), 61 (1986).

5. A. P. Thakoor, John Lambe, and A. Moopenn (To be published).

A LAYERED NEURAL NETWORK MODEL APPLIED TO THE AUDITORY SYSTEM

Bryan J. Travis
Los Alamos National Laboratory, Los Alamos, NM 87545

ABSTRACT

The structure of the auditory system is described with emphasis on the cerebral cortex. A layered neural network model
incorporating much of the known structure of the cortex is applied
to word discrimination. The concepts of iterated maps and attractive fixed points are used to enable the model to recognize words
despite variations in pitch, intensity and duration.

INTRODUCTION

Neurobiological systems embody solutions to many difficult
problems such as associative memory, learning, pattern recognition,
motor coordination, vision and language. It appears they do this
via massive parallel processing within and between specialized
structures. The mammalian brain is a marvel of coordinated
specialization. There are separate areas for each sense modality,
with massive intercommunication between areas. There are
topographic maps, many specialized neuron types, and quasi-regular
small scale structure (columns and layers) which vary from area to
area to accommodate local needs, and plasticity in connections
between neurons. Feedback occurs on many levels. This complexity
is apparently necessary for the kind of multi-mode processing that
brains perform but it's not clear how much structure is necessary
to perform isolated tasks such as vision or speech recognition, nor
do we know if nature's solutions are optimal.

Our knowledge of neural systems originates with experimental
studies. Experimental data lead to conceptual models, computational theories, which should be implemented through well-defined
algorithms.[1] These provide a tool for high-level information
processing whose worth can be evaluated explicitly. If one is
interested in their biological validity, model predictions can be
tested with further experimentation. This usually leads to new
models and algorithms.

In the area of learning and pattern recognition, a fruitful
approach has been the study of neural networks. Much of the work
in this field has focused on the mathematical properties of nets
with varying degrees of realism.[2] Recently, the minimum energy
viewpoint[3] has led to a breakthrough in the ability of networks to
perform cognitive functions. Capacity has increased as structure
increases from symmetric neural connections[4] to nonsymmetric[5] to
high-order correlation[6] models. Another viewpoint that promises to

be very powerful is the application[7,8] of concepts from dynamical systems theory to neural networks.

The goal of the project described in this paper is to build a biologically valid model of a sensory system, the auditory system, which integrates what is known both structurally and physiologically at several levels, from the peripheral sense organ (cochlea) to the midbrain nuclei to the auditory cortex. Despite the complexity of the system, there is a great deal of data available at each level to permit a fairly realistic model, although simplifications must be made especially at the cortex level to remain within computer memory and speed limits. Attractive features of such a model are that it: 1) includes more structure (and presumably more processing capacity) than previous models, 2) provides each subsystem with the kind of input that the biological counterpart receives, 3) emphasizes the dynamic or time dimension.

STRUCTURE OF THE AUDITORY SYSTEM

The comprehension of sound and speech begins with transduction of sound waves into transient motions of the basilar membrane of the cochlea in the inner ear. Displacements of the basilar membrane lead to discharge of hair cells in the organ of Corti and generation of action potentials in auditory nerve fibers. A particular frequency will cause peak displacement at a characteristic location along the basilar membrane. The dynamics of the organ of Corti is still a subject of great interest.[9] Details of cochlear structure and function are given in Ref. 10.

The auditory nerve bundle feeds into several nuclei in the lower brain and midbrain regions before the cortex is reached. Some of these nuclei use auditory input to initiate behavior such as head turning and other reflex actions. Others serve to sharpen auditory input and to integrate input from each ear. Fibers from all these nuclei converge onto the auditory cortex. More details are given in Ref. 11.

A great deal of experimental work over the last hundred years has slowly revealed the intricacies of the structure of the cortex of the mammalian brain. Although our knowledge of the cortex is still incomplete, many features are well established. The cortex consists of several (generally six) layers of neurons, each layer having a characteristic composition of specific neuron types.[12] The number and thickness of layers varies in different regions of the brain, presumably reflecting differences in function. The upper layers of the cortex are fairly recent in the evolutionary development of the mammalian brain. These layers are especially rich with neurons in man. Neural activity follows a generally vertical pattern; signals from the periphery enter mainly in layer 4, pass to the upper layers and then to the lowest layers, and then back up. Neurons in the upper layers also project to other cortical areas and neurons in the lower layers also project to subcortical nuclei. In addition to the layered structure a vertical columnar organization has been established through physiological measurements. These columns are roughly 0.5 mm in diameter and contain several thousand cells. Cells within a column generally exhibit the same sensitivity to sensory stimuli. Considerable lateral interaction between columns is now known[11] to occur.

In the auditory system, the cortex surface is organized into several tonotopic maps. At least four are known in the cat brain; two are very specific maps, the other two are diffuse. This multiplicity of frequency maps also appears in various nuclei of the medial geniculate body. The tonotopic map is conserved throughout the auditory system, but mappings onto isofrequency strips are both divergent and convergent. Association fibers connect isofrequency curves from each map onto the corresponding isofrequency curves in the other maps. Along an isofrequency curve, each ear is represented by alternating clusters of neurons. It is believed[13] that this second dimension of the auditory cortex map is used with the frequency map to accomplish spatial localization of sound.

Little is known about processing of auditory data at higher levels of the brain, except that neural activity extends outward from the primary auditory area to associational areas and is integrated with other sense modalities. There is of course a strong interaction with the speech generation areas (Wernicke and Broca). Most of the experimental work has used cats, with some measurements in owls and monkeys.

THE MODEL

A three-part mathematical model of the auditory system has been developed. It includes a hydromechanical model of the cochlea with hair cell innervation for studying transduction of sound into basilar membrane motion and then into neural signals. This model is similar to others[14,15] that have been studied, but attempts to treat the temporal dimension more realistically. Output from this portion of the auditory model can be and is being compared to auditory nerve recordings. The second portion applies neural networks to simulate auditory processing in nuclei such as the dorsal cochlear nucleus. Several of these nuclei have a relatively simple structure (two or three layers with feed forward only) and, based on recent studies,[16,17] use lateral inhibition to sharpen the auditory signal train. The third part of the model attempts to simulate neural activity in the auditory cortex. Because of the extremely large number of neurons and interconnections between neurons in the brain, only a very small portion of the cortex (roughly a few square millimeters) can be modeled in great detail, and even then certain aspects such as connection to other cortex areas will be difficult or impossible on present computers. Alternatively, a coarser-grained model can be used to approximate an "average" structure of larger cortex regions. Even these crude models may contain enough structure to have cortex-like dynamics.

The emphasis in this paper is on the application of this third part of the model to auditory processing.

A general neural network model is used to simulate cortex-like structures. The model, called LCM[18], operates in discrete time. It allows a variable number of layers, number of neurons and neuron types. The composition of each layer is determined by the user. For each layer, the types of neurons are specified, as well as the spatial density (number/unit area) for each type, radii of influence (distance over which a neuron's axonal projections and dendrites extend in each direction in each layer), refractory periods and the excitatory or inhibitory character of each type. Neurons can send axonal projections into multiple layers and dendrites can extend beyond a single layer. Within radii of influence, actual connections between neurons are determined from a user supplied probability. Two or three-dimensional structures can be created. Input stimuli to a network can be simple or time-dependent and can be fed in as a topographic map (frequently referred to as a tonotopic map in the auditory literature) and in a "dual" manner to simulate input from multiple sources, e.g., binaural input. Super networks can be created which consist of neural networks that communicate with one another. The state of each neuron at each time step is determined by summing inputs from all other neurons, subtracting its threshold value and then, if the neuron is not in a refractory state, using the Fermi probability to calculate whether the neuron fires an action potential. Plasticity in connectivity is governed by a Hebbian rule in which connection strength depends on previous activity (increases with use, decreases with disuse), with the restrictions that the sign (excitatory or inhibitory) of a connection cannot change and that connection strength is bounded.

In the present application a six layer two-dimensional model containing the neuron types[11,12,19-21] described in Tables I-III was constructed. Other details are given in Ref. 18. It has been assumed that neuron types in the auditory areas are approximately the same as those found in other cortex areas. Input stimuli are directed into layer 4. The flow of neural activity is then up into layers 2 and 3 (and 1), then down to layers 5 and 6. Activity flows from layers 5 and 6 back up to 4 and the upper layers, creating an iterated loop. Pyramidal cells in layer 6 are assumed to represent output units. Input vectors were constructed to represent approximately the frequency and temporal patterns seen for words and syllables in auditory nerve recordings[22] in animals and voice spectrographs in humans.

Table I Distribution of dendrites and axons of primary
excitatory neurons

Neuron Type	Layer Containing Cell Body	Layers for Apical Dend.	Layers for Basal Dend.	Layers for Axonal Proj.
Pyramidal	2	1	2	2,3,5,6
Pyramidal	3	1	3	2,3,5,6
Spiny Stellate	4	1,4	4	2,3,5,6
Pyramidal	5	1,4,5	5	2,3,5,6
Spindle	6	1,4	6	2,3,4,5,6

Table II Distribution of dendrites and axons of
two excitatory intracortical neurons

Neuron Type	Layer Containing Cell Body	Layers for Dendrites	Layers for Axons
Horizontal Cell	1	1	1
Martinotti Cell	2	2	1
Martinotti Cell	3	3	1,2
Martinotti Cell	4	4	1,2,3
Martinotti Cell	5	5	2,3,5
Martinotti Cell	6	6	2,3,5,6

Table III Common inhibitory neurons in the cortex

Neuron Type	Layer Containing Cell Body	Layers for Dendrites	Layers for Axons
Golgi II	1	1	1
Golgi II	2	2	2
Chandelier	2	2	1,2
Golgi II	3	3	3
Basket	3	3	3
Chandelier	3	3	1,2,3
Golgi II	4	4	4
Basket	4	4	4
Double Bouquet	4	4	1,2,3,4,5,6
Golgi II	5	5	5
Basket	5	5	5
Golgi II	6	6	6

CORTEX DYNAMICS AND ITERATED MAPS

An iterated function $f^n(x)$ is defined as

$$f^n(x_0) = x_n = f(x_{n-1}), \tag{1}$$

$$x_{n-1} = f(x_{n-2}), \ldots, x_1 = f(x_0), \text{ i.e.,}$$

$f^n(x) = f(\ldots f(f(x))\ldots)$ where f is applied n times.

Nonlinear functions allow a rich variety of behavior under iteration.[23,24] Stable points may exist which act as attractors. The iterated function maps a set of points X(i), the basin of attraction, into the attractor point y. For a system evolving in time, the membership of basins of attraction may change as well as the attractor set.

Our multi-layered model of the cortex is an iterated function. The looping from middle to upper to lower to middle layers provides the iteration and the all-or-none firing of model neurons provides the nonlinearity. Plasticity in the connection strengths allows for modifications of attractor points and basins of attraction.

In the auditory system, a word can be recognized despite differences in intensity, frequency shifts, and time duration. In the model this behavior is approximated when

$$f^n\left(ax^j_{-i+m}(bt)\right) \rightarrow f^n\left(\underline{x}^j_i(t)\right) \tag{2}$$

for an appropriate range of (a, b, m) where $x^j_i(t)$ is the rate of firing of input neurons corresponding to frequency i (or location i because of the topographic mapping) for input pattern j, subscript m represents a shift in frequency, a corresponds to a change in intensity or rate of firing, b effects a time dilation. The underbar indicates a reference point, i.e., $\vec{\underline{x}}^j(t) = (x^j_{-1}, x^j_{-2}, \ldots,$ $x^j)$ is the input vector at a reference intensity, time duration and frequency. The arrow in Eq. (2) means that the iterated map of the transformed pattern reaches the same limit point as the iterated map of the reference pattern after a sufficient number of iterations.

APPLICATIONS

The layered neural network model is being used in several ways:

1) Model behavior of individual neurons are being examined to see if they have qualitative behavior as observed in auditory cortex single neuron microelectrode recordings (e.g., monotone or nonmonotone response vs. frequency and intensity ranges, width of best frequency range, etc.)

2) The model proposed by Edelman and Finkel[25] for columnar organization and behavior has been simulated. Regular columnar structures develop spontaneously when connection strengths are allowed to vary. The input layer becomes strongly inhibited after an input is presented constantly for a long period of time. Connection strengths constantly change; some oscillate in time, while others show a more erratic behavior.

3) The model can be taught to recognize individual words over limited ranges of frequency shifts, intensity changes and time durations by simply presenting input vectors for versions of each word in sequence several times in a rehearsal phase. Pyramidal cells in layer 6 receive input from other cells over a very wide range. These pyramidal cells appear to serve as attractor points, or act as abstractions of words invariant to pitch and intensity variations. Much more work needs to be done to systematize this behavior.

4) A three-dimensional model is being constructed to test a model[13] for spatial localization of sound. This would provide a mechanism for separating simultaneous sound inputs.

SUMMARY

A layered neural network model having some of the structure of brain cortex has been developed. Applications for limited word sets show that fixed points can be created and merged very rapidly, allowing recognition of words despite changes in pitch, intensity, and duration.

REFERENCES

1. D. Marr, Vision (W. H. Freeman and Co., San Francisco, 1982), Chapter 1.
2. J. W. Clark, J. Rafelski, and J. V. Winston, Physics Reports, 123, 215 (1985).
3. J. J. Hopfield, P.N.A.S., 79, 2554 (1982).
4. D. H. Ackerly, G. E. Hinton and T. J. Sejnowski, Cognitive Science, 9, 147 (1985).
5. A. S. Lapedes and R. M. Farber, Los Alamos National Laboratory report LA-UR-85-4037 (1985) submitted to Physica D.
6. T. Maxwell, H. H. Chen, G. Z. Sun, Y. C. Lee, P. Guzdar and H. Y. Lee, this proceedings.
7. T. Hogg and B A. Huberman, Phys. Rev. A32, 2338 (1985).
8. T. Hogg and B. A. Huberman, P.N.A.S., 81, 6871 (1984).
9. A. J. Hudspeth, Science, 230, 745 (1985).
10. J. O. Pickels, Introduction to the Physiology of Hearing (Academic Press, Orlando, 1982).
11. T. J. Imig and A. Morel, Ann. Rev. Neurosci., 6, 95 (1983).
12. R. C. Truex and M. B. Carpenter, Human Neuroanatomy (Williams and Wilkins, Baltimore, 1969), Chapter 22.

13. M. M. Merzenich, W. M. Jenkins, and J. C. Middlebrooks, in G. H. Edelman, W. E. Gall, and W. M. Cowan, eds., Dynamic Aspects of Neocortical Function (John Wiley and Sons, New York, 1984), Chapter 12.

14. J. B. Allen, J. Acoust. Soc. Am., 61, 110 (1977).

15. M. H. Holmes, J. Fluid Mech., 116, 59 (1982).

16. S. A. Shamma, J. Acoust. Soc. Am., 78, 1612 (1985).

17. S. A. Shamma, J. Acoust. Soc. Am., 78, 1622 (1985).

18. B. J. Travis, Los Alamos National Laboratory report LA-UR-86-1004 (1986).

19. C. Gilbert, Ann. Rev. Neurosci., 6, 217 (1983).

20. B. Kolb and I. Q. Whishaw, Fundamentals of Human Neuropsychology (W. H. Freeman and Co., New York, 1985), Chapter 8.

21. G. M. Shepherd, The Synaptic Organization of the Brain (Oxford University Press, New York, 1979.)

22. D. G. Sinex and C. D. Geisler, J. Acoust. Soc. Am., 76, 116 (1984).

23. J. Feigenbaum, Los Alamos Science, 1, 4 (1980).

24. H.-O. Peitgen and P. H. Richter, Frontiers of Chaos (Forschungsgruppe Komplexe Dynamik, Universität Bremen, 1985).

25. G. H. Edelman and L. H. Finkel, in G. H. Edelman, W. E. Gall, and W. M. Cowan, eds., Dynamic Aspects of Neocortical Function (John Wiley and Sons, New York, 1984), Chapter 22.

EPSILON CAPACITY OF NEURAL NETWORKS

Santosh S. Venkatesh

California Institute of Technology

Abstract

It is shown that the capacity of neural networks for storing associations under error-tolerant conditions is linear in n, where n is the number of neurons in the network. Error-tolerance is introduced into the retrieval mechanism by specifying components in the retrieved memory which are to be ignored (i.e. are treated as dont-cares) by means of a binomial distribution of choice. The epsilon capacity $C_\varepsilon(n)$ is defined to be the largest rate of growth of the number of associations that can be stored such that with high probability the retrieved memory after one synchronous step differs from the desired associated memory in no more than (essentially) a fraction ε components. It is shown that for large n, and with $0 \le \varepsilon < \frac{1}{2}$, the epsilon capacity $C_\varepsilon(n)$ is at most $2n/(1-2\varepsilon)$. The result is universal in the sense that any other mode of choice of dont-care components with essentially a fraction ε of components being dont-care will have an upper bound of $2n/(1-2\varepsilon)$ on capacity. The results are not tied to any particular algorithmic formulation of neural networks, but are based on a consideration of the universe of possible networks.

I. ASSOCIATIVE MAPS

Mathematical idealisations of neural networks proliferating in the literature have been demonstrated to be very adept at the storage of associations and memories. Recently the memory storage capacity of some popular algorithms has been characterised [1],[2]. The relative efficacy of various algorithms, however, can best be gauged if the ultimate storage capacity of the neural network model itself is determined. This will be our focus in this paper. In particular, we will provide answers for questions such as: What is the maximum number of associations that can be stored when all possible (McCulloch-Pitts) neural networks are allowed for consideration? What gains can be achieved in capacity if there is some tolerance to errors?

The rest of this section describes the neural network model under consideration, and sets up the framework of error tolerance. A rigourous definition of capacity is also provided. The capacity results are quoted in section 2. Finally, the distribution of errors is tackled in section 3. In the interests of brevity, the theorems will be stated without proof for the most part in this exposition. The formal proofs of the quoted results are fairly involved and can be found in Ref. [3].

A. McCulloch-Pitts Neural Networks

We consider a network of n labelled neurons. Each formal neuron (modelled after McCulloch-Pitts) is a threshold gate characterised by a vector of real weights $(w_{i1}, \ldots, w_{in})$, and a real threshold (which we take to be zero). The neurons accept real n-tuples $\mathbf{u} \in \mathbb{R}^n$ as input, and return binary scalars $v_i \in \mathbb{B}$ as output according to the threshold rule $v_i = \mathrm{sgn}\,(\sum_{j=1}^{n} w_{ij} u_j)$. Given an input $\mathbf{u} \in \mathbb{R}^n$ the neural network under advisement yields as output a binary n-tuple $\mathbf{v} \in \mathbb{B}^n$, whose components v_i are the outputs of each of the individual neurons.

Our concern will be with the storage of prescribed associations of the form $\mathbf{u} \mapsto \mathbf{v}$ in the neural network by suitable choice of weights w_{ij}. We tacitly assume a synchronous mode of operation for simplicity; the inputs to the network are presented simultaneously to each neuron, and each neuron returns an output binary variable in concert. We will further restrict our attention to single step synchronous transitions of the form $\mathbf{u} \mapsto \mathbf{v}$. Our results on capacity will bound cases where more complex associative behaviour (such as soft error correction in distorted memories) is achieved through the medium of feedback and dense neuronal interconnection.

Let m denote the number of associations of the form $\mathbf{u} \mapsto \mathbf{v}$ to be stored in the network. Specifically, we require to store m associations $\mathbf{u}^{(\alpha)} \mapsto \mathbf{v}^{(\alpha)}$, $\alpha=1,...,m$. We call the input probe vectors $\mathbf{u}^{(\alpha)}$ the *fundamental memories*, and the desired resultant vectors $\mathbf{v}^{(\alpha)}$ the *associated memories*. The specified m-set of fundamental memories $\{\mathbf{u}^{(1)},\mathbf{u}^{(2)}, \ldots, \mathbf{u}^{(m)}\} \subset \mathbb{R}^n$ is assumed to be chosen independently from any probability distribution invariant to reflection of coordinates in real n-space. The corresponding m-set of associated memories $\{\mathbf{v}^{(1)},\mathbf{v}^{(2)}, \ldots, \mathbf{v}^{(m)}\} \subset \mathbb{B}^n$ is also a randomly specified set, with components $v_i{}^{(\alpha)} \in \{-1,1\}$, $i=1,...,n$, $\alpha=1,...,m$, chosen from a sequence of Bernoulli trials with equal probabilities of success and failure.

We will also refer to the components $v_i{}^{(\alpha)}$ of the associated memories $\mathbf{v}^{(\alpha)}$ as *decisions* (made by neuron i w.r.t. fundamental memory $\mathbf{u}^{(\alpha)}$). This is motivated by the fact that if we consider neuron i as an isolated threshold gate, then $v_i{}^{(\alpha)}$ is simply the decision made by the threshold gate when pattern $\mathbf{u}^{(\alpha)}$ is the input.

B. Error Tolerance

For error free associative maps we require that $v_i{}^{(\alpha)} = \text{sgn} \left(\sum w_{ij} u_j{}^{(\alpha)} \right)$ for each component i and for each associative map. Under error tolerant conditions, however, some of the components of the retrieved states could be allowed to be in error. We now prescribe a mechanism which determines the allowed error distribution in the components of the retrieved states.

It is clear that if for a neuron (threshold gate) we specify a certain number of don't-care decisions, then the number of don't-care decisions determines the maximum number of decision errors made by the neuron. Similarly, if we specify don't-care components for each associated memory, then the number of don't-care components in each associated memory determines the maximum number of errors made in retrieving each associated memory from the corresponding fundamental memory. Our approach to introducing error tolerance is hence to specify don't-care decisions in the associated memories by a suitable distribution of choice such that the expected number of don't-care decisions coincides with (twice) the allowable fraction of errors. Specifically, we perform a sequence of mn independent, and identical experiments to determine whether each (random) decision $v_i{}^{(\alpha)}$, $i=1,...,n$, $\alpha=1,...,m$, is to be labelled a don't-care decision, with the probability that any particular decision $v_i{}^{(\alpha)}$ be labelled a don't-care decision given by twice the allowable fraction of errors.

Formally, let $0 \le \varepsilon < \frac{1}{2}$ denote the allowable fraction of decision errors made by a neuron for the case of threshold gates, and let ε also denote the allowable fraction of component errors in the retrieved associated memories for the case of neural networks. Let $V_i{}^{(\alpha)}$, $i=1,...,n$, $\alpha=1,...,m$, be the outcomes of mn identical, and independent experiments whose outcomes are subsets of $\{-1,1\}$ such that

$$V_i{}^{(\alpha)} = \begin{cases} \{v_i{}^{(\alpha)}\} & \text{with probability } 1-2\varepsilon \\ \mathbb{B} & \text{with probability } 2\varepsilon . \end{cases} \qquad (1.1)$$

If the outcome $V_i^{(\alpha)} = \{v_i^{(\alpha)}\}$, then we will require that neuron i produce decision $v_i^{(\alpha)}$ as output whenever it receives fundamental memory $\mathbf{u}^{(\alpha)}$ as input. If, however, the outcome $V_i^{(\alpha)} = \mathbb{B}$, then we associate a don't-care decision with component $v_i^{(\alpha)}$ of the associated memory $\mathbf{v}^{(\alpha)}$, so that neuron i can result in either -1 or 1 as output when $\mathbf{u}^{(\alpha)}$ is input. For obvious reasons we call $V_i^{(\alpha)}$ the *decision set* associated with decision $v_i^{(\alpha)}$. Clearly, once the fundamental memories $\mathbf{u}^{(\alpha)}$, the associated memories $\mathbf{v}^{(\alpha)}$, and the decision sets $V_i^{(\alpha)}$ have been specified, we need to find neural networks for which the neurons yield correct decisions only for the restricted set of decision sets which are not don't-care.

Definition. Let $\mathbf{w}_i \in \mathbb{R}^n$ be the vector of interconnection weights associated with neuron i, for each neuron $i = 1, ..., n$ of a neural network. The event:

$$\text{sgn}\left[\sum_{j=1}^{n} w_{ij} u_j^{(\alpha)}\right] \in V_i^{(\alpha)}, \quad i = 1, ..., n, \ \alpha = 1, ..., m, \quad (1.2)$$

is described by saying that *the neural network stores m associations with tolerance epsilon.*

Note that by virtue of the random decision sets $V_i^{(\alpha)}$ being drawn from independent, and identical experiments, the actual distribution of errors is spread independently across the components $v_i^{(\alpha)}$ of the associated memories. Furthermore, the conditional distribution of the decision sets $V_i^{(\alpha)}$ given the associated memories $\mathbf{v}^{(\alpha)}$ is binomial. Hence, the expected number of don't-care decisions attributed to each neuron is $2\varepsilon m$, and the expected number of errors in each of the associated memories is εn. Thus the modified definitions reflect (at least in an average sense) a tolerance of up to a fraction ε of errors in the decisions.

C. Definition of Capacity

Let a tolerance $0 \le \varepsilon < \frac{1}{2}$ be fixed. In querying whether there exists a choice of interconnection weights for which the neural network maps the fundamental memories to the associated memories with at most εn errors (on average), the event of interest is described by the following attribute:

Event E. "$\exists$ a neural network which stores m associations with tolerance ε."

Note that the event E is defined on the sample space obtained as the product of the probability spaces over which the components of the fundamental memories $u_i^{(\alpha)}$, the components of the associated memories $v_i^{(\alpha)}$, and the decision sets $V_i^{(\alpha)}$, ($i = 1, ..., n$, and $\alpha = 1, ..., m$,) are defined.

Definition. A sequence of integers $\{C_\varepsilon(n)\}_{n=1}^{\infty}$ is a *sequence of epsilon capacities* for neural networks iff for each $\lambda \in (0,1)$, the following implications hold:

(1) $m \le (1-\lambda)C_\varepsilon(n) \ \Rightarrow \ P\{E\} \to 1$ as $n \to \infty$,

(2) $m \ge (1+\lambda)C_\varepsilon(n) \ \Rightarrow \ P\{E\} \to 0$ as $n \to \infty$.

The above definition is arrived at formally by a very general consideration of lower and upper estimates of capacity, and subsumes within it many commonly used notions of capacity. The only requirement that we might wish to impose is that the probability of the event E be monotonic with the number of associations

m, and this does indeed turn out to be the case. Note that the situation is rather reminiscent of sphere hardening. Note also that in our definition, zero-capacity corresponds to the maximum number of associations that can be stored under conditions of perfect recall, i.e., with the tolerance epsilon being identically zero.

The following result demonstrates that if sequences of epsilon capacity do exist, then they are not very different from each other. The proof of the assertion follows solely from the definition.

Proposition 1. If $\{C_\varepsilon(n)\}$ is a sequence of epsilon capacities then so is $\{[1 \pm o(1)]C_\varepsilon(n)\}$. Conversely, if $\{C_\varepsilon(n)\}$ and $\{C_\varepsilon(n)'\}$ are any two sequences of epsilon capacities, then $C_\varepsilon(n)' \sim C_\varepsilon(n)$ as $n \to \infty$.

The proposition establishes that if sequences of epsilon capacity do exist, then: (1) they are not unique, and (2) they do not differ significantly from each other. In light of the above result, we define an equivalence class of sequences of epsilon capacities $[\{C_\varepsilon(n)\}]$ with equivalence relation defined as follows: if $\{C_\varepsilon(n)\}$ and $\{C_\varepsilon(n)'\}$ are members of this equivalence class of epsilon capacities, then they must satisfy the equivalence relation $C_\varepsilon(n)' \sim C_\varepsilon(n)$. Henceforth, if a sequence of epsilon capacities $\{C_\varepsilon(n)\}$ exists, then we shall say without elaboration that $C_\varepsilon(n)$ is the *epsilon capacity* of neural networks; by this we mean that $\{C_\varepsilon(n)\}$ is a member of the equivalence class $[\{C_\varepsilon(n)\}]$ of sequences of epsilon capacities.

II. EPSILON CAPACITY

In this section we quote our main result on the ultimate storage capacity of neural networks.

Theorem 1. Let ε be a given error tolerance, $0 \le \varepsilon < \frac{1}{2}$. Then, for every fixed λ in the open interval $0 < \lambda < 1$, the following implications hold:

(a) $m \le \dfrac{2n(1-\lambda)}{(1-2\varepsilon)} \Rightarrow P\{E\} \to 1$ as $n \to \infty$,

(b) $m \ge \dfrac{2n(1+\lambda)}{(1-2\varepsilon)} \Rightarrow P\{E\} \to 0$ as $n \to \infty$.

Corollary 1. $C_\varepsilon(n) = \dfrac{2n}{1-2\varepsilon}$.

Thus, with an error tolerance of $\varepsilon \in [0, \frac{1}{2})$, we can find neural networks to store almost all choices of $2n/(1-2\varepsilon)$ associations. The proofs of the above assertions utilise large deviation Central Limit Theorems, very large deviation estimates, and function counting theorems in combinatorial geometry.

For the case of perfect associative recall, $\varepsilon = 0$, a capacity result due to Abu-Mostafa and St. Jacques [4] gives n as an upper bound for the capacity of neural networks, while we anticipate a capacity of as much as twice n. The capacity result of [4], however, required that for *every* choice of m fundamental memories with $m <$ capacity there must exist at least one neural network in which the chosen m-set of vectors can be stored as memories. This leads to the upper bound of n on capacity. In our definition of capacity on the other hand, we require that asymptotically as $n \to \infty$, for *almost all* choices of m fundamental memories with $m <$ capacity, there exist some neural network in which the chosen m-set of vectors can be stored as memories. This leads to the upper bound of $2n$.

Thus, as long as we are willing to eschew *all* of the possible choices of m fundamental memories in favour of *most* of these choices, then we can potentially gain by as much as a factor of two in capacity. Another consequence of the above two statements on capacity is that if $2n > m > n$, then there are guaranteed to be choices of m fundamental memories which cannot be stored in *any* neural network; such choices of fundamental memories will however constitute an asymptotically negligible proportion of the total number of choices.

III. DISTRIBUTION OF ERRORS

The binomial distribution of choice of don't-care decisions ensures that each associated memory has an average of $2\varepsilon n$ don't-care components—half of these on average resulting in errors, and the other half resulting in correct decisions. However, we might query whether, when all the associated memories are considered jointly, there is a significant probability that there are much more or much less than $2\varepsilon n$ don't-care decisions for many of the memories. An alternate problem of importance is whether the capacity results are strongly tied to the choice of a binomial distribution for don't-care decisions—natural though it be in some sense.

A. *Strong Convergence to Mean Error Rate*

By the Strong Law of Large Numbers, *each* individual memory has $2\varepsilon n$ don't-care decisions attributed to it with probability one. As m increases very rapidly with n, however, it is not clear whether the joint probability of essentially $2\varepsilon n$ don't-care decisions for each associated memory also approaches one. In fact, the following rather stronger assertion holds.

Theorem 2. As $n \to \infty$, the number of errors in each retrieved memory approaches εn; further, for each neural site $i = 1, ..., n$, the number of erroneous decisons made approaches εm.

The proof utilises standard Borel-Cantelli lemmas, and is similar in statement to the Strong Law of Large Numbers.

B. *Universality of Capacity Bounds*

Thus far we have considered a particular mechanism for the introduction of error tolerance into the decision process. The specification of the decisions to be labelled don't-cares was through the agency of the random decision sets $V_i^{(\alpha)}$ assumed to be drawn from a binomial distribution corresponding to a sequence of mn Bernoulli trials, each with probability ε of resulting in a don't-care decision.

This particular mode of choice of don't-care decisions is natural and intuitively appealing as it outlines a method of specifying don't-care decisions in a random and independent fashion. Further, for any two particular choices of sequences of don't-care decisions, the choice with fewer don't-cares is more likely in accordance with our preference for more accurate recall. However, as we saw from the theorem of the last section, *typical* sequences of don't-cares are essentially $2\varepsilon n$ in number for each associated memory. It is hence reasonable to say that the resultant upper epsilon capacity of $2n/(1-2\varepsilon)$ indeed specifies a tolerance of up to εn errors in decision for each associated memory.

The notion of introducing the random decision sets $V_i^{(\alpha)}$ to specify decisions which we treat as don't-cares is much more general, however. In the general case we could specify the values taken by $V_i^{(\alpha)}$, $i = 1, ..., n$, $\alpha = 1, ..., m$, to be taken from some product space, with probabilities of points in the joint ensemble given by any suitable distribution. The only requirement we impose on the distribution of choice is that (at least in an asymptotic sense) the number of don't-care decisions approaches $2\varepsilon n$ for each associated memory with high probability, where $\varepsilon \in [0, \tfrac{1}{2})$ is the prescribed error tolerance.

With a plethora of possible distributions of don't-care decisions confronting us, we might wonder if by some suitable choice of distribution we could obtain a significant increase in capacity over the $\dfrac{2n}{1-2\varepsilon}$ result obtained earlier using a binomial distribution of choice. The answer is furbished by the following

Theorem 3. For any choice distribution of don't-care decisions resulting in essentially εn or fewer errors asymptotically in each associated memory, the epsilon capacity is bounded from above by $\dfrac{2n}{1-2\varepsilon}$.

The proof essentially rests on the following result wherein best and worst case capacities are constructively obtained for suitable deterministic error spreads.

Proposition 2. For any particular choice of exactly $2\varepsilon mn$ don't-care decisions, the following hold:

(a) The maximum upper epsilon capacity of $\dfrac{2n}{1-2\varepsilon}$ is achieved iff each associated memory has $2\varepsilon n$ don't-care decisions attributed to it, and each neural site has $2\varepsilon m$ don't-care decisions attributed to it.

(b) The maximum achievable upper epsilon capacity is $2n$ iff there is at least one neural site which has no don't-care decisions attributed to it.

(c) The maximum achievable upper epsilon capacity lies between $2n$ and $\dfrac{2n}{1-2\varepsilon}$ for any choice of $2\varepsilon mn$ don't-care decisions.

Conclusion

Recapitulating, the ultimate associative storage capacity of neural networks is $2n/(1-2\varepsilon)$, where ε is the allowed fractional component error in the retrieved memories, and particular algorithms can, at best, achieve this capacity. Note that we require that almost all choices of associations be permissible. For a binomial distribution of errors, the capacity is achievable, and in fact, the number of errors approaches εn for each associated memory. The universality result indicates that further gains in capacity are not possible for other choices of error distribution. Extensions can be readily made to more complex network structure.

References

[1] R. J. McEliece, E. C. Posner, E. R. Rodemich, and S. S. Venkatesh, "The capacity of the Hopfield associative memory," submitted to the *IEEE Trans. Inform. Theory*; presented at conference on *Neural Networks for Computing*, Santa Barbara, May 1985.

[2] S. S. Venkatesh and D. Psaltis, "Information storage and retrieval in two associative nets," submitted to the *IEEE Trans. Inform. Theory*; presented at conference on *Neural Networks for Computing*, Santa Barbara, May 1985.

[3] S. S. Venkatesh, PhD Thesis, California Institute of Technology.

[4] Y. S. Abu-Mostafa and J. St. Jacques, "Information capacity of the Hopfield model," *IEEE Trans. Inform. Theory*," vol. IT-31, pp. 461-464, 1985.

AIP Conference Proceedings

		L.C. Number	ISBN
No. 1	Feedback and Dynamic Control of Plasmas – 1970	70-141596	0-88318-100-2
No. 2	Particles and Fields – 1971 (Rochester)	71-184662	0-88318-101-0
No. 3	Thermal Expansion – 1971 (Corning)	72-76970	0-88318-102-9
No. 4	Superconductivity in d- and f-Band Metals (Rochester, 1971)	74-18879	0-88318-103-7
No. 5	Magnetism and Magnetic Materials – 1971 (2 parts) (Chicago)	59-2468	0-88318-104-5
No. 6	Particle Physics (Irvine, 1971)	72-81239	0-88318-105-3
No. 7	Exploring the History of Nuclear Physics – 1972	72-81883	0-88318-106-1
No. 8	Experimental Meson Spectroscopy –1972	72-88226	0-88318-107-X
No. 9	Cyclotrons – 1972 (Vancouver)	72-92798	0-88318-108-8
No. 10	Magnetism and Magnetic Materials – 1972	72-623469	0-88318-109-6
No. 11	Transport Phenomena – 1973 (Brown University Conference)	73-80682	0-88318-110-X
No. 12	Experiments on High Energy Particle Collisions – 1973 (Vanderbilt Conference)	73-81705	0-88318-111–8
No. 13	π-π Scattering – 1973 (Tallahassee Conference)	73-81704	0-88318-112-6
No. 14	Particles and Fields – 1973 (APS/DPF Berkeley)	73-91923	0-88318-113-4
No. 15	High Energy Collisions – 1973 (Stony Brook)	73-92324	0-88318-114-2
No. 16	Causality and Physical Theories (Wayne State University, 1973)	73-93420	0-88318-115-0
No. 17	Thermal Expansion – 1973 (Lake of the Ozarks)	73-94415	0-88318-116-9
No. 18	Magnetism and Magnetic Materials – 1973 (2 parts) (Boston)	59-2468	0-88318-117-7
No. 19	Physics and the Energy Problem – 1974 (APS Chicago)	73-94416	0-88318-118-5
No. 20	Tetrahedrally Bonded Amorphous Semiconductors (Yorktown Heights, 1974)	74-80145	0-88318-119-3
No. 21	Experimental Meson Spectroscopy – 1974 (Boston)	74-82628	0-88318-120-7
No. 22	Neutrinos – 1974 (Philadelphia)	74-82413	0-88318-121-5
No. 23	Particles and Fields – 1974 (APS/DPF Williamsburg)	74-27575	0-88318-122-3
No. 24	Magnetism and Magnetic Materials – 1974 (20th Annual Conference, San Francisco)	75-2647	0-88318-123-1